Gustavus Weissenborn

American Engineering

Vol. 1

Gustavus Weissenborn

American Engineering
Vol. 1

ISBN/EAN: 9783337413330

Printed in Europe, USA, Canada, Australia, Japan

Cover: Foto ©berggeist007 / pixelio.de

More available books at **www.hansebooks.com**

AMERICAN ENGINEERING,

ILLUSTRATED BY

LARGE AND DETAILED ENGRAVINGS

EMBRACING VARIOUS BRANCHES OF

MECHANICAL ART,

STATIONARY, MARINE, RIVER BOAT, SCREW PROPELLER, LOCOMOTIVE,

PUMPING and STEAM FIRE ENGINES, ROLLING and SUGAR

MILLS, TOOLS, and IRON BRIDGES,

OF THE

Newest and Most Approved Construction.

BY

G. WEISSENBORN,

CIVIL AND MECHANICAL ENGINEER,

181 FULTON STREET.

NEW YORK:

1861.

INDEX.

INTRODUCTION.

THE publication of this work has been conceived in the hope of obviating some very great difficulties, which practical observation has suggested to our mind as lying in the way of scientific men, inventors, young engineers, students, mechanics, and machinists, who may not be practical draughtsmen; to set before them clear views of mechanism and mechanical relations, which by practice have become part of the habits of their minds; and to enable them to apply those principles as well to the structure as to the laboring force of machinery.

Our workmen far excel the artisans of other nations, in genius to invent, in skill to fashion, and in assiduity to finish, whatever they undertake: but they chiefly excel in contriving tools with which to complete their designs, and to confer on the workmanship all the perfection of which it is susceptible. To this truth every observer will assent who has calmly contemplated the specimens of mechanism in our manufactories, or the giant steam-engine by the shaft of a mine, on a railroad, or in a ship.

The knowledge of our various artisans and mechanics is the result of an education of a very high order.

In the workshop, the manufactory, or the steamship, we see them enter on their task with the most thorough understanding of the various intricacies of the operations in which they are engaged.

We see their well-trained hands subjecting to the dominion of their experience all the materials requisite for giving existence to some new and untried forms and arrangements of mechanism, which shall be applicable to a desired purpose.

All the artistic dexterity shown in the various manipulations by which the task is accomplished, appear to be the result of a species of intuition, while, in point of fact, it is only the development of that scholarship which every master in his art, which every man learned in his calling, carries about with him as a matter of course; constituting him the owner of a property of which neither fortune nor locality can deprive him.

The application of mechanics to steam navigation, railways, manufactures, and all branches of mechanical art, has acquired that importance in this country which authorizes correct drawings of all important inventions of machinery, in mechanical combination, arrangement, and proportion, to be published on such a scale, with such full description of the parts, as will enable the student or machinist to fully understand their properties and adaptation, and also afford aid to inventors in their researches. As the construction of the machinery which we are representing and illustrating, has been after the designs and drawings of the most eminent engineers in the country, we are encouraged to hope that our publication may take its stand at once as a national work.

We are under obligations to gentlemen at the head of different establishments, who have generously honored our enterprise with their confidence; and we can only say, that if we should not be remunerated for our labor and expense, in the commencement, it will not be from any want of promptness on their part in affording us all facilities for the selection of the newest machinery as subjects for our publication, and we trust that by furnishing finely executed engravings, after the most correct drawings, we shall very soon find ourselves in the agreeable position of having our efforts substantially appreciated by our subscribers.

Similar publications are well sustained in European countries, and we know of no good reason why one should not be in this. At all events, it will be our constant endeavor to win the approbation of the public.

THE EDITOR.

AMERICAN ENGINEERING ILLUSTRATED.

BEAM ENGINE

OF THE NEW YORK STEAM SUGAR REFINERY.

GENERAL DESCRIPTION.

We commence our first number, with a description of a steam-engine constructed by Messrs. Hogg & Delamater (foot of Thirteenth street, New York) for the use of the New York Steam Sugar Refinery.

The following are the dimensions of the engine:

Cylinder, 24 inches in diameter,
Stroke, 6 feet,
Vacuum pumps, 26 inches in diameter,
Stroke, 3 feet,
Cold water pumps, 18 inches in diameter,
Stroke, 18 inches.

This engine, when in operation, makes an attractive appearance. It is beautifully proportioned, and the highest style of workmanship is visible in all its parts. It is designed, as we have stated, for a sugar refinery, and is operated in connection with two vacuum pumps, *G G*, which are placed on each side of the engine, for the purpose of producing the vacuum in the sugar pans, and for condensing the steam from the cylinder. The beam *L L* communicates by the connecting rod *M* with the crank *N*, on one end, and on the other communicates with the engine, by means of the usual appliances, as seen in Plate I. The crank is connected with a shaft, by gearing, as seen in Plate II., by means of which the power of the engine is applied throughout the building, in the usual manner.

The reader, by a careful examination of this engine with respect to general arrangement and combination of the different parts, will find that much economy of labor and solidity of the engine, have been exhibited in its construction. A point chiefly worthy of attention, is the combination (Plate I.) of the bedplate *V V* and the columns of this engine *T T T*. The column *T* and the lower part T^1, are cast together, by which arrangement, the parts of the bedplate *V V*, are necessarily bolted to the side of the base T^1. The bedplate and the columns *T T T'* are thus made to form one solid piece. The front pieces of the bedplate are also bolted to the side of the columns, as is represented in the top view, (Plate III., Fig. 23.) The column is shown in the same plate, (Fig. 29.) The entablature *W W* is bolted to the top of the six columns. It will be at once perceived, that by means of this combination, not the least vibration on the engine frame is perceptible—the whole standing as firm as if it were of a single piece of iron.

The mechanical arrangements for unhooking and disconnecting the air and cold water pumps with the engine, are worthy of reference. In Plate I., representing the side elevation of the frame, the coupling cases E^4 are made of cast iron, and keyed to the lower rods *F F*. If the air-pumps are not required to be in operation, the keys by which the coupling cases E^4 are fastened to the rods E^3 are taken out, and then the rods E^3 will slide loose in the cases E^4. The rods E^3 are of such length as to reach about 3 feet 6 inches into the cases E^3 when coupled with the rods *F*; so that when the buckets of the vacuum pumps are at the bottom of it, the rods E^3 will be about six inches within the cases E^4.

The upper keys on the parts E^4 can be removed or returned to their places, while the engine is in motion, and are consequently a great convenience to the engineer; whereas, they would otherwise be a source of annoyance and interruption.

Plate I. The governor *Z Z* is driven by a belt from the main shaft of the engine, and two bevel wheels, one connected with the governor's spindle, and the other with a small shaft communicating by a belt with the main shaft of the engine to receive its turning motion. A throttle valve, bolted to the flange A^2, in communication with the governor by a gearing, regulates the admission of steam into the cylinder.

The motion of the steam and cut-off valves is performed through the medium of two eccentrics in connection with two rods. The one for the steam valve has an eccentric hook on the end, and that for the cut-off valve has a strap end, by which means it is attached to a cast-iron lever keyed to the rockshafts *C* and C^1. The valve and cut-off rods *B* and B^1 receive their motion by means of two cast-iron levers that are fitted to each rockshaft.

The cold water pump studs I^2 are fastened by pins to the crossheads to which the cold water pump rods are connected.

Plate II. The parallel motion is represented (Plate II., Fig. 1) with the beam of the engine at the highest position in its stroke. The piston and air-pump rods connected with it move in a vertical direction, parallel to the centre line of their pumps and parallel to each other; the principles of which are well known, and are found in almost every work on mechanics. A point to be taken into consideration is, that the air-pump centres must always be equi-distant between the main and end centres of the beam; and the length of the main and air-pump links, which must be equal in distance from centre to centre, but may be made in any suitable length.

Plate III. The sectional view of the cylinder, valves, steam-chests, and cut-offs, is represented in Fig. 10. The following rule, with respect to the position of the valves in communication with the piston, must be followed: When the piston is at the end of the stroke, whether up or down, the valve in communication with the entrance of steam for the cylinder, must remain open about one-eighth of an inch. This is mainly necessary to prevent any vibration or noise on the part of the piston or piston rod. It also has the effect to counterbalance the pressure on the piston, before it reaches the dead point, thus facilitating the return motion of the piston, and preventing any noise or derangement in the mechanical parts. Our own experience, however, is such, that we recommend that valves for an engine of this size be so arranged that the steam shall exhaust and enter the cylinder about four inches from the dead point. This is about the rule that ought to be observed, in proportion to the size of the cylinder, in the construction of all valve arrangements.

Plate IV. The sectional views of the vacuum pumps *G*, the cold water pumps *H*, the condenser G^2, and the waste water reservoir G^1 are shown in Plate IV.

A very singular and beautiful design for a water pump, originated by Messrs. Hogg and Delamater, and constructed by them in connection with this engine, and driven by two spur wheels, is represented in Plate IV., showing the two pumps in sectional views, their front and side elevations; also the ground plan and some details.

DETAILED DESCRIPTION.

Plate I.

A Bottom plate of cylinder bolted to the top of the engine bedplate, A^1 cylinder of cast iron, A^2 nozzle for the entrance of steam into the steam-chest. A^3 steam-chest, A^4 exhaust pipe cast to each side of the cylinder. *B* Rods for steam valves. B^1 Rods for cut-off valves, made of wrought iron, to be all finished. *C* Rockshaft for steam valves. C^1 Rockshaft for cut-off valves made of cast iron, to be turned. C^2 Pillow-blocks for the two shafts. *D* Piston-rod of common steel. D^1 Socket of wrought iron. *E* Main links, straps of wrought iron, and centre piece of cast iron; E^1 back links; E^2 links of parallel motion; E^3 vacuum pump rods, all of wrought iron; E^4 coupling cases of cast iron. *F* Vacuum pump rods (wr. iron;) F^1 vacuum pump bucket rods of brass or covered with copper $\frac{1}{8}$ of an inch thickness. *G* Vacuum pumps; G^1 waste water reservoirs; G^2 condensers; G^6 bedplates for pumps and condensers; *H* Cold water pumps; H^1 flanges cast to the pumps to which the ascending pipes are bolted; H^2 flanges for the suction pipes; H^3 pump bucket rods of brass; H^4 pump rods, wrought iron. *I* Coupling cases cast iron; I^1 pump-rods; I^2 cold water pump-straps, wrought iron. *K* Main pillow-blocks. *L L* Beam (cast iron.) *M* connecting-rod (wrought iron;) *N* crank; *O* shaft pillow-block (cast iron;) *P* eccentric for steam valve gearing; *Q* hub of fly-wheel; *R* fly-wheel (cast iron;) *T T T* columns of engine frame; T^1 T^1 T^1 bases of columns, cast solid to them; *V V* parts of bedplate bolted to the side of the bases of the columns; *W* entablature; *Z Z* Governor.

Plate II.

Fig. 1 represents a side elevation of parallel motion; Fig. 2 front view; Fig. 3 top view; Fig. 4 front and side view of connecting rod (wrought iron;) Fig. 5 piston rod, common steel or hard iron; Fig. 6 coupling case for cold water pumps, and Fig. 7 coupling case for vacuum pump rods; Fig. 8 is the governor's spindle (cast iron;) Fig. 9 governor stand (cast iron) and a Front View of the engine, showing the cylinder, frame of the engine, eccentrics, bedplate, fly-wheel, and a part of the gearing for the main building.

Plate III.

Fig. 10 is a longitudinal section of the cylinder and cover, steam chest, steam and cut-off valves, all of cast iron; valve stems *A* and *B*, made of common steel; Fig. 11 top view of the cylinder and steam-chest; Fig. 12 side view of the cylinder and steam-chest; Fig. 13 front view of steam-chest; Fig. 14 air-pump centre (wrought iron;) Fig. 15 end centres (wrought iron;) Fig. 16 socket keyed to the piston rod and fitted to the crosshead on part *c*; Fig. 17 guard for governor, represented in top and side view; Fig. 18 top and side view of steam and cut-off valve stem cross-head; Fig. 19 governor ball fastened to a wrought-iron lever, represented in a side and front view; Fig. 20 rockshaft stands, shown in a side and top view—journal seats *d d*, made of brass; Fig. 21 shaft pillow-block; Fig. 22 shaft pillow-block near the crank—both of cast iron—and journal seats of brass; Fig. 23 top view of engine bedplate, showing the bottom plate of cylinder marked *E*, and foot of governor stand marked *a*, giving a clear idea of the bolting of the different parts of the bedplate to the columns; Fig. 24 is a side view of a part of the bedplate, to which the base of the column marked T^1 is bolted. Fig. 25 beam pillow-block—represents a side elevation and top view; Fig. 26 cold water pump buckets, represented in section and top views, made of composition; Fig. 27 vacuum pump buckets, shown in section and top views, made of composition; Fig. 28 engine beam shown in side and top views, made of cast iron; Fig. 29 columns represented in section and outside views.

Plate IV.

Fig. 30 represents a longitudinal section through the vacuum-pump G, waste water reservoir G^1, condenser G^2, cold water pump H, and bedplate G^3. The cold water and vacuum pumps are made of cast iron and lined with brass through the whole length of the stroke of the buckets $\frac{1}{4}$ inch in thickness. The segments are about 4 inches wide, and planed on the side, to leave about $\frac{1}{32}$ of an inch clearness on the inside surface of the joints, which will be filled up afterwards by hammering the joints, by which operation the segments will keep firm to each other. The same can be done with one brass cylinder turned inside and outside $\frac{1}{4}$ inch in thickness, but the cast-iron cylinder would necessarily require to be heated before the brass cylinder can be put in, to keep the two cylinders firm to each other. However, segments are preferable, as it will spare some labor, at the same time will give greater compactness to the composition and firmness with the cast-iron cylinder, when it is hammered. G^4 valve and valve seats of composition, fastened on two sides by means of keys to their seats. Fig. 31 top view of bedplate, pumps and condenser. Fig. 32 top view of cover for vacuum-pump. Fig. 33 top view of cover for cold water pump.

DETAILED DESCRIPTION

OF THE

LIFTING AND FORCE PUMPS FOR THE NEW YORK STEAM SUGAR REFINERY.

Plate IV. Fig. 35 represents a front view: A is a column cast to the shaft pillow-block, and bolted to the bedplate C by 4 bolts; the top part of the column is connected with the bedplate by two braces B B made of wrought iron, all to be finished; C is the bedplate with strong ribs on the bottom, bolted on each side to the brick wall by 1-inch bolts. D is the lifting pump (cast iron) lined with brass on the inside, $\frac{1}{8}$ inch thick through the whole length of the stroke of the pump bucket, E air-vessel, F connecting rod (wrought iron,) G crank (cast iron,) H crosshead (wrought iron,) I pump rod, K square guide rods, L guard for guide rod, M braces, N main shaft, O spur wheel. Fig. 36 represents a side elevation of the pump; P force-pump with valve chamber; Q plunger; R crosshead; K K square guide rods; M M braces; F F connecting rods; G G cast-iron cranks; N main shaft; O spur wheel; A columns, D lifting pump, I pump rod, H crosshead. Fig. 37 represents a top view of the bedplate and the mechanical parts of the pumps; Fig. 38 top view of braces and guard for the pump rods, showing the columns in section; Fig. 39 braces in detail (cast iron) bolted on each side to the columns, for the purpose of securing the two guards by bolts to the braces, to guide the pump rods in a parallel direction to the two pumps. Fig. 40 represents a section through the lifting pump bucket and air-vessel. Fig. 41 is a top view of the lifting pump; Fig. 42 side view of manhole plate of the same; Fig. 43 section of forcepump and plunger—pump cast iron, plunger brass; Fig. 44 top view of the same; Fig. 45 pump bucket represented in section and top view; Fig. 46 guard to guide the pump rod represented in side and top views; Fig. 47 forcepump rod crosshead; Fig. 48 lifting pump rod crosshead.

PRACTICAL TREATISE ON TOOLS AND MATERIALS.

Iron is the material almost exclusively employed in mechanical engineering. Iron ships, buildings, pavements, and railings,—iron bridges, aqueducts, cars, and statues,—machinery of a thousand kinds, tools, domestic and mechanical, hardware, stoves, gratings, and columns,—and the noblest offspring of the iron manufacture—the Railway and Steam Engine,—all involving the employment of millions of capital and tens of thousands of strong and skilful laborers,—serve to show how truly is iron the great agent of modern art and industry. And as new wants have arisen, and new fields of invention and industry have presented themselves, the iron manufacture has been adapted, in direction and detail, to supply the requisite fixtures and machinery.

Iron is rarely found in a pure or native state. In Canaan, Conn., iron has been found pure, in a vein or plate two inches thick, and sufficiently ductile to be wrought into nails by a blacksmith. In almost all parts of the world there are also some few deposits of native iron.

Iron exists naturally as an ore,—in the form of a rusty, metallic stone. The most common kind of ore used—the hematites—may be described, in common words, as iron rust solidified or concreted by water. Chemists call it hydrated oxide of iron, or red or brown oxide of iron,—and chemically, it is composed of iron, oxygen, and water.

In the United States, the earliest iron works (and they made iron in Massachusetts previous to the year 1700) were supplied with "bog ore," dug in neighboring swamps. Perhaps the oldest iron *mine* is the Warwick mine, near Phœnixville, Pa. This was opened a few years before the Revolution, and is yet worked with much success It is 150 feet deep, and has been mined over sixteen acres of surface. A considerable portion of the ore used in the great rail mills of Reeves, Buck & Co., at Phœnixville, is obtained from this mine.

The extent and abundance of the deposits of iron in this country are remarkable. We have the common reputation of existing iron furnaces in nearly every state in the Union; while in others, explorations or genealogical analogy prove the existence of vast bodies of iron. The Pembroke iron of Maine, the Franconia of New Hampshire, the Adirondack, Hudson, Sterling, and Clinton of New York, the Salisbury of Connecticut, the many beds in New Jersey, the wholesale deposits in Pennsylvania, Maryland, Virginia, Alabama, and in every Western state, are all as notorious as the wheat, corn, or cotton of either of those districts. Take Pennsylvania, for instance: the Schuylkill, Susquehanna, Alleghany, Juniata, Monongahela, and smaller valleys, are literally filled with iron. In Virginia, the Shenandoah, the James, the Kanawha, and other valleys, are underlaid continuously with iron. In Missouri, the "Iron Mountain" and "Pilot Knob" are towering masses of iron—*millions* of tons in extent. It has been estimated that the "Iron Mountain" alone —containing over one hundred millions of tons of iron—would supply the entire probable wants of the *whole world* for that material for at least *ten centuries*. In Wisconsin, and on Lake Superior, some of the purest known ores exist in the same lavish profusion of nature.

We can apply no figures to the extent of the supply of iron now opened and in store for the future generations of mankind; for the present, the supply is exhaustless.

American iron is stronger and softer than most of the iron imported. It is in general considerably superior to English iron, but not to Russian and Swedish bars. Scotch pig is a kind of cast-iron much used on account of its superior fluidity when melted, but it is not as tough nor as desirable in any other respects for machinery as ordinary American iron.

The following results relative to the strength of American irons were obtained by Professor W. B. Johnson, who experimented at the expense of the American government.

				Strength in lbs. per sq. inch.
Iron from	Salisbury, Connecticut, by means of	40	trials	58,000
"	Sweden, "	4	"	58,084
"	Centre County, Penn., "	15	"	58,400
"	Lancaster County, Penn., "	2	"	58,061

				Strength in lbs. per sq. inch.
Iron from McIntire, New York, by means of	4 trials	.	.	58,912
" England, (cable bolt,)	" 5 "	.	.	59,105
" Russia,	" 5 "	.	.	76,069
" Carp River, Lake Superior, determined by Major Wade			.	89,582

Every one is familiar with the practical difference between cast and wrought iron. Yet it is difficult to describe, strictly, in what the difference consists. It is hardly a chemical difference. It may be called, perhaps, a mechanical difference in the arrangement of the atoms of the iron. In pig or cast iron, the atoms are united in homogeneous masses, with no intervening bond except the natural cohesion of the particles themselves. In wrought iron the atoms are disposed in distinct fibres, united each with the other by a cementing principle contained in the cinder of the iron.

The conversion into wrought iron is effected, not by any mixture of the pig metal with other matter, but simply by an additional heating, which heat is prolonged for some time at just above the melting point, and during which the iron is stirred up until every particle has been brought under the cementing action of the heat.

There are various distinctions arising from the precise nature of the materials and processes employed, which exert a great influence on the nature and value of the product. Thus, if charcoal

* ANALYTICAL VIEW OF THE MANUFACTURE AND USE OF IRON.

Natural *Iron Mines* and *Iron Beds* contain ☞	Magnetic Ore, or Black Oxide of Iron, used in U. S., Russia, Sweden, India, etc. The richest ore known. Specular Ore, or Red Oxide of Iron, used in U. S. Hematite Ore, or Brown Oxide of Iron, used in U. S. Clay Iron Stone, used in England, Scotland, etc. Black Band Ore, a compact Carbonate of Iron, used in England, Scotland, etc. Sparry Carbonate of Iron, used in Germany.
These undergo in the *Blast Furnace* ☞	Digging or Mining, Roasting,* SMELTING.
and assume the form of *Pig Iron*, rated as ☞	White Iron, Mottled Iron, Bright Iron, Gray Iron, Foundry Iron, No. 1, flows very readily, worth $32 per ton. Foundry Iron, No. 2, worth $28 per ton. Foundry Iron, No. 3, worth $27 per ton. Forge Iron, No. 1, worth $26 per ton. Forge Iron, No. 2, worth $25 per ton. Forge Iron, No. 3, worth $28 per ton.
They next in *Iron Works* and *Foundries* undergo ☞	Refining,* PUDDLING, Squeezing or Hammering, Rolling. — Melting, MOULDING. Cleaning.
And are put *in use* as ☞	Flat Bar Iron, Round Bar ", Boiler ", Sheet ", Band ", Scroll ", Angle ", Tee ", Railroad ", Shafting, Wire, Anchors. — Hollow Ware, Stoves and Furnaces, Machinery Castings, Gratings, Fences, Statues, Architectural Ornaments, Girders, Buildings, Bridges, Cradles, Coffins, Pavements.
And when *worn out* become ☞	Wrought Scrap Iron. — Cast Scrap Iron.

* Those processes marked by a Star are sometimes omitted.

fuel is employed in the smelting process, the product is charcoal iron, and is generally esteemed tougher and better for making into wrought iron, especially for steam boilers, where great strength and little thickness are required. The product of such furnaces, when blown with air previously heated, is considered inferior to "cold blast" iron; and the peculiar ores and processes employed in Sweden and Russia, give "Swede's Iron" and "Old Sable" a peculiar character, which can only be described by thus designating the location of its manufacture. So, too, in practice, most of the irons in commerce are conventionally designated by the name of works, as Stirling, Sligo, Juniata, etc., the character of processes, etc., at each being supposed to be well understood, so far at least as the workable qualities of the product are concerned. In the broadest general view, however, the ores and smelting processes may be considered as nearly identical, but producing, by laws as yet imperfectly understood, six distinct qualities of pig metal, ranging from the "Gray," which contains the most carbon and oxygen and flows the most readily in the mould, to the "White," which contains the least of these substances, and is extremely hard, but brittle. These distinctions, which are represented by numbers, have been already alluded to; and in the subjoined note is exhibited, in one general view, a connected outline of the whole process of manufacture. The distinctions between forge and foundry iron must be considered as general, and not absolute; foundry No. 3 being frequently used in the puddling furnace, and forges Nos. 1 and 2 being sometimes employed with complete success in mixing with others for large and not very intricate castings.

One great reason for the universal employment of iron in the mechanic arts is its cheapness. The prices per ton given in the analytical view on the previous page were designed chiefly to illustrate the *comparative* values of the various qualities, but may serve approximately to indicate the absolute cost of the ordinary cheap qualities of anthracite iron in its crude condition. The cost of iron as compared with other metals may be realized by inspecting the following figures:—

PRICES PER POUND OF METALS IN QUANTITY.

Crude iron, (pigs)	$1\frac{1}{2}$ cts.	Copper	25 cts.
Wrought iron	$4\frac{1}{2}$ "	Tin	26 "
Cast-steel	15 "	Mercury	44 "
Lead	6 "	Silver	$17.00 "
Spelter	6 "	Platinum	128.00 "
Zinc	7 "	Gold	270.00 "

But, interesting and important as is the subject of iron, and valuable as might be made a full treatise both on it and its rival, copper, we must hasten very lightly over it to spend more time on its more perfected form, iron in the state of steel.

Cast-steel is the metal now universally employed as the master metal, or that which is most available in giving shape to others. In the form of hammers and anvils, dies, swages, and a host of kindred tools, it gives to any malleable metal when skilfully manipulated an approximation to almost any form desired. In the form of chisels and files it removes superfluous parts even of the most rigid castings, and gradually reduces an unshapen mass to the finely proportioned pieces required in scientific modern engineering. The ancients are affirmed to have possessed the art of hardening copper, and it is probable that many of the finely carved stones in the buildings and ruins of the old world were executed with copper tools, but the art of hardening this metal (except to a slight degree by admixture with other metals) has long been lost, and the superior cheapness of excellent steel makes it undesirable to expend any considerable time in the apparently hopeless task or rediscovering any other hard materials for tools.

The cause of the great hardness of properly prepared steel is yet unknown. The only agents necessary to its development are calorific. All steel, and indeed many kinds of iron, both cast and wrought, when heated to a red heat and suddenly cooled, assume a much harder condition than when allowed to cool slowly, and nearly or quite all forms of iron and steel are capable of being annealed or considerably softened by a long and general cooling in an oven or a pile of ashes. Except for a comparatively few purposes, however, the softening of steel is rarely of much interest, and we will confine our attention in this chapter entirely to the proper production of the opposite effect. We shall, in the few pages devoted to this important subject, quote liberally from the "Machinist's and Engineer's Assistant," as also from Mr. E. P. Freedley's recent work on "Leading Men and Leading Pursuits," in the United States, without further acknowledgment. The treatise on this subject in the former standard work, is in fact the basis of which this chapter may be considered but a very thorough revision.

One of the operations which naturally seems the simplest to be effected by the mechanical engineer is the production of plane surfaces of metal. But, however simple it may seem to enunciate, it is practically quite difficult to effect. Strictly speaking, there are no perfectly plane surfaces. All bodies are composed of particles so connected together as to leave pores or cavities between them, and all artificial surfaces are necessarily irregular and uneven to a degree which can be readily seen with the aid of a microscope. Even the most highly polished razors or needles appear rough and jagged under a high magnifying power, and although there are surfaces formed by nature in the wings and shells of insects, etc., which appear perfectly smooth under the most powerful microscope which can be brought to bear on them, we may safely assume that even these exquisite substances are really porous and irregular, but only to a less degree than artificial objects. When surfaces are in contact, we cannot assume that the whole faces are together, but only certain most prominent points. It is theoretically required, in a surface for mechanical purposes, that all the bearing points should be in the same plane, that they should be equidistant from each other, and that they should be sufficiently numerous for the particular application intended. Where surfaces remain together in fixed contact, the bearing points may, without disadvantage, be fewer in number, and consequently wider apart; but in the case of sliding surfaces, the points should be numerous, and in close approximation. In the use of emery for producing plane surfaces, by grinding, very many serious difficulties are involved. The surfaces are usually made adhesive with oil, and the emery powder is sprinkled thereon as evenly as may be, after which the faces are pressed together and rubbed upon each other. Were it possible to rub the metals together in every possible direction, bringing successively every part of one in contact with every part of the other, a close approximation to actual plane faces would be produced on both; but this is rarely the case in practice, and the abrasive effect is usually far less perfect, producing ridges and grooves of a greater or less extent and regularity according as the process is conducted. Ordinarily also the powder collects in greater quantity about the edges of the metal than upon the interior parts, producing the well-known effect of the bell-mouthed form. This is particularly objectionable in the case of slides, from the access afforded to particles of dust, and the immediate injury necessarily occasioned thereby. Another circumstance materially affecting the durability of ground slides, is, that a portion of the emery becomes fixed in the pores of the metal, and can never be entirely eradicated therefrom; causing a rapid and irregular wear of the surface.

In short, there can be little chance of a multitude of points being brought to bear, and distributed equally under a process, from which all particular management is obviously excluded. To obtain any such result, it is necessary to possess the means of operating independently on each point, as occasion may require; whereas, grinding affects all simultaneously. It is subject neither to observation nor control; there is no opportunity of regulating the distribution of the powder, or of modifying its application, with reference to the particular condition of the different parts of the surface. An appearance, indeed, of beautiful regularity is produced, to which, no doubt, we may trace the universal prejudice so long established in favor of the process; but this appearance, so far from being any evidence of truth, serves only to conceal error; and under this specious disguise, surfaces pass without examination, which, if unground, would be at once rejected.

In addition to what has been stated, it must be remembered another great evil of grinding is, that it takes from the mechanic all sense of responsibility and all spirit of emulation, while it deludes him with the idea that the surface will be ultimately ground true; hence, he slurs his work over in a slovenly manner, trusting to the effect of grinding, being conscious that it will efface all evidence either of care or neglect on his part, and thus it appears that the practice of grinding has seriously impeded the progress of improvement.

We have shown above, that absolutely true and perfectly plane surfaces are entirely unattain able by human skill. The remark propagated by some writer that the phrase "near enough" should never be used in the machine shop, and that nothing but absolute perfection should be tolerated, is extremely unwise. It is possible to aim at a degree of perfection far beyond the necessity of the particular case, the difficulty of which would more than counterbalance the advantage, and every adjustment, every touch given, is but a compromise. It is found that an error exists, and the blow is struck with the knowledge that it may introduce another, but it is hoped a smaller one. The extensive class of machinery denominated machine-tools, affords an important application of the subject; here every consideration combines to enforce accuracy.

The tool employed for scraping is not only simple, but easily made. It should be of the best cast-steel, and carefully sharpened to a fine edge on a Turkey-stone, the use of which must be frequently repeated. Worn-out files may be converted into convenient scraping-tools. A flat file, with a broad end bent and sharpened, will be most suitable in the first instance, and afterwards a three-angled file, sharpened on all the edges. The process of scraping is equally simple, requiring rather care than skill on the part of the workmen, whilst it affords a certain and speedy means of attaining any degree of truth that may be deemed necessary, thus tending to the gradual establishment of a higher standard of excellence, the influence of which cannot fail to affect beneficially all mechanical operations, opening at the same time to the mechanic himself a new field in which he will find ample scope for the exercise of skill, both manual and mental.

We are now in a condition to proceed with the matter more immediately under consideration. The value of every cutting instrument depends upon the excellence of the steel of which it is made, the care bestowed during the several processes of forging, hardening, and tempering, and the just adaptation of the angle or bevel which forms its edge to the work it is intended to perform. Generally speaking, this angle is determined by the hardness of the substance to be operated upon. Thus, we see chisels for cutting soft woods are thinner than those used for the harder species, and these, again, are more acute than chisels employed for cutting metals, or, in other words, the greater the resistance offered by the material to be cut, the more obtuse must be the angle of the tool. This definition is not propounded as rigidly correct in all cases, although it is susceptible of abundant practical illustration; for example, in hand-turning, the workman is enabled, by raising or lowering the *T* of the rest, to vary the direction and limit the cut of the tool employed, according to circumstances. This one fact, amongst a multitude of others equally palpable, that could be adduced, might have been expected to induce inquiry and investigation. On the contrary, we have to state that the form of tools, more especially those used in turning and planing, iron, brass, etc., has not hitherto received that attention which the importance of the subject calls for, nor has any attempt been made to reduce it to plain and general principles, of which it is highly susceptible, and if so treated, would be of much service to those in whose hands the management of such tools is for the most part entrusted. So many considerations of a practical nature are inseparable from this subject, that the quality, as well as the quantity of work producible from turning-lathes and planing machines, depends entirely upon the skill of the workman in giving to his tools the proper form.

The general principle of the working tool, which is equally applicable whether the motion be horizontal, circular, or vertical, is deduced from a consideration of the direction in which the metal is to be cut or penetrated. With regard to the first case, as in the planing-machine, (it is immaterial, so far as the principle here contended for is concerned, whether the tool be stationary and the table of the machine movable, or the reverse, since the action is precisely the same in either case,) it is manifest that if, while the axis of the tool stands at right angles to the plane of the material to be cut, its face be bevelled, and consequently, if its point or cutting edge be made in the form of a very obtuse angle, it will possess little or no penetrating quality; such a tool would not cut, but rather abrade, or probably crush off the particles of metal. Again, if we resort to the other extreme, and give to the cutting edge the shape of an extremely acute angle, we shall find, however sharp it may appear, a total absence of penetrating quality, or, at all events, in the required direction, and what is equally objectionable, the point being weak, would snap off, incapable of resisting the least applied force.

From an investigation of these and other obvious facts, a tool of the form shown in the annexed cut, Fig 1, fulfils the requisite conditions, as it combines a high degree of acuteness, with sufficient strength—the former in the direction of the cut, and the latter behind the point or cutting edge, where it is most needed. Hence, the following principle may be established—namely, that in forming and setting a tool to cut any surface, it is essentially necessary so to place it that the end shall form but a small angle with the surface to be cut, or, in other words, be as nearly parallel as practicable, and whatever degree of acuteness may be deemed necessary must be obtained by hollowing out the face *E C* on which the shavings slide. An apt and very familiar illustration of the principle may be drawn from the common plane of the joiner. An artificial end being given to the plane-iron, which is here the cutting tool, by means of the sole of the plane, this necessarily limits the penetrating quality in all directions, except that in which it is required to remove the material. Further, it can scarcely have escaped observation that the bevelled surface next to the wood itself shall form the least possible angle.

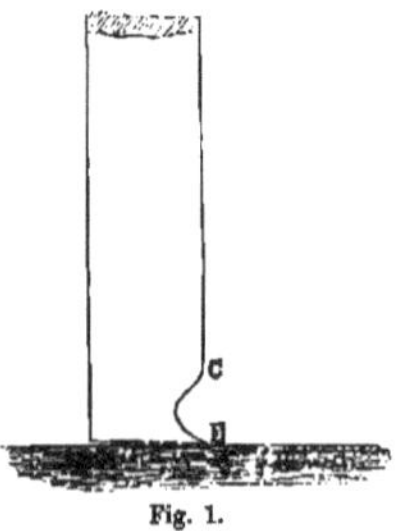

Fig. 1.

In practice it is frequently more convenient for the smith to so forge the tool that the point *E* shall project forward beyond the line *C*. This form facilitates the subsequent grinding, which may be necessary, as the edge becomes dulled, and is in several respects superior; but the point must not be projected too far so as to spring, and the effect, so long as the proper angle is given to the acting face, is evidently the same as in the illustrations here given. The forms represented show the principle under construction in a very strong light, and it is evident that it applies with equal force in the case of turning-tools, and, indeed, in every tool, from the smallest and most delicate of the clock and watch maker's, up to the largest and most powerful tool in any engineer's lathe or planing-machine.

Fig. 2.

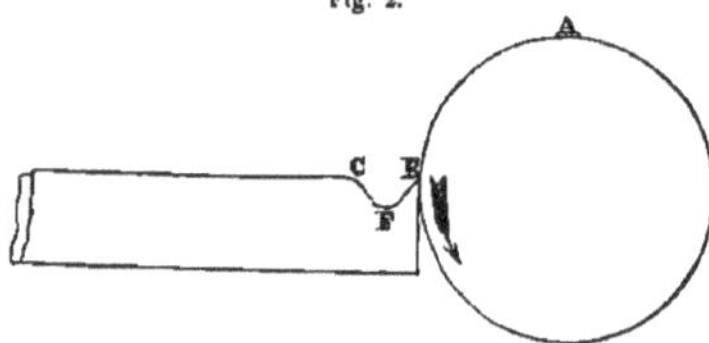

As regards circular motion, we have a clear exemplification of the principle by merely considering the tool already described as a turning-tool. Here *A B*, Fig 2, shows a section of a cylindrical bar in the lathe, and *E F*, so placed as to be, as nearly as circumstances will permit, a tangent, that is, at right angles to the radius of the curve—the requisite acuteness being obtained, as before, by hollowing out the face *E G*.

The same principle applies to drills. Thus, *H* being the end view of a drill, the edge *O P* should be in the least degree prominent or out of the plane of the surface, of which the bounding lines are the edges, *O S* being slightly less prominent than *O P*, so that the penetrating quality at the edge *O P* may be limited as much as possible. An adherence to this rule will produce a drill that shall cut a smooth and equal hole, without chattering as is commonly the case when the edges are bevelled very much back, as shown at *R*. The necessary acuteness to the cutting edge of the drill is easily obtained by merely observing the principle laid down in respect to turning tools—that is, by hollowing a groove at *X*, at each cutting face.

Fig. 31.

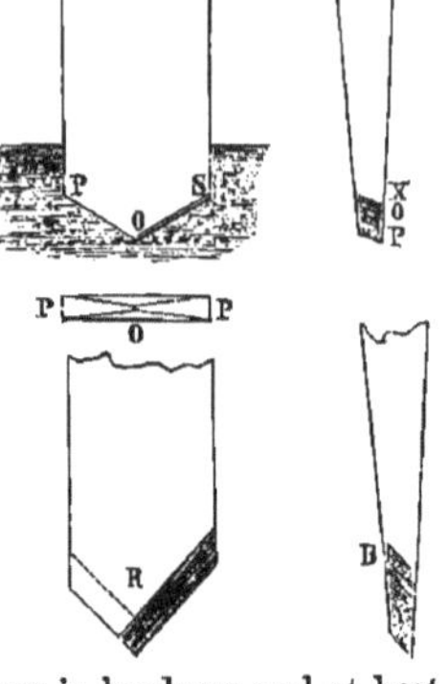

Every mechanic is sensible of the value of good tools as necessary appliances to the performance of his work more quickly, with less exertion, and more accurately than can be done with inferior ones; yet how few are in a position to answer this apparently simple question, what constitutes this quality denominated goodness? The excellence of cutting tools is generally decided by their relative degrees of endurance; but how many incidental circumstances may, and frequently do, interfere to vitiate any accurate comparison? As regards hardness, nearly the only test is the resistance the objects offer to the file, a mode extremely fallacious, because files differ among themselves in hardness, and at best only serve to indicate in a very imperfect manner, to the touch of the individual, a vague notion without any distinct measure.

Take, for example, two chisels for turning iron, both of cast-steel, and from the hands of the same maker, and although precisely alike in outward appearance, the one may be absolutely worthless and the other equally valuable. Nay, one portion of the same chisel may be good and the other bad. If, during the process of fabrication of two cutting tools, the same treatment be practised with similar care, we should naturally expect, all things being coincident, that the one would correspond with the other. Experience shows the fallacy of this mode of reasoning. Nearly every metal turner makes his own tools, and for this plain reason—that he cannot place any dependence on those he purchases.

In order to elucidate a subject confessedly of the highest possible importance, but unfortunately obscured, beyond any other connected with mechanical art, by conflicting opinions, we now propose to investigate the nature and properties of steel, and to introduce a variety of practical examples, collected from authentic sources, in preference to any speculative theories. These examples will be found, in most cases, to confirm, but in some to confute each other, thus leaving to each individual the means of testing, by his own experience, the value of the information offered.

Steel, as is well known, is made by combining carbon with iron. Bar steel is made by a process called *cementation*, which consists in placing bars of the purest malleable iron in alternate layers, with powdered charcoal, in a proper furnace; air is carefully excluded, and the whole kept at a red heat for several days. By this process carbon combines with the iron, and alters its texture from fibrous to granular crystalline. In these furnaces, twelve tons of bar iron may be converted into steel at each charge. This has been called *blistered* steel, from the air-bubbles which cover its surface, derived apparently from the formation of carbonic oxide during the process of cementation. The action of the carbon causes cavities and fissures, which render the steel unfit for use until it has undergone the operation of *tilting*, which is performed by beating it under tilt-hammers, weighing usually two hundred weight, until it acquires a very uniform structure. *Shear* steel is made by binding together several bars of blistered steel, by means of a steel rod, and heating them to a welding heat, the surface being covered with sand or clay to prevent oxidation. It is then drawn out into a bar, by means of a tilt-hammer, and rolled. In this state it is more tenacious and malleable, and susceptible of a finer polish. *Cast*-steel is made by melting blistered steel, broken into small pieces, in fire-clay crucibles, closely covered, by means of a coke fire. It is then cast into "ingots" and rolled into bars. In this condition it has a much more close, fine-grained, and uniform texture. *Sheet* steel is made by being rolled between revolving cylinders.

Damascus steel is made directly from the iron ore, principally a red oxide of iron. *Wootz* is the name given to the steel, the most ancient known, derived from India, celebrated for the toughness and durability of the cutting edges made from it; it is made from magnetic iron ore, such as is found in abundance in New York, New Jersey, and Pennsylvania. It is highly probable that the genuine Damascus blades were made from "wootz;" the imitation blades, of modern make, though they have a damasked appearance, have none of the superior qualities of the true Damascus swords. It has been said that the peculiarities of this steel depend on the presence of a small quantity of aluminum; but this is by no means settled. *German* steel derives its name from the manner in which it is manufactured. It is made of pig or white-plate iron, in forges where charcoal is used for fuel. To this last, and to the character of the ore, (bog-iron or the sparry carbonate,) its properties may be attributed.

The most important element in making steel is the iron ore. Bog-ore, impure hematites, impure magnetic and sparry ores, will make steel, but at great expense and of an inferior quality. According to Mr. Overman, the only iron ores in this country, which can be profitably used for the manufacture of natural steel, are the ore of the Missouri iron-mountain, the recently-discovered deposits near Lake Superior, and the specular ores of Pennsylvania and New Jersey; either carbonates or peroxides of iron. A low heat, an abundance of coal, (well-charred pine charcoal is the best,) and good ore, will produce good steel; and in no other way can it be produced. Many attempts have been made to make good cast-steel directly from the iron, on a large scale, without resorting to the converting process into blistered steel, but without success.

The question whether steel contains any thing besides iron and carbon is purely chemical, the consideration of which would form, did space allow, an interesting theoretical illustration of the present inquiry. The researches of the French academicians, Monge, Berthollet, and Vandermande, indicate the distinction between cast-iron and steel to be, that the former is charged with a superabundant, the latter with a minute yet sufficient dose of carbon; wrought iron, on the contrary, if pure, is free from all heterogeneous matter. It is to be regretted that the constituent proportions of steel have not been accurately determined.

Vauquelin assumes the average amount of carbon to be 1-150th, and Clouet places it as high as 1-32d. Mr. Parkinson considers the quantity of carbon necessary for making of steel to be very small—indeed, the actual amount seldom exceeding 1-200th or 1-300th, and perhaps never more than 1-100th, the remaining portion of charcoal flying off at the time of cementation in the form of gaseous oxide of carbon.

Dr. Thomson analysed some specimens of cast-steel, and the general results of his trials gave the constituents as follows:—

Iron, ...	99
Carbon, with some silicon, . .	1
	100

Now this approaches—

Iron, 20 atoms, . .	70.00
Carbon, 1 atom, . .	0.75
	70.75

And this Dr. Thomson considers as likely to be the constitution of cast-steel. He did not, in like manner, attempt the analysis of blistered steel, but concluded the proportion of carbon in it to be rather less. It is well ascertained that iron and carbon are capable of combining together in a variety of different proportions; when the carbon exceeds, the compound is carburet of iron; when the iron exceeds, the compound is steel, or cast-iron in various states, according to the proportion: all these compounds may be considered as sub-carburets of iron. The most satisfactory deduction from experiments on these compounds which has yet appeared is, that the hardness of iron increases with the proportion of charcoal with which it combines, till the carbon amounts to about 1-80th of the whole mass. The hardness is then a maximum, the metal acquires the color of silver, loses its granulated appearance, and assumes a crystallized form. If more carbon be added to the compound, the hardness diminishes in proportion to the quantity, as appears from the following tabular arrangement on iron and steel.

Iron, semi-steelified, contains	1-150th	of carbon.
Soft steel, capable of welding, ...	1-120th	"
Cast-steel for common purposes, ...	1-100th	"
Cast-steel requiring more hardness, ...	1-90th	"
Steel capable of standing a few blows, but quite unfit for drawing, ...	1-50th	"
First approach to a steely granulated fracture, .	1-30th, 1-40th	"
White cast-iron, ...	1-25th	"
Mottled cast-iron, ...	1-20th	"
Carbonated cast-iron, ...	1-15th	"
Super-carbonated crude iron, ...	1-12th	"

But even in the present advanced state of chemical and microscopic science, it is impossible to say how far the peculiar properties of steel are due to the crystalline structure, rather than to the chemical constitution of the metal. Steel is undoubtedly to a great extent a mechanical production of the forge-hammer, which tears the molecules of certain species of white cast-iron out of their original positions, into which the forces of attraction, in respect to the centres as well as to the position of the molecules, had arranged those molecules by the slow action of heat.

Perfectly pure iron, cemented in equally pure carbon, would doubtless produce steel free from blisters; but as in practice these blisters are unavoidably evolved, it is needless to inquire into their origin more minutely than we have already done, especially as it seems to be admitted that blistered steel is unequally carbonized, the outside retaining the larger portion. It is, therefore, rendered fit for the market by doubling and welding several times, by which means the parts are more intimately blended together, and the carbon more equally distributed; in this state it is called spur steel. These repeated weldings, although they tend to condense metal, are apt to produce flaws, 1st, by imperfect union; 2d, by the carbon burning out of the commingling surfaces, thereby interposing a stratum of iron or imperfectly converted steel, and this being softer than the surrounding particles, would give way during the extension of the steel. To what cause such defects are to be attributed, must necessarily remain a matter of conjecture; but that they do very largely accompany this description of steel is certain; and it is a question whether any process short of actual fusion can totally remove them; nevertheless, it is ascertained that long-continued forging essentially conduces to soundness or homogeneity.

An excess of carbon renders steel harder and more brittle, therefore an inequality is liable to occur. Good steel, hardened by sudden immersion in cold water, when at a red heat, will rapidly return to a soft state by slow cooling from such heat, and more equally so if the external atmosphere be carefully excluded; but chilled cast-iron will not; it requires to be exposed for many hours to an intense heat, and must not be smothered by fuel to prevent the escape of the superabundant carbon with which it is charged. The air, too, should be allowed free access as a means of disengaging some portions of the carbon, while the remainder has a tendency to equalize itself; then, if slowly cooled, the mass will be found to be sufficiently annealed.

The reduction of iron from the condition of cast-iron to that of wrought is a partly chemical and partly mechanical process, never perfectly understood *either in theory or practice*. Very slight changes (frequently no apparent ones) in the processes produce very widely varying results. Absolutely pure iron is not as strong as that with just enough carbon, silica, and the like. Cast-iron can be found containing a little more, a little less, and exactly the same, amounts of carbon as the best cast-steel. It is believed that wrought iron consists of fibres surrounded by cinder, while cast-iron has the cinder chemically combined, and that steel resembles and yet differs from both in its internal structure.

Most of the steel used in the world is manufactured in Sheffield and Warrington, in England, and the British manufacturers have, to a great extent, secured a monopoly of the product of the Danamora mines in Sweden, which, it is affirmed by these parties, have thus far proved the most suitable for the purpose. There are about ten manufactories in the United States, manufacturing steel of various kinds from American ore, but although very successful in producing the cheaper varieties, they have so far failed to establish a reputation for their cast-steel, as good as that of similar material from the mammoth British works. The difficulty does not lie in the materials, but in the skill and care with which the operations are performed, or rather in the smallness of the scale on which the works are conducted. The large concerns of Great Britain have a great number of workmen, each trained in his especial department for many years, always working with the same metals and fuel, and with similar machinery. As a consequence, they become enabled, almost intuitively, to distinguish and assort the products at different stages of the manufacture, so as to produce given and reliable qualities of steel. There is really a great diversity in cast-steels, even among those kinds which are sold at the same price. The hard steel, suitable for a pick or cold chisel, is treated differently, and is made from material which, at an early stage, presented a different appearance from that suitable for surgical instruments. This is not generally known, but is affirmed by the agents of English manufacturers to be strictly true, and the same parties affirm that much of the trouble arising from the failure of steel, arises from an improper selection. We would recommend to parties using large quantities, to give their orders to an intelligent and active man, with full instructions in relation to the purposes for which it is to be used, and always thereafter to obtain from the same works an article marked in precisely the same manner. We are unwilling to do any thing which may seem to discourage the home manufactures of our country, but justice to the subject, with our present information, compels us to acknowledge that at this date, the cast-steel of no American works is as uniform in quality as that from the English. A lot which is intended to be precisely similar to another previously used, may prove in practice to require very different treatment.

Steel, at a red heat, when suddenly plunged in cold water, becomes both brittle and hard; but even in this state its toughness greatly exceeds that of any other brittle substance. This characteristic hardness cannot be given in part, but always in full, and to its highest limit. So true is this, that in a piece of steel, a portion of which is hard and a portion soft, no gradation of hardness can be detected, the parts adjacent to the hard portion being quite soft, or, as some think, softer than if slowly cooled. This singular fact has been thus accounted for:—Suppose a rod of well-hammered steel to be heated at one end for hardening, there will be a gradation of temperature from the coldest to the hottest extremity, and the annealing or reduction of that hardness which it has received will be in proportion to the heat, consequently, the rod will be softer and softer towards the end where the heat is applied. On plunging the bar into cold water, that portion which has become sufficiently hot to harden, is rendered quite hard, but that part immediately adjacent to it will be found to be most annealed, and will endure more twisting and bending than any other. Although this hardness may be imparted in its full extent, it may nevertheless be lowered in any assignable degree—that is, a portion of its brittleness may be removed by the application of moderate heat, a greater portion by more heat, and so on, as the purposes may require. This is called tempering. If hard steel be brought to a red heat, and then suffered to cool slowly, it will become as soft as if never hardened. This is called softening, and is distinguished from annealing, which is a similar process of slow cooling, but applied to steel, iron, or brass, merely to remove all mechanical condensation, whether by hammering or otherwise; for if metal has been altered in shape by the hammer or any other process, as much as it will bear without breaking, then by annealing, it will be softened, and may again be altered in form as often as requisite. Now, as different degrees of heat remove different degrees of condensation received from the hammer, and a white heat removes all, it is of great importance to harden steel from the lowest possible degree of heat, in order to retain as much condensation as practicable; and it is a fortunate coincidence, that the greater the condensation, the lower is the heat from which steel will harden, and the stronger and tougher it will be. But should this condensed metal be once overheated, it will then no longer harden from that lower degree, but only from a heat nearly approaching that to which it was originally raised. In this case, the condensation, with all its attendant advantages, can only be restored by rehammering. The lowest heat at which steel will generally harden, is a dull or cherry-red, just visible in daylight; therefore, to be safe, the same test—that is, a dull, red heat, just perceptible in the dark, is chosen for the process of hammering; it offers, too, the advantage of coating the article with carbonaceous matter, thereby securing instead of losing by the action of the fire a due supply of carbon, which is of particular consequence. Different modes of performing this part of the process may be adopted.

A paper on the constitution of steel, recently read before the Manchester Society of Arts, by Mr. C. Binks, presents facts which indicate a small quantity of nitrogen to be an essential element in addition to the carbon. Mr. Binks, in that paper, gave an account of some analyses made by himself, which proved that the best kinds of steel contain about one-fifth per cent of nitrogen, and the general results of his experiments tend to show that the substances which change pure iron into steel all contain nitrogen and carbon, or that nitrogen has access to the iron during the operation. He concludes that steel is a triple alloy of iron, carbon, and nitrogen. With regard to improvements in the present system of manufacture, he was of opinion that the most extensive use of cyanogen compounds, such as ferro-cyanide of potassium, was highly important, and he drew particular attention to the fact that these compounds might be economically formed in the ordinary operations of the blast furnace, so that these operations, properly conducted, might serve the double purpose of purifying the metal and converting it into steel.

We will here remark, that it will be found advantageous if the forge is prevented from coming under the influence of a strong light; the fire should be deadened with fine coal, just kept alive by the use of the bellows, in order that the gas may not be allowed to take fire and produce a flame, while, as near as may be under existing circumstances, the anvil should join the flat bed of the forge. The bars of steel that may be undergoing this process, when placed in the partially kindled and half-burning fuel, and enveloped in its smoke, imbibe carbon to a certain extent; this the heat proper for hammering is insufficient to expel. Next, they are successively transferred to the anvil from the forge, and of course returned when cooled. In hammering, every part of the steel is subjected to the force of the hammer, which is moved very quickly. In repetitions of this operation, which are as numerous as may be requisite, the position of the articles is somewhat altered, until they have undergone a good hammering on all sides. The art of discerning when the proposed end is arrived at, and the steel has acquired a sufficient degree of condensation, can only be acquired by the experience of the workman, in enabling him to judge by the feel produced by the diminished effect of the hammer, and the detonation consequent upon the blow. The expansion of the metal is greatly diminished at a heat so low, as compared with the proportion of that quality under the influence of a higher temperature, yet it must not be forgotten that the expansion which is caused by the effect upon the outer portion, by the inner coating, unless it should be fastened in proper moulds or recesses, will be to such an extent as to destroy the value of the exterior band. The failure of steel, especially if it be subjected to very hard usage, it is apparent from what we have already said, may be attributed to two chief causes. First, the metal is liable to fracture, in consequence of the different degrees of tension arising from the brittleness of some parts, and the toughness of others. This arises from an unequal combination of carbon with the iron, which is the case with all steel in a greater or less degree, until it is hammered in a proper manner. In the second place, (as it must be admitted,) however much steel may be hammered, workmen often leave it in a condition in which its several parts are violently conflicting—some almost breaking with strains, and others compressed and girded. Bad and imperfect hammering is therefore a fruitful source of failure.

When thoroughly hammered, the least temperature is requisite to produce the necessary degree of toughness; it will lose all the condensation from the hammer previously acquired, if hardened at a temperature too high, and at the same time so large a proportion of strength as to destroy its tenacity—less strength, with equal hardness, produces the quality of brittleness. Therefore, to obtain a practical degree of toughness, steel thus incautiously treated needs more letting down by the process of tempering. Such tools do not stand long for any purpose, being much too soft and weak to be capable of turning iron or steel. To us, it appears reasonable to suppose that there is a certain degree of cold to which, in a given time, steel must be brought, to produce hardness; there evidently is, also, a given degree, below which it will not harden, and that the metal is only weakened by any additional increase in heat; while a diminution in heat only has the effect to harden a greater proportion of a large mass by cooling it within the time required. If steel is overcharged with carbon to any great extent, it will be found too brittle or harsh to be susceptible of all that improvement which otherwise is imparted by hammering; but, by choosing a sound quality of the most malleable steel, it will, if hammered at a heat so low that it will retain a coat of carbonaceous matter, imbibe the carbon in quantities so minute and so slowly during each hammering, that the workman will be enabled, so far as is compatible with sound hammering, to enforce the fullest charge. Beyond this limit, carbon must produce brittleness, when the steel will not receive any compression from the hammer, by which alone the quality of toughness is acquired; but to the limit referred to, carbon undoubtedly improves the strength. As we should be under the necessity of resorting to extreme letting down in tempering to prevent breaking, it is plain that it would be useless to obtain hardness, if it is not accompanied by toughness; yet many suppose that by coming up to this brittleness, the hardness is advantageously increased; and consequently, in order that extreme hardness may be combined with toughness, dies, taps, and some tools, used in turning, are allowed, while the tenacity of the interior metal remains unaltered, to imbibe a small quantity of carbon in addition on their surface, by being placed in red-hot carbonaceous matter. Burnt leather is generally used for this purpose, and is said to be superior to wood charcoal in the celerity with which it imparts the carbon. "This," says the Dictionary of Arts and Manufactures, "is clearly a modification of case-hardening. Steel may alike be softened or entirely decarbonized. According to Dr. Ure, the common way to soften is to put it into an iron case, surrounded with a paste made of lime, ox-gall, and a little nitre and water, then to expose the case to a slow fire, which is gradually increased to a considerable heat and afterwards allowed to go out, when the steel is found to be quite soft." It is stated that a celebrated English mechanic first decarbonized the metal, and then reimparted a proper portion of carbon by a process which resembled case-hardening.

Steel, as is well known, is composed of iron and carbon. The latter substance becomes *volatile* under the influence of a high degree of heat. It is important to remember this in conducting the actual process of forging. When the carbon is volatilized from being too powerfully heated, steel assimilates itself to iron; it loses its characteristic properties, and therefore it cannot be expected that tools constructed from it will possess either durability of edge or keenness. It is impossible to use too much precaution, more especially in working cast-steel, since it is a fact, that for the evil there is no reliable cure, although it is contended by some that steel which has been overheated may be restored, and a few pretend to prefer steel so "restored" for use in cutting-tools. According to Holtzapffel, in his work on Mechanical Manipulation, cast-steel which has been injured in consequence of overheating, may be recovered to some extent, by four or five reheatings and quenchings in water, each being carried to an extent progressively a little less than the first excess, and lastly, the steel must have a good hammering at the ordinary red heat. But it is obviously doubtful whether hammering in any degree—evidently the leading requisite in this process—will restore the loss of tenacity. Notwithstanding, therefore, the additional expenditure of time and labor required, it cannot be enforced too strongly upon the understanding, that steel should be worked at the *lowest possible temperature, and with the fewest number of heats.*

It is a very difficult matter to regulate the temperature, in a forge with a fire not covered, particularly with articles either thin or small, and in this case much advantage, in acquiring and sustaining the proper degree of heat, may be found in the use of a muffle, which prevents the steel from coming in immediate contact with the coal, as in the common forge fire. Muffles may be constructed of cast-iron or of earthenware, which are, however, liable to crack or open, and it is said that an old gun-barrel, with one end closed, may be advantageously used for this purpose. In consequence of the well-known fact that bituminous coal emits a sulphurous gas during combustion, which is extremely prejudicial to steel, coke and cinders are used to a great extent in our best foundries both for hardening and forging; we believe, however, pit coal is most generally in use, although the Sheffield workmen prefer coke, both for hardening and forging.

The process of hardening and tempering steel involves the consideration of three important points. We shall take them up in their order.

I. *How the requisite degree of heat may be communicated.*
II. *How the heat may be abstracted with rapidity.*
III. *How a reduced or partial heat, for the purpose of tempering or letting down, may be applied.*

I. *How the requisite degree of heat may be communicated.*

All articles that are either too small to be exposed conveniently to the naked fire, or are too large to be heated by the blow-pipe, and in all cases, in fact, where it may be desired to harden tools of more than ordinary length, some protective contrivance is considered prudent, in order to diffuse, uniformly and correctly, the temperature required. A safe and easy apparatus is afforded by a sheet-iron box or an iron tube. Medium sized drills may be heated in the flame of the blow-pipe, directed a little below the point, and very obliquely. Watchmakers' tools, and small tools of that description, are occasionally supported upon charcoal, but are generally heated in the blue part of the flame of a candle. To obtain a perfectly uniform temperature in hardening such articles as gravers for artists, and other delicate instruments of a long and thin description, there have been suggested a great variety of methods. An effectual yet simple plan by which the difficulty encountered in communicating a uniform temperature to many articles in a fire of ordinary character, may be avoided, consists of a bath of molten lead, which shall contain little or no admixture of tin, heated to a moderate redness and well stirred; the steel is plunged into this for a few moments, and when brought near to the surface, that part does not appear less luminous than the rest;—the steel is then stirred rapidly in the bath, quickly drawn out, and thrust into a large quantity of water. A plate of steel may be so hardened in this manner as to be quite brittle and at the same time be very sonorous. We doubt whether similar results could be accomplished by any other means. The use of melted metal, whose point of fusion is a trifle below the proper degree of heat, has been approved to some extent, and it is admitted that the result will prove quite satisfactory, if the melted metal be of sufficient quantity, and the temperature accurately designated by a pyrometer. For the same purpose, pans of charcoal dust, heated to redness, are sometimes brought in requisition.

It is of the first importance that all the parts requiring to be hardened, should receive a uniform temperature, whatever method be adopted, and this can only be insured by subjecting all parts equally to the action of the fire, by moving the article continuously to and fro. A large proportion of tools in general use are heated in the fire of a common forge, which ought to consist of charcoal, cinders, or coke. Sometimes, hollow fires are made use of; but the blast from the bellows should in all cases be applied to a very limited extent, principally to aid in getting the fire up and ready for the introduction of the work, which requires sufficient time to "soak," as it is technically termed—that is to get hot. It will be found better to err on the side of deficiency rather than excess of heat, and a safe rule is, to harden little if any above the state which the tempering is designed to produce; a definite point will depend to some extent on the quality of the steel, and will be found extremely difficult to determine. This point involves, it will be seen, considerations as to the quality of the steel, and a suitable heat. We find, in Gill's Technical Repository, some valuable suggestions concerning these considerations, the substance of which we here present; it is not an ordinary occurrence to find two bars of the best cast-steel of similar quality. It is desirable that some method or proof should therefore be adopted, by which the workman may be enabled to assort the bars, as they respectively indicate their qualities during the process. For however dissimilar, relatively considered, any number of bars may be, it is probable that every part of a bar of cast-steel will be of the same quality as that portion which was subjected to proof. The process of proof is thus described in the work we have alluded to: the bar must be heated carefully at one end, and drawn down to dimensions suitable for bending; then heated to the proper degree for hardening; quenched, leaving the thick portion of the bar still hot; blazed off to spring temper, and again quenched. Then about half an inch of the hardened and tempered portion of the bar is firmly screwed in a horizontal position, in a strong vice; the bar is then used as a lever and moved in any direction, bending the steel. If it snap suddenly, it is evidence of *hardness;* if it endure bending a quarter of a circle, breaking then quietly, a *mild* or *middling* quality is indicated; and *softness* suitable for springs is shown, when it bends a full semicircle, and then separates like lead.

Cast-steel should never be heated above cherry-red, visible in daylight. But with regard to heat, an exact limit cannot be defined. There is in other qualities, such irregularity in purity, and they are so unequally carbonized, that a degree of heat will be found inadequate in one, which will readily bring out the required hardness in another. The only mode of determining, then, this point, is by experiments, commencing at a heat certain to be below that required, gradually increased until the exact heat is ascertained, which, when once known, should be carefully noted. With many it is a favorite theory, that by heating pieces of steel to different degrees of heat, before plunging them into the water, the one heat attains full hardness, the next the temper of a tool fit for metal, another of a tool fit for wood, and the fourth for a spring, and so on; that, in fact, the different degrees of heat suitable for tools of various kinds, may be ascertained with a sufficient degree of exactness for all practical purposes, without resorting to the ulterior process of tempering. A celebrated mechanic of Great Britain, in making the cutters or dividing points for his dividing engine, hardened the end of a larger piece of steel than was required, and sharpened the point upon a grindstone, exactly at that point where the temper suited, without the steel being let down at all. This experiment in a measure sustains the view last stated. It was more elaborately put in practice by another celebrated workman, whose method was, to carefully heat the point, and quench it after shaping the tool and condensing it by hammering; then he made a trial by filing, with the edge of a file, along from the soft and unhardened part to that part where it became hard, when the cutting edge was formed by grinding and whetting that particular part to shape. The tool thus did not require tempering in the usual manner, and the maker was assured that its quality was the best that could be produced from the steel employed.

But the skill required and the time employed are serious drawbacks to the adoption of hardening without tempering, and to the working mechanic it is not available, except in a few instances —so few, indeed, that it is practically of very little value.

After being hardened, some tools require tempering; others do not. Among the latter are many tools for metal which are left of the full hardness, for greater durability, and are used with a uniform and quiet pressure; while the former class includes tools which are liable to be exposed to violent strains or blows, such as hatchets, screw-taps, cold chisels, etc. Another consideration —namely, the quality of the steel, must not be forgotten; for cast-steel let down to a straw-color is softened nearly as much as other kinds of steel let down to a purple or a blue. If a tool will stand without tempering, so much the better—and this may be stated as a general rule; but in the first instance the ignition must be carefully regulated; that degree of heat which is just sufficient to effect the purpose, produces the most useful hardness; and it is clear that we must have recourse to tempering if we once pass the precise limit.

We conclude this point of our remarks, with again enjoining prudence, in regard to heat. It must always be remembered that steel, which has been overheated or burned, suffers injury by the pores of the metal becoming open and expanded, the fineness of its texture being destroyed, (in which state it is quite incapable of sustaining a cutting edge,) and it is superficially injured by its surface becoming covered with scales. Any attempt to remedy the injury done by overheating, by the process termed *letting* down, is simply the introduction of one error to overcome the results of another. As we have before remarked, the lowest possible heat at which steel can be hardened, is unquestionably the best; and in reference to the opinion that if steel be overheated previous to immersion, an extra amount of heat is necessary to reduce it to the degree of hardness requisite to produce a good cutting edge, we repeat that the properties of which steel has been deprived by being overheated, cannot be restored by any degree of *temper* which may be imparted to it.

II. *The means of rapidly abstracting the heat.*

The process of hardening is necessarily completed with the rapid extraction of the heat by the employment of some cooling medium. Simple water at a temperature of 40 deg. Fahrenheit is recommended by Mr. Stodart, but water in various states, currents of cold air, flat metallic surfaces, immersion in oil or wax, freezing mixtures, and mercury, have been resorted to, with a success proportionate to the power of conducting heat respectively possessed by them. Mercury, it is generally conceded, imparts the greatest degree of hardness; salt or acidified water ranks next; simple water next, and lastly, oily substances. The superiority of mercury is doubtless owing not only to its strong conducting power, but to the fact that it produces no steam, which, where water is used, prevents the metal coming in close contact with the water. The superiority of salt water over pure water is supposed to be attributable to its greater density; but it has the disadvantage of imparting to the metal a tendency to rust. This, however, may be corrected by the addition of some neutralizing agent, lime water, for example.

Files are hardened by methods peculiar to the manufacturers, which have never been fully divulged. One process is said to be through a medium composed of water saturated with salt, with a small quantity of white arsenic added, and that an extraordinary degree of hardness is contributed by a limited proportion of the arsenic entering into combination with the steel. A celebrated file of Parisian manufacture, said to be of unrivalled hardness, is supposed by a writer in Gill's Technical Repository, "to receive its hardness from the following composition:

Mutton suet, *not rendered*, but chopped small, . .	2 lbs.
Hog's lard, ...	2 lbs.
White arsenic in powder, ...	2 oz.

These ingredients being put into an iron vessel, with a cover fitted to it, must be boiled until all moisture is driven off. It is advisable to perform this operation, as well as that of quenching any article in it, under the hood of a smith's forge, in order to avoid the noxious arsenical fumes which are disengaged. In using this compound, it must be first melted, and any thin article may be placed upon a red-hot plate of iron to receive the proper heat, and then immersed in the liquor."

Water that has been in constant use for the purpose for a number of years, is preferred by many workmen, provided it be free from grease. It is advisable, when the steel is hard, to take the chill off plain water, as a means of lessening the risk of cracking. A reduced yet sufficient degree of hardness is imparted to springs and other thin articles, and with less risk of cracking, by the use of oily mixtures, and a thin film of oil is spread over the water in some cases, as an expedient to reduce the suddenness and to assuage the effect of the transition from heat to cold. A contrivance was invented by a late engineer to the Bank of England, for the purpose of hardening the rollers used in transferring the impressions of bank-notes to steel plates. He caused a vessel to be made with a true and false bottom—the latter perforated with holes. A copious supply of water was admitted beneath the partition or false bottom, which ascended through the holes and escaped at the top, carrying off that portion which becomes heated by use. It is denominated, with reference to the principle upon which the water is admitted and escapes, the *natural spring*. Mr. Adam Eckfeldt employed a somewhat analogous process in hardening steel dies. In the upper part of the building, forty or fifty feet above the apartment in which the hardening process was performed, a vessel holding about three hogsheads of water was placed, from which the water was conducted through a pipe, and directed in a jet to the centre of the heated die by means of nozzles of different sizes.

Some mechanics are accustomed to use a thin layer of loam moistened with water, as a coating for light steel articles while in the process of hardening. It is applied by dipping the articles into the coating substance and gradually drying them before the fire; and it is said to prevent the scaling or oxidizing of the steel. Provided it be cooled with the requisite celerity, steel may be hardened when covered as well as when the surface is exposed, and in the use of the loam there is no difficulty, because the covering is removed the moment the article is plunged into water. And delicate articles are sometimes inclosed in a perfectly air-tight metal box, and subjected to the heat of the fire and the cold of the water without being brought in direct contact with either.

The adjoining parts of a piece of machinery are required, in some cases, to be hard and soft respectively; and lampblack, from its slow conducting power, has been found a useful agent, during the process of hardening one portion, in protecting the other from the absorption of heat. The Engineers' and Mechanics' Assistant gives the following example: "Suppose we wish to harden the neck of a lathe mandrel, but not the outside screw, lest it should break when roughly used; this may be done by means of an iron tube, fitted a little way on the neck of the mandrel, and ramming the intervening space between the screw and the inner surface of the tube full of lampblack, so as completely to envelop the screw; the orifice of the iron tube or case must then be closed with a metal plug; the mandrel being hardened in the usual way, the exposed parts will be hard, and the protected portions will be soft."

Holtzapffel's Mechanical Manipulation remarks (vol. i. p. 244) upon the difficulties, dangers, and uncertainties which attended the process of hardening, arising principally from the suddenness of the change from heat to cold, which sometimes causes the work to crack, and often to twist or become otherwise distorted. The danger of the former calamity, it is remarked, "is much increased with thick massive pieces, which appear to be hardened in layers, or, at all events, the outer crust may be perfectly hard, while the inner portions are less and less hard, as they approach the centre, which in many instances is found to be so soft as to yield easily to the file. The most rational explanation of this fact is, that the steel under the influence of heat becomes expanded, but when plunged in water, the outer crust is made to contract from the abstraction of the heat; but this loss of heat goes on less rapidly at the centre than at the exterior parts, whence it may be assumed that the inner bulk continues to contract after the outer case is fixed, and thus tends to tear the two asunder, the more readily if there happen to be any flaw or defect in the steel itself. Nor is it by any means unusual to find an external flake shell off in hardening; and it sometimes happens, that work which is apparently sound when removed from the water, will eventually give way, perhaps after the lapse of some hours, to the unequal tension of the hardening process, and crack with a loud report. As regards both cracks and distortions, their avoidance depends, generally speaking, upon the *manipulation*, or the successful management of each step; first, the original manufacture of the steel, its being forged and wrought so that it may be equally condensed on all sides with the hammer; otherwise, when the cohesion of the matter is lessened, from its becoming red-hot, it recovers in part from any unequal state of density in which it may have been placed.

It must not be forgotten, we may add, that steel is in its weakest condition, and consequently, liable to injury from two causes, when at a red-heat; namely, from becoming cooled irregularly, and from careless handling. The safest practice, it is therefore considered, is, to plunge articles vertically, as preferable to immersion obliquely or sideways in the water; in the former case all parts are similarly exposed to equal circumstances, and there is less risk of their disturbance. But at the same time, different treatment is required by some articles—for instance, swords, and implements of similar form—which it is found preferable to dip exactly as in making a downward stroke with a sabre; but this point does not properly belong to the present inquiry.

We conclude this branch of our subject, with an extract from the Engineers' Assistant. "*In all cases*, the thick, unequal scale left from the forge should be removed. The vicious and most defective process of hardening direct from the anvil cannot be too strongly deprecated, since it must, we apprehend, be sufficiently obvious to any person who will take the trouble to disabuse his mind of an old custom which has nothing but antiquity to recommend it, that steel hardened with the scale adhering to its surface must naturally receive a very imperfect impression either of heat or cold, because this scale, which is produced by the act of forging, varies in thickness according to the degree of heat employed in forging; it is also a bad conductor of heat; consequently, the sudden transition from heat to cold, on which the success of the operation depends, is not uniform in respect to time, the results of which are shown by one piece of steel being hard, while another is scarcely hardened at all, owing to the resistance presented by the scale to the cold temperature of the water. Hence, when such articles are *drawn down*, scarcely any two out of a large number will be found alike in temper, a fact which can scarcely have escaped the observation of any one in the habit of using edge-tools. As a means of avoiding this slovenly and very imperfect mode of hardening, all that is requisite is, to pass the metal from the anvil to the grindstone, a slight application of which will remove the whole of the scale or coating, and render the steel fit for the operation of hardening, with an accuracy which cannot otherwise be obtained; for it is undoubted that steel thus cleansed, heats in the fire with great regularity, and is equally sensible to the immediate action of the water on its body, when immersed so as to receive a similar degree of hardness from one extremity to the other."

III. *Tempering or letting down, by means of a partial or reduced heat.*

There are two modes of hardening steel, in common practice; but before describing them it may be remarked, generally, that it is a fact, singular as it may appear, that the means usually employed in hardening steel, should be applied in the process of softening. We have never seen a satisfactory solution of the problem, why a certain temper is imparted through a certain temperature. The tempering or reduced hardness is given by exposing the metal to a particular temperature, which, for the reason named, cannot be determined by any well-defined principle, but which must be chosen and tested by the workman from such indications as his own experience may have pointed out. One mode of tempering is termed *blazing*, from the use of oil or tallow, which takes fire. The article is first smeared with one of these substances, and then held over the fire until a thick smoke rises, which is supposed to indicate a temper equivalent to

450 deg. Fahrenheit: when the oil or tallow takes fire, a still lower temper is given. This method of drawing down or lowering the temper is used exclusively in some branches of hardware manufacture, and finds a very general application; but from the fact that no accurate means is afforded by it for estimating the effect produced on hard steel of different and varying quality, by certain degrees of heat, it is regarded as very defective and indefinite. Another and convenient mode, (based upon the same general principle,) of tempering a number of articles, and heating them uniformly, even if irregular in their shape, all at the same time, is, by placing them, covered with oil or tallow, in a vessel of proper shape and dimensions. This vessel being put over the fire, the following are the indications for determining the temper: a similar temper to that called a straw-color, is denoted when the oil or tallow first begins to smoke; it will equal a golden brown, when smoke of a darker color becomes more abundant; a temper equivalent to a purple is shown when the tallow throws off a heavy black smoke; a blue is indicated when the oil or tallow becomes so hot that it will take fire on a candle or lamp being brought in conjunction with it, and yet is not hot enough to continue to burn after the lamp is withdrawn; and the temper generally selected by clockmakers for their work, is imparted if the tallow be allowed to burn dry, or entirely burn away. A trifling degree of heat beyond this point, which is equivalent to the lowest grade of red heat, would just be visible in a dark place, consequently a further supply of oil or tallow would have no beneficial effect.

That the color produced on the surface of steel, says Brande, is the effect of oxidation, is proved from the circumstance, that when steel is heated and suffered to cool under mercury or oil, none of the colors appear, nor do they when it is heated in nitrogen or oxygen; therefore it is, that the second method we have to present under this head, has reference to the color or film of oxide, which is developed by the application of heat, on the bright surface of hardened steel. By this criterion, the workman is enabled to judge, with a near approach to correctness, of the intermediate grades of steel, between the two degrees of hard and soft; and for all practical purposes it is regarded as the most direct and simplest. But if the nicest accuracy is required, resort must be had to the bath, which may be of mercury or of any fluid whose boiling point is not less than 600 deg., and the thermometer. The articles to be tempered must be put together into the bath, with the bulb of a thermometer graduated to near the boiling point of mercury. The temperature at which the different colors make their appearance upon hardened steel, has been the subject of many experiments; we subjoin the results arrived at, some years ago, by Mr. Stodart.

1.	Very pale straw yellow,	430 deg. Fahr.	Tools for metal.
2.	A shade of darker yellow,	450 "	
3.	Darker straw yellow,	470 "	Tools for wood screw-taps, &c.
4.	Still darker straw yellow,	490 "	
5.	A brown yellow,	500 "	Hatchets, chipping-chisels, and other percussive tools, saws, etc.
6.	A yellow, slightly tinged with purple,	520 "	
7.	A light purple,	530 "	
8.	A dark purple,	550 "	Springs.
9.	Dark blue,	570 "	
10.	Paler blue,	590 "	Too soft for the above purposes.
11.	Still paler blue,	610 "	
12.	Still paler blue, with a tinge of green,	630 "	

At 430 deg. the tint is so faint that it cannot be discerned, except by comparison with a piece of polished steel, that has not been tempered. The French nomenclature of colors differs in some respects from Mr. Stodart's arrangement. The following shades are specified in the "*Dictionnaire Technologique*:" first, yellow, which also has three distinct shades—straw-color, canary-color, and lemon-color; second, orange; third, red; fourth, violet; fifth, blue; sixth, grey. When steel arrives at this last color, however little the heat may be increased, it returns to its first state; that is, as it was before being hardened and tempered. It must be remembered, too, that the tempering colors differ slightly with the various colors of steel, and also that the intermediate shades produced on steel by heat between 500 deg. and 580 deg., are frequently exhibited irregularly on different parts of the surface. "As I had," observed Mr. Nicholson, "noticed this irregularity, particularly on the surface of a razor of wootz, and had found in my own experience that the colors on different kinds of steel do not correspond with like degrees of temper, and probably of temperature; and in their production, I was desirous that some experiments might be made upon it by Mr. Stodart. Four beautiful polished blades were therefore exposed to heat in a bath of fusible metal. The first was taken up when it had acquired the fine yellow, or uniform deep straw-color; the second remained till the part nearest the stem had become purplish, at which period a number of small round spots of a purplish color appeared in the clear yellow of the blade. The third was left till the thicker part of the blade was of a deep ruddy purple, but the concave face still continued yellow. This also acquired spots like the other, and a slight cloudiness. These three blades were of cast-steel; the fourth, which was made out of a piece called styran steel, was left in the bath till the red tinge had pervaded almost the whole of its concave face. Two or three spots appeared upon this blade, but the greater part of its surface was variegated with blue clouds, disposed in such a manner as to produce those waving lines, which in Damascus steel are called the water. Two results are more immediately suggested by these facts: first, that the irregular production of a deep color upon the surface of brightened steel, may serve to indicate a want of uniformity in its composition; and secondly, that the deep color, being observed to come on first at the thickest part, Mr. Stodart was disposed to think its more speedy appearance was owing to those parts not having been hardened."

It follows, therefore, it being thus known at what degrees of heat the several colors are produced, that the only remaining requisite is, that the bath be heated to the proper point; the heat being equally applied, the workman has but to note the indications of the thermometer, without attending to the colors. The advantages of this process are obvious; although according to Brande's Manual, it is highly probable that, for many uses, steel may be sufficiently tempered in a range so extensive as 212 deg. to 430 deg.; and it is quite true, that by some comparatively recent experiments, it is proved that steel, for certain uses, is sufficiently tempered long before it is heated to produce any change of color, thus promising to give additional value to this process by a thermometer. A knife edge attached to the pendulum described by Captain Katen, in Philadelphia Transactions, 1818, p. 38, was forged by Mr. Stodart from a piece of fine wootz. It was carefully hardened and tempered in a bath at 430°. On trial, it was found too soft. It was a second time hardened and heated to 212°. The intention was to increase the heat from that point, trying the temper at the advance of about every ten degrees. In the present instance this was not necessary. The heat of boiling water proving to be the exact point at which the knife edge was admirably tempered.

A process introduced by Mr. David Hartley must take precedence of the above in point of time and was undoubtedly an important step—perhaps the first—towards improvement, whether as regards dispatch or precision, on the basis of color. This process was, to prepare a bath of oil heated to a regular temperature, measured by a thermometer, and immerse the articles of hardened steel therein. Respecting this process, the Mechanics' Assistant remarks, that it is "evident that a bath of any of the soft metals, whose fusible points are just above that required for tempering, may be used instead of oil, and alloys of those metals might be so proportioned as to obtain points of fusion at the exact degree of heat required. In these cases, however, to guard against oxidation, it would be advisable to keep the fluid metal covered with grease. Oil is preferable to the fusible mixture for several reasons. It is cheaper, and admits of work being seen during the immersion, by reason of its transparency."

Approved Recipes for Hardening Steel.

We give herewith a few additional examples for the hardening of steel, which are practised at some of the largest iron works, for the purpose of refining the steel for tools, to be used upon the very hardest cast-iron—a purpose for which the best English steel, hardened in the usual manner, is not suitable. Those examples have been successfully applied to English or Huntmen's steel.

NO. I.

For refining and hardening of English steel of superior quality, take

Pitch,	2 lbs.
Whale-oil,	1 lb.
Tallow,	½ lb.

The pitch is to be pulverized, sifted, and slowly heated. The tallow must be molten so that the mixture will form a semi-liquid, in which the steel is hardened in the usual manner. If besides this, the steel is subjected to a second hardening, in No. III., the hardening becomes still more perfect.

NO. II.

Tools for turning or chipping the hardest kind of cast-iron, such as rollers for rolling-mills; or for chipping hard iron. Take

Sal-ammoniac,	1 oz.
Borax,	½ lb.
Water,	2½ pts.

The sal-ammoniac and borax must be pulverized and mixed with the water. Then the steel is hardened in this mixture, which is suitable for English steel.

NO. III.

Tools for turning lathes or other purposes, which were hardened in mixture No. I., and are hardened a second time in this preparation, receive still greater hardness. Take

Sal-ammoniac,	1 lb.
Tartaric acid,	⅓ lb.
Water,	4 qts.
Red wine,	1 pt.

The sal-ammoniac and tartaric acid must be pulverized and mixed with the water and wine. The steel is hardened in this in the usual manner. This mixture is particularly good for boring and turning tools.

NO. IV.

For hardening turning-tools, for turning hard rollers for rolling-mills, on which steel hardened in cold water would not cut; take

Tallow suet,	1 lb.
Black rosin,	¾ lb.
Sal-ammoniac,	¼ lb.
Prussiate of potash	¼ lb.

The prussiate of potash and sal-ammoniac must be pulverized, and then mixed with the tallow suet and black rosin, which is previously slowly melted on a fire—the whole forming a semi-liquid, and the steel hardened as usual.

The above recipes may be prepared in larger or smaller quantities, but the proportions of the ingredients must be the same.

V.

Preparation to restore overheated steel to the same quality as before, and at the same time, to make its hardness of long duration. Take

Tallow,	1 lb.
Sal-ammoniac,	¾ lb.
Prussiate of potash	¼ lb.
Black rosin,	½ lb.
Pepper,	1 oz.
Shaving powder,	1 oz.

The tallow and black rosin are molten in a stone or earthen vessel till they form a liquid; the other ingredients are pulverized and mixed with the liquid. Heating the steel twice or thrice and cooling it in this preparation, will give to it a superior and durable hardness.

This preparation is recommended for all kinds of steel.

Recipes for Welding Cast-steel with Wrought Iron.

Borax,	2 oz.
Sal-ammoniac,	⅓ oz.
Spirits of wine	90 drops.

Boil them together in a close iron vessel until the composition is red-hot; then pour the mixture out so as to cool quickly, and pulverize it. The parts to be welded are then to be well dressed and heated to a red heat; sprinkled with this powder and subjected to a strong red heat, by heavy blows they can be welded together. Instruments of pieces of wrought iron, which, on account of their scant dimensions, cannot stand a regular welding heat, may be welded in the same way.

Consequently, to harden the steel to make it tough, take

White arsenic,	1 oz.
Litmus (or azure stone) blue stone,	1 oz.
Vinegar,	2 qts.

Mix them well together, and keep it in a covered vessel. The cast-steel heated as usual, is cooled in this mixture. Pieces which in cooling are apt to warp, are, after being made slightly red-hot, sprinkled with tartar, then heated again, and cooled in the above mixture.

Cement for connecting Cast-iron Pipes or Tubes, which will remain perfectly Air-tight.

Cast-iron filings, or turnings, (dust,)	10 lbs.
Pulverized sal-ammoniac,	1 lb.
Pulverized sulphur,	½ lb.

Mix them most thoroughly; take of the whole a quantity sufficient for cementing two or three joints in a separate vessel, and add water slowly while continually stirring it with a wooden spoon, the mixture thus will become heated and begins to make steam. Within half an hour from this the filings, sal-ammoniac, and sulphur, will be completely dissolved, and the mixture appear like a fine dough. In this state the mixture is applied to both parts of the joints, about half an inch thick, and the joints then screwed together. Twenty-four hours after this the cement is so hard as to be water and air tight, and when heated red-hot, will be as substantial as the cast-iron itself, and last equally as well as cast-iron constantly exposed to a red heat.

All the recipes for hardening steel, and others given, have been practically tested and proved to be of great importance for the practical machinist, as it will save him not only time in fixing his tools, but he can perform more and better work. To make steel of a very superior hardness, and at the same time very tough, is, it is perhaps needless to say, a most valuable acquisition to any workman, and, therefore, we have no hesitation in demanding more attention to this subject than has hitherto been paid to it.

SINGLE OSCILLATING ENGINE

OF THE STEAMSHIP KNOXVILLE,

BUILT BY STILLMAN, ALLEN AND COMPANY,

At the Novelty Iron Works, New York.

THE Steamship Knoxville was built for Samuel Mitchell's Savannah and New York Steamship Line. Her engine and boilers were built at the Novelty Works, Stillman, Allen & Co., proprietors; the hull by W. H. Webb.

The following are her dimensions:	Feet.	Inches.
Length on deck,	220	8
Breadth of beam,	35	4
Depth of hold,	21	1
Tonnage,	tons 1262	
Average draft of water,	13	0
She is provided with an oscillating engine, of the following dimensions:		
Diameter of cylinder,	7	1
Length of stroke,	8	0
Diameter of air-pumps,	3	10
Length of stroke,	3	8
Diameter of force-pumps,	0	11
Length of stroke,	2	5
Diameter of paddle wheels,	32	0
Length of paddles,	9	10
Depth of paddles,	1	8
Number of paddles in each wheel	28	
Average dip of wheel,	6	6
" Number of revolutions per minute,	15	
" Pressure of steam in steam boilers, per square inch,	lbs. 25	
Cutting off at	4	6

She is provided with two iron boilers, each 30 feet long, and 11 feet 6 inches in diameter.

Whole amount of fire surface in each boiler,	2816 square feet.
" " Flue surface,	2000 "
" " Grate "	89 11 "
Capacity of cylinder,	367 cubic feet.

Ratio of fire surface to a cubic foot of cylinder,	$15\frac{1}{3}$ to 1.
" " " Grate surface, .	$63\frac{1}{15}$ to 1.
Area of upper (or) return flues for one boiler,	1995 inches.
" Lower flues, for one boiler, . .	2288 inches.
" Chimney, ...	4266 square inches.
Consumption of coal per hour, . .	$1\frac{3}{8}$ tons.
Water evaporated by 1 lb. of coal, . .	$6\frac{1}{4}$ lbs.
Coal per hour to a square foot of grate, .	$40\frac{1}{4}$ lbs.
Cubic feet in steam boilers to 1 cubic foot of cylinder,	18 to 1.

GENERAL DESCRIPTION.

The *Knoxville*, for which the engine represented in the accompanying engraving was constructed, is a very superior vessel, of great speed. She has made the run between New York and Savannah in fifty-four hours, which is the shortest time recorded, and this, with the regularity of her trips, proves her to be a steamship of no ordinary qualities.

Her speed may not only be attributed to the superior design and workmanship of her machinery, which is visible in all its parts, but to many valuable improvements which have attracted our attention, and which will be found represented in our large, correct, and detailed engravings, showing minutely, the nicest mechanical combinations and arrangements applicable to marine engines. Prominent among these is the arrangement by which the steam and exhaust chests on the cylinder A, are separated from each other, as shown in Plate V., which represents a longitudinal view of the engine, and a sectional view of the boilers. The steam and exhaust chests $A^1 A^1$ and $A^2 A^2$, are cast solid to the cylinder A. It will be seen that the exhaust chests are constructed considerably larger than usual in proportion to the steam chests. This, it must be perceptible by all who are concerned in the application of steam, is a great advantage, from two considerations. First, it overcomes the difficulty which arises from the partial condensation of the steam which passes through the steam chest into the cylinder; and second, by further increasing the surface of the exhaust valves and chests, it facilitates a quick exhaust of the steam which passes to the condenser P.

The arrangement of the valve motion is a very ingenious and simple piece of mechanism, which performs all the movements with great perfection. Connected with it are such appliances, that the engineer, by simply turning a screw, may cut off the steam in the cylinder at any point in the stroke, while the engine is in motion. It is called the variable cut off motion.

The steam cylinder is provided with safety valves, both in the cover and bottom, the construction and intent of which will be fully explained below.

We take a special pleasure in presenting to our readers the full drawings and details of this engine. It is one of the latest, and is generally acknowledged among engineers, as one of the best specimens of American marine engineering. Without any unusual ornamental portions, it is in the highest degree neat and elegant in design, and has proved itself capable of the very highest degree of efficiency. The method of working the air-pump by means of a lever suspended overhead, and connected by a link to the head of the piston rod, is free from many of the objections urged against many other methods of operating this important adjunct. Beam engines present great facilities for procuring every possible variety of reciprocal motion for pumps of less stroke than the piston, as the various feed-pumps, bilge-pumps, and air-pumps are very conveniently attached to the beam at any point from the centre, outward, and each, consequently, is moved with a length of stroke proportional to such position. It is common to give the air-pump about one-half the throw of the main steam piston, an end which is accomplished in this engine by constructing the light but strongly trussed lever represented, connecting it at one extremity directly to the crank, or rather to the head of the piston rod, and attaching the other extremity to a fixed point, while the air-pump rods are secured at or near the centre. The common methods of working air-pumps when not analogous to this, may be generally comprised under one of three classes,—either the motion is derived from the oscillating of the cylinder, or from a crank, or large eccentric in the main water wheel shaft, (or in the centre shaft when double engines are employed;) or the air-pump is worked by a separate engine independently.

The framing of the Knoxville has been very justly admired. In comparison with the mammoth castings employed in the framing of English engines, the frames of all American marine engines are extremely light, strong, and graceful. In nearly every instance the pillow-blocks are of cast-iron, and the supports thereof are stout wrought-iron rods, nicely finished. The pillow-blocks of the Knoxville employ but four such supports, two on each side, inclined at a considerable angle as represented, so that the wrenching and diagonal strains thrown on the pillow-blocks in the powerful revolution of the crank, are all received in the most direct and easy manner, by these substantial fixtures, and thus transmitted directly to the bed-plate. The boilers are at just a sufficient distance from the engine to prevent crowding, and in all the details, whether large or small, a high degree of skill and of scientific proportioning is exhibited. Many engineers have been particularly anxious to secure drawings of this particular engine, and we shall endeavor, both in the engravings and description, to give the fullest and clearest exposition of her construction, which is consistent with the limits of our work.

DETAILED DESCRIPTION.

Plate V.

A is the main steam cylinder. A^0 is the water-wheel shaft, by which the wheels are revolved. A^1 A^1 are the upper and lower steam chests, enclosing the balance puppet valves, through which, at the proper time, steam is admitted, or shut off from the cylinder. A^2A^2 are upper and lower exhaust chests, in which are enclosed the balance puppet exhaust valves, which, when open, allow the steam to escape into the condenser *P*.

B is the steam pipe through which the fresh steam is received from the boiler. B^2 simply represents prolongations of the same. The passage of steam through this is regulated by the stopcock or screw valve *B* and B^1.

C is the pipe through which the steam escapes to the condenser.

C^1 is a broad flange on the end of the exhaust pipe, to allow it to be bolted to the pillow-block at the points represented.

C^2 is simply a kind of pedestal to support the weight of the exhaust pipe. It is bolted upon the top of the bed-plate *D*.

D is the bed-plate, cast with deep flanges at each side, to support it as rigidly as possible.

D^1 D^1 are the trunnion pillow-blocks cast on *D*.

D^2 is the cot or binder which surmounts the trunnion pillow-block, and receives, through the aid of the bolts and nuts represented, all the upward strain which is at every revolution powerfully thrown on the trunnion.

E E are the pillow-blocks, extended as represented, which allow of securing by bolts to the cross braces E^3 E^4, which enable these heavily strained castings to brace and support each other, and also to be connected through these by bolts, as represented, to the deck-beams E^5.

E^1 E^1 are caps or binders which hold down the main shaft by the aid of the stout bolts and nuts E^2.

F F are the inclined wrought-iron rods, which receive alternately the tensile and crushing strains due to the action of the steam on the piston. When the piston is on its upward motion, with the full pressure of the steam below, it lifts with immense force on the pillow-blocks *E*, and this is a force which must be resisted entirely by the rods *F*. When the action is reversed and the piston commences to descend, an equally powerful downward force is imparted to *E*, which tends to crush *F*, and the highly inclined positions of the piston rod when the engine is nearly at half stroke, gives an additional straining tendency to this force, by causing it to act in oblique directions. The rods *F* are inclined toward each other, the better to resist the compound action of these varied forces; and except for the facility of removing the cylinder cover, would be inclined so as to stand in lines ranging directly toward the centre of the water-wheel shaft A^0. The sockets in which these rods are fastened, both in the bed-plate *D* and the pillow-blocks *E*, are very nicely fitted, and the parts are secured therein by stout keys F^1 F^1. These keys are capable of resisting forces in both directions, but are set to draw the collars on *F* tightly into contact with the castings. The method of inserting these mammoth rods in such diagonal positions is somewhat peculiar, and consists in dropping one through the bed-plate, for which purpose it must be constructed without a collar on its lower extremity, and the keelsons must be correspondingly arranged at that point, to allow of its being lowered several feet through the bed-plate. The other rod *F* may be fixed rigidly in position, and there is then no difficulty in lowering the pillow-block *E* into its place on the other rod, and securing it firmly, after which the rod previously dropped may be hoisted in its proper position, and firmly secured. The necessity for the two key seats F^1 and D^3, in the bed-plate is thus explained: one being set with its draught inclined to force the rod *F* downward, and the other acting in a contrary direction, to supply the rigidity which would be afforded by the collar represented, were such present.

D^5 is a brace or tie which passes transversely across, above the bed-plate, and aids in stiffening and supporting this portion.

G is the head of the piston rod, which embraces the crank pin. The construction of this portion will be more fully explained in the succeeding plates.

H is a single link, H^1 being the upper centre on the beam *I*, and H^2 the lower centre, which is carried in lugs on the top of *G*. H^2 consequently describes the motion of the crank at each revolution, and gives *I* a reciprocating or slightly circular curved linear motion in the dotted line represented, which is described from the fixed centre I^0.

I is the air-pump lever, trussed by the addition of the stout cast-iron strut I^2 and the strap I^1, passing under the strut, and secured to both extremities, as represented.

$K\,K$ are stout uprights which support the fixed extremity of I. They stand on the deck-beam E^5, which is connected by the casting K^1 to the lower deck-beam E^6, and this latter, in turn, is connected to the keelsons by the upright rod K^2, so that the whole forms in effect a continuous column leading from the bottom of the ship, to support the centre I^0.

$L\,L$ are diagonal braces extending from the pillow-block to aid K in resisting lateral fore-and-aft strains.

M is the single air-pump link which descends from the centre M^1 in the air-pump lever, through the top of the hot well, and is connected in the ordinary manner to the air-pump rod within.

N is the feed-pump rod, which descends from the point N^1 on the air-pump lever, and which works the feed-pump O, which, with the arrangement of valves in the box O^1, will be more fully represented in the succeeding plate. The centre N^1 in the air-pump lever is supported by the slight braces I^3 I^3.

P is the condenser, a casting of sufficient strength to resist the powerful external pressure of the atmosphere, and to support all the strain of the air-pump, hot well, etc. It is firmly bolted to the keelsons, as represented. The condenser receives through the large pipe C all the steam which is worked in the cylinder A; but this is all condensed by the constant influx of large quantities of cold water. The office of the air-pump is to maintain this vacuum, by continually pumping out the hot water resulting, and which would otherwise flood the condenser. The term *air-pump* is derived not so much from its occasionally pumping small quantities of air which are mingled with the steam, as from its being obliged to pump out the water in opposition to the pressure of the exterior air; a work which requires considerable force. The arrangement of the air-pump will be more fully represented in the details.

Q is the hot well, a simple reservoir in which the water pumped from the condenser P is received. From Q a portion is pumped by O into the boilers, but the principal portion is discharged through Q^1, a large pipe which leads directly from Q through the side of the vessel to the external atmosphere or the ocean without.

We assume in this description, that our readers are all familiar with the general principles of the steam-engine, and with the nature and effects of the various parts; but as a large number, especially in the interior and western portion of the country, are little accustomed to low pressure or condensing engines, we are particular in the description of every important portion. The parts now described, with the addition of the crank, and a few other portions not distinctly seen in this drawing, constitute the principal features of the engine. The parts remaining to be described may be considered simply as auxiliary to, or aiding in the action and control of these parts.

R is a hand lever socketed, as represented in the lever R^1, which is hung to a fixed point in the framing.

R^2 is a rod or link which connects R^1 to the extremities of the levers R^3 and R^4, on the side of the cylinder. To the extremity R^0 of the lever R, may be communicated at will, by the hand of the engineer, a reciprocating vertical motion, which motion is communicated through the whole train. The motion of R^3 is imparted to the exhaust valves: that of R^4 to the steam valves, so that R is the starting bar of the engine.

S is an eccentric fixed on the water-wheel shaft A^0, and communicating its motion through the eccentric rod S^1, which is braced by the side braces S^2 S^2, to the lever S^5.

T is a hand lever which works the unhooking gear. It is located very convenient to the engineer.

T^1 is a rod operated by T. It leads down to the lever

T^2, which latter has a curved slot to allow the corresponding pin on the short bent lever

T^3, to play back and forward as the cylinder oscillates, without giving any motion thereto. When T is depressed by the hand, T^1 rises, and raises T^2, which turns the bent lever T^3, and thus lifts the eccentric rod and unhooks it.

U is a nicely revolved and convenient hand wheel, directly in front of the engineer. This is capable of being revolved with great ease, in either direction, but may be set in any position by tightening the small hand-wheel and nut U^1 immediately above it.

U^2 is the hollow stand on which U is mounted.

U^3 is the shaft on which U is keyed. U^3 leads down through the engine-room to the point U^4, where it carries a pinion which gears into a larger wheel, by which is operated the screw-valve or stop-cock U^5. This stop-cock controls the communication from the exterior of the ship to the interior of the condenser. The powerful suction (as it is sometimes termed) on the interior of P draws the cold water with great violence through the passage thus controlled. The engineer by turning U controls the quantity thus admitted, carefully shutting it off, or diminishing the quantity, when the engine is slowed, and opening it to condense the greater volumes of steam when working at full speed.

V is the boiler, a very common and favorite form of this important feature for low pressure purposes. It is termed a return flue or ascending flue boiler.

V^4 is the furnace, entirely enclosed within the shell. The dotted line represents the grate on which the fuel is laid.

V^1 V^1 are flues leading the intensely hot products of combustion through the water to the back connection V^2.

V^3 V^3 are return flues in which the same is discharged into the uptake or stack V^6.

V^5 is the steam chimney, an attachment on the top of the boiler in which the steam is dried and delivered in a pure state, free from particles of water, through the pipe B.

W is the blow-off valve which regulates the escape of dirty and concentrated brine from the boiler through the pipe W^1, into the sea.

X X^1 X^2 form the bilge-pump and connections, through which the water is raised from the hold, and discharged overboard, by the aid of the pump-rod X^3 worked by hand on the main deck, at X^4.

Z is the feed-pipe. It leads from the feed-pump O to the check-valve, imperfectly represented behind W. Through this is received the water to be evaporated in the boiler.

Before proceeding to describe the succeeding plates, we will introduce a few suggestions for the benefit of those not familiar with the subject, in regard to how drawings should be studied. Every separate piece of an engine or large machine is a separate study to the engineer, and every modification of the form or proportion is, or should be, made with a view to favor some particular end. The learner should ask himself the effect of every angle, and every convolution. He should inquire what would be the effect, were each individual part made longer or thicker; and although, in many instances, it will be found that little difference would be produced by slight changes, in others, very interesting facts may thus be developed. The answer, in a large number of cases, with regard to the effect of increasing the size or strength, will be simply that it involves a necessary increase of expense; but there are cases, very numerous in the aggregate, where lightness is an almost indispensable requisite; and, in general, all machinery is proportioned in such manner, in regard both to economy and efficiency, as to secure the greatest possible amount of strength, with the least material. This problem, which can be enunciated in very few words, is the great general object in view in designing machinery. Drawings, and especially large and strictly correct drawings like these, may be studied with as much profit as reading matter. And in many respects, the instruction is far more positive, practical, and both directly and indirectly advantageous. The best method of studying under different circumstances may be various.

The most full and perfect means of studying any drawing is to copy it. Let the learner take off the dimensions singly, and transfer them to another sheet, carefully inquiring, at each step, the reasons therefor. One unaccustomed to drawing machinery, or only accustomed to do so from the instruction of others, and without any training of the mental faculties in this direction, may find it very difficult to answer these questions; and, in fact, all will find many points in machinery in which the answer to this question must be waived at the moment of its suggestion; but it will probably be developed in the course of the succeeding studies. Having finished the tracing of the outlines, they may be inked in, and the parts all shaded, more or less fully, with India ink or colors.

The time consumed by this process is too great to justify most persons in commencing it. The second method is to trace the drawing in full, on tracing paper, or on transparent muslin. Having stretched the plate on a suitable board or table, lay the transparent material over it, and confine the whole with suitable pins or tacks. In the absence of these latter, proper weights or any other convenient means may be employed as fastenings. Thus prepared, the learner may commence to trace, either first with pencil, and afterwards with ink—or with ink in the first instance, every line found in the plate. Care should be taken in both instances to shade, or make heavier, those lines which are on the under and right side of every object. This, although termed shading, is not strictly so, but serves greatly to relieve the heaviness of a simple outline drawing, and to make very clear every part thereof. It causes the object drawn to, as it were, stand out from the paper, in relief.

The third method, and one which can be practiced with but a comparatively small expenditure of time, and one of great profit to almost every class of readers, is to simply trace the lines on the plate, direct, with a quill or splinter. Spreading the plate on a suitable table or board, without any fastening, travel slowly and carefully over every line in the drawings, in the same manner as when tracing as directed in the last paragraph. It may seem an idle process, but a few experiments will develop its efficiency. Hundreds of points, not before observed, will be discovered by this means.

Scores of questions will arise from time to time, in either of these methods of study, which, as before suggested, can only be answered at a later period. The first method brings out these with the most distinctness. The last named method is liable, if carelessly conducted, to become a merely mechanical operation—one in which the hand traces every line without reference to what it is, or why it is so drawn. This must be carefully guarded against.

A fourth method, and one which is equally efficient and probably more agreeable than the last, is to tint the plate. There are colors which have come to be recognized as conventional means of representing different materials. India ink, or China ink as it is sometimes termed, prepared from lamp black and gum, is employed to represent wrought iron; neutral tint, a mixture of black, blue and red, is used for cast iron; indigo blue, for steel; gamboge, for brass; burnt sienna, for copper; yellow ochre, for wood; etc. These tints should all be used very much diluted, and unless the operator is accustomed to tinting, he would do well to practice, for some time, on other pieces of paper, before commencing on the plate. Remember, that if too faint a tint is given, the dose can be repeated. Blue is frequently used for all forms of iron and steel, and black, in different shades, is frequently used for all substances. To some eyes, drawings simply and neatly shaded in black, look better than in tints; but for engineering purposes, tints are preferable. In either case, the exercise of laying them on will prove a very wholesome one.

It is important to have good sets of instruments, but these are not as essential as many suppose; a very few will serve if they are kept in good order. Whatman's is esteemed the best drawing paper. India ink is valuable in proportion as it dissolves easily and does not settle to the bottom of the water in standing. Lead pencils should be black and firm, and free from both grit and gum. Faber's have the best reputation, but we use and prefer Rehbach's for all purposes. No. 1 is softest, and No. 4 the hardest; we recommend Nos. 3 and 4 for general engineering drawings.*

While on this subject, we may be excused for digressing somewhat, to suggest the importance of study to the young mechanic, as also to advance a few hints with regard to how it should be conducted.

The late lamented Hugh Miller was for a long time engaged in a quite sharp discussion, with regard to the relative superiority of self-made, as compared with liberally-educated men. We do not recollect which got the best of the argument, Mr. Miller—the student of quarries and nature—the hard-fisted working man—or the gentlemen of the satchel and gown; but in either case, the fact is undisputed that in the present age, with all the facilities afforded by the printing-press, and the general diffusion of science, it is possible and practicable for the unaided workman, who commences early, or even if rather tardy in the commencement of the enterprise, to compete successfully for many—perhaps most—positions of honor and profit in this country. Scores of examples might be adduced of men who have accomplished this; and the absolute and imperative necessity for study, even to maintain one's position with credit, is every year becoming more obvious. A century—yes, two-score years ago, our fathers acquired, by long and patient drilling, that degree of mechanical skill in the use of certain tools which was termed "learning a trade;" and properly relying on the trade as the best capital with which life could be entered upon, they pushed boldly forth into the mingling currents of the world. But labor-saving machinery has introduced serious innovations on this mode of doing business. Every few years, or oftener, some tool is manufactured, or machine invented, which gradually or rapidly supersedes hand labor, and compels a large class of mechanics either to lower their wages below a respectable living point, or change to some more lucrative occupation. It has, therefore, become necessary in preparing to enter "the world's broad field of battle," that more varied acquirements be made, and that the judgment and business capacity, as well as the intellectual faculties generally, be developed and matured. If machinery is introduced to supersede hand labor, some persons must construct and attend the machines, and every new modification which practical science assumes, calls for a set of new hands—in fact, more frequently than it dispenses with the help of old. We will assume, without further preface, that proper culture and storing of the mind is desirable. The next questions are—What is proper culture? and how can it be obtained?

We know, from practical experience, some of the difficulties with which young workmen and apprentices are surrounded, even in this free and enlightened country. There is often, perhaps

* The manufacturer of these recently introduced pencils is "L. J. Rehbach, in Regensburg." There are three qualities;—the ordinary, (which is intended for common use, and is equal to the best of other makers,) the second quality, for the use of merchants and mechanics, and the very finest, for artist draftsmen. Willy Wallach, 181 William street, New York City, is sole agent.

generally, a jealousy on the part of older workmen by their side; and even, and very stupidly, in their employers, to prevent the youthful aspirant from learning any more than what is esteemed his proper part. We have felt the inconvenience of attempting to study in the chamber of a boarding-house, and we know, to some extent, the disadvantages of learning from antiquated and incorrect books, by a very slow and tedious process, what was subsequently discovered to be not only useless, but positively erroneous. We are aware that men so situated have little opportunity to procure, or judgment to select books; and more than all, they have not sufficient inclination, except in very rare cases, to contend, alone, against the powerful influence of their surroundings. There are strong inducements in most localities to idleness if not to dissipation.

Without attempting now to dictate what branches should be pursued, and what books procured, we will simply add a few words on *how* to study. And first in practical importance, let, if possible, two or three unite, or rather, pursue a study cotemporaneously. Let them discuss the principal points during the intervals of labor; and, although it will generally follow, that one is so much in advance of his neighbor as to become a teacher rather than a fellow-student, yet the exercise of comparing notes, even in this manner, will be profitable to both. It will certainly benefit the indolent man, and will refresh the memory, and confirm the opinions of the more advanced. Do not attempt to unite more than three, as the chances of confusion and final abandonment of the enterprise are increased by the addition of each new member. But all this depends, after all, on the force of the student. There is a quality, so termed, almost or quite as important in real life as intellectual development. If you will look abroad among your acquaintances, particularly those filling, with credit, important and responsible situations, a close analysis will show you that they derive their best qualities as much from force as from sense. If you have force enough, when fully summoned, to propel alone, spend no time in soliciting companions. There are always those who will be pupils if no labor is required. Explain to such occasionally some prominent point in your newly-acquired knowledge. If it does not benefit the listener it is no fault of yours. The improvement of the lecturer is the main point, and you will frequently find much truth in the very obstinacy with which they will object to the propositions advanced.

Every text book is, or should be, adapted to a certain progress in the learner. There are generally several books on the same subject, some very simple, and others very deep. You will very likely—procuring your books by accident or chance—find points difficult to master. Overcome this by procuring, if possible, two or three different text books. Whether the subject be mathematics or chemistry, electricity or law, hydraulics or book-keeping, two or three authorities will settle every difficulty. The explanations of one writer will cover the "joints" left by the other. And even if one book is very old and musty, both covers eaten off by mice, and the whole stained as if fresh from the slop-pail, skim through what it has to say, after you have faithfully studied a section of the more modern one. Two books are sufficient—four are too many.

Morning is the best time for close deep thinking: the evening best for charging the memory. In the long days with warm mornings, arise at five, and study mathematics; practice drawing from supper till dusk; then spend the remaining hour in general social enjoyment. But if you wish simply to remember an array of terms, or acquire simple "word knowledge," a matter of great importance in chemistry, read over what you wish to fix in your memory at night, just before retiring. In short, go to work every morning full of new and strong *ideas;* retire at night with new *words* stamped in the memory. By following the suggestions we have advanced, without losing an hour's wages or diminishing your usefulness in the shop, you may progress about half as rapidly as you would were you at a popular school and devoting your whole time to study.

DETAILS OF THE STEAMSHIP KNOXVILLE.

Plate VI.

In lettering the drawings of this work, we have studied to, as far as possible, employ the same letters to indicate like parts in all the figures. The immense number has rendered this impracticable in some instances, but in the main we have preferred to attach numbers to the letter rather than to complex the subject and confuse the student by a different course.

Fig. 1 is a side view of the cylinder; Fig. 2 a top view; Fig. 3 an enlarged section of the lower portion, and Fig. 4 a front view of the same important casting. The casting is made with the bottom attached, or rather a large portion thereof, a circular opening, four feet in diameter, represented very distinctly in Fig. 2, being left to be subsequently covered.

A^1A^1 and A^2A^2, respectively the steam and exhaust valve chests, have been already briefly described, and will be further detailed below.

A^3 Fig. 4, is the cylinder bottom, a casting of sufficient size to stop the orifice left in the bottom of the cylinder, with stout flanges to allow its being secured thereon by bolts. The whole diameter is four feet seven inches. (This part is lettered A^6, by mistake, in Fig. 1.)

A^4 is an attachment to the bottom of the cylinder, to provide for the stout nut which helps to secure the piston on the piston-rod. The interior diameter of A^4, which is 15½ inches, is designed to be amply sufficient to allow the entrance of the nut at each stroke, without a liability of contact thereof with its sides.

A^6 is a crossbar, two views whereof are represented. The object of this will appear below.

A^7 is a housing bolted to the cylinder bottom, to contain a stout coiled spring, as represented. This spring allows the descent of A^6, if impelled by any sufficiently great force.

A^9 is a safety valve, which fulfils a very important office, and merits a tolerably full explanation. Steam is water expanded into a gaseous form by heat. But it almost invariably contains more or less fine particles of water in its ordinary form; and although every effort is made in steam engineering to procure dry steam, or steam as free as possible from such particles, it is sometimes the case that the boilers "prime," "foam," or "work water," very seriously, in other words, water comes out in large quantities with the steam. The precise nature of this action is not understood, as it is of course not subject to inspection under such circumstances as occur in practice. There are some reasons for supposing that the water, under the influence of some unknown law, ascends the side of the boiler, in a thin sheet, and thus flows out with the steam, from the steampipe; but there are, we think, stronger reasons for supposing it to rise in the form of thick spray, mingled intimately with the vapor. This theory derives much support from the fact that the simple deflecting plates of iron placed in the boiler in such a manner as to deflect in splashes and throw them against the side of the boiler, to descend by gravity, have, in some cases, almost or entirely remedied this evil. Some boilers are far more liable to work water than others; and the reason cannot always be satisfactorily assigned for this difference; but in general a large area of water surface in the boiler, or, in other words, a liberal provision for the escape or disengagement of the steam from the water, so that it does not rise therefrom in any considerable velocity, tends very greatly to prevent priming. The steam domes, added on the top of many high-pressure boilers, and the steam chimney, on low-pressure boilers, are both intended for the same purpose, i. e., to take the steam at a considerable elevation, so as to avoid the commotion of the water as far as possible. The steam chimney, however, in the last-named example, contains the heated smoke stack or up-take, which tends very considerably to dry the steam by evaporating all the particles of water which come in contact with it. But with all precautions, engines are always liable to receive a greater or less quantity of water, to which may be added an allowance for the quantities, sometimes very considerable, which are condensed by contact with the cold metal of the cylinder in commencing to work. As water is incompressible, except to a very small degree, and as the piston at each stroke comes into almost absolute contact with all parts of the cylinder end, it follows that a quantity of water sufficient to more than fill the small space remaining before the piston at the end of the stroke, must necessarily compel either a stoppage of the engine or a fracture of some portion of the machinery, unless means are provided for its escape. In ordinary small engines, with slide valves, a sufficient vent for such water is allowed by the ability of the slide valve to rise slightly from its seat and allow the escape, partially into the steam chest, and partially into the exhaust port; but with poppet valves such as those here employed, there is no possibility of producing such effect, and extra provision must be made, both at the top and bottom of the cylinder, for the escape of water. A^9 is the valve applied for this purpose, at the bottom of the cylinder. It is constructed substantially like the ordinary safety valves employed on boilers, but stands in an inverted position, and is held up to its seat by the aid of the bent crossbar A^6, which, as before described, is suspended by powerful coiled springs, enclosed in the housings A^7A^7. The action is precisely similar to that of the ordinary safety valve, the tension of the springs being such as to hold the valve A^9 firmly in its seat, unless the pressure within the cylinder exceeds a certain proper amount. The moment the pressure becomes excessive—which can never occur except from the presence of water, or the breaking of some of the valve gear—the valve A^9 is urged downwards with such force as to carry with it the crossbar A^6, despite the tension of the coiled springs.

A^{10} are nuts which fit on the screw-bolts represented, and by means of which the tension of the coiled springs may be graduated to any desired amount. If, for example, the coiled springs become somewhat set, or partly lose their force, so that the safety valve A^9 is no longer able to hold its seat against the ordinary full boiler pressure of the steam, it will follow that the valve will open somewhat at the commencement of each stroke, when the pressure is greatest, and steam will escape, on observing which the engineer has simply to turn the nuts A^{10} a little, and thus increase the force with which the valve is held to the proper extent.

The seat of the safety valve A^9 is of brass, and it is common to make A^9 also of similar material, to prevent a possibility of either of the surfaces rusting so as to cause adhesion to the other. The valve seat has a flange on its lower edge, as represented, and the body is fitted tightly into a hole bored in the casting A^4.

The manner in which the safety valve A^9 is guided is worthy of especial note, as the space for guides of the ordinary character either above and below the valve is very limited. The opening in the seat is crossed by a radial bar cast therein, and from the centre of this projects a short rod or spur, nicely turned, as represented, and only ¾ inch in diameter. The safety valve is cast with a thicker stem projecting downwards, the centre of which stem is bored with a corresponding hole to fit easily on the projection from the seat described. It follows that the motion of A^9 is simply vertically downwards and directly back to its seat, as it is perfectly guided by the central rod described, in its interior cavity. A^{11} represents hand holes to allow access to the valve or seats. A^{12} is the opening in which the exhaust rock shaft is carried. A^{13} is a similar opening for the steam rock shaft. Both openings are, in effect, hollow pipes or tubes cast in the cylinder and extending across the passages. A^{14} is a lug under the exhaust chest. A^{15} is a corresponding lug under the steam chest. Both these are merely fastenings for brackets or hangers to be described below. A^{16} are bosses to centres in which are attached the rods of the bilge pumps. A^{17} is a projection to which is bolted a part of the valve motion as seen in Plate V. A^{18} is a boss on the opposite side of the cylinder. It is fixed to a centre from which a motion is obtained which lowers the valve to its seat.

Fig. 5 is a plan view, and Fig. 6 a vertical section of the cylinder cover. This casting is fitted into the upper end of the cylinder, making a tight joint therewith, at the extreme upper edge, and is bolted thereon by a series of bolts which are inserted through the flange, plainly represented in Fig. 6. The holes for the bolts are shown in the flange of the cylinder, in Fig. 2, and these fastenings must be sufficiently near together to prevent any considerable springing of the casting, when subjected either to the full pressure of the steam, or the greatly increased strain due to the occasional presence of water, as before explained. This casting must, in short, form a rigid and perfectly tight top to the cylinder, while at the same time, it must be easy of removal, in order to allow access to the piston within.

The cover of an oscillating engine is necessarily deeper at its central portion than the like parts of other engines, and, in fact, it is generally desirable to make the ends of a cylinder as thin relatively as consistent with the proper strength of the parts, and the provision for a suitable stuffing box; but with oscillating engines the case is different. The strains developed by the oscillating motion of the heavy cylinder induce a great amount of side pressure in alternate directions, on the piston rod, at the point where it passes through the cylinder cover. The cylinder cover is a hollow casting in a form approximating that of a cone, but with an aperture through its axis of a diameter greater than that of the piston rod. Through this hole the piston rod must play tightly and easily, while provision must also be made for receiving the lateral strain due to the oscillating of the cylinder. This is effected by lining the aperture for a considerable part of its length with brass, and making it in another portion an ordinary stuffing box.

This construction is represented distinctly in Fig. 6. The central aperture referred to, is bored truly and smoothly to a size of one inch greater diameter than the piston rod. About one half the distance is now bored from the outside to a still larger diameter, as represented. The brass sleeve A, having a flange at its upper extremity, is next fitted tightly and permanently in its place in the smallest part of the hole, and this makes a suitable bearing to receive the side pressure. Above this is left an annular space around the piston rod, which receives packing or soft material of any ordinary character. This packing is compressed and compelled to keep a tolerably tight joint with the piston rod, by the aid of the gland D, which is fitted loosely within the bore, as represented, and driven down to a suitable extent by screw bolts, not represented, passing through its flange, and tapped into the thicker flange on the cylinder cover. The cup-like space above the flange is somewhat contracted near its upper edge, as shown, to prevent the spilling of the fluid employed in lubricating the piston rod.

The durable fitting of this portion of an engine in a manner just sufficiently tight and capable of easy adjustment, is a matter of great importance, and one of no little difficulty. The stout piston rod which moves outward and inward through this stuffing box, is a medium through which all the power of the engine is transmitted. The packing must be of soft material, and not too tightly compressed, or it will heat and abrade the necessarily finely polished surface of the rod. It is also necessary to maintain a steam and air tight contact with the rapidly moving surface. On the inward stroke of the piston, there is a pressure of steam endeavoring to force a passage outward through the annular opening around the rod, and on the outward stroke there is a pressure of air on the exterior, endeavoring to force a passage through into the vacuum which now exists within. Any leaking of steam outward renders itself readily visible to the eye; but the leaking of air inward is not equally plain, and it is found in practice that air will leak quite rapidly through spaces too narrow to allow the passage of steam. All the air which leaks through this or any other portion of the engine on the exhaust side of the piston, tends to vitiate the vacuum, and thus detract from the power of the engine. In the endeavor to diminish as far as possible, this source of loss, it was at one time very common for English engine builders to insert what they termed a "lantern brass," a hollow ring, fitting the piston rod at about the centre of the stuffing box, and to supply this ring with steam at full boiler pressure by means of a small tube. Such a device insures that in case any inward leakage occurs through the stuffing box into the interior of the cylinder, the fluid drawn inward is certain to be steam and not air, and as steam is readily disposed of by the cold water in the condenser, the evil accruing is very slight unless the leakage be very excessive. This is one object attained by the use of the lantern brass; but another, and perhaps more important one in practice, is the tell-tale effect produced by the escape of steam so soon as the stuffing box becomes to any considerable extent leaky. It is easier for steam to leak outwardly, while the motion of the piston rod is outward, than while the surface of the piston rod is moving inward. As ordinarily arranged, the suction or vacuum within the cylinder induces an inward leak at that movement, so that a considerable imperfection might exist without detection; but with the lantern brass a slight escape of steam is observed at every outward movement of the piston rod, and this escape becomes very sensibly increased whenever the packing becomes defective, and thus attention is immediately invited to the fact and the adjustment is at once attended to. Such brasses are, however, rarely or never employed on American engines of any kind.

B is an opening plainly represented both in Figs. 5 and 6. It serves the purpose both of a manhole to allow occasional access to the interior of the cylinder end, without removing the whole cylinder cover, and of a suitable aperture on which to mount the upper safety valve. *CC* etc., are circular openings in the lower side only of the cylinder cover. These are provided to allow the removal of the core from the casting, and are subsequently stopped by screw plugs inserted as tightly as possible. *D* and *E* are respectively the inclined channels which allow the free influx and escape of the steam through the steam and exhaust ports. *FF* etc., are vertical webs which connect the upper and lower portions of the cylinder cover, and extend radially from the centre to the periphery. *GG* represent an annular recess on the under surface near its outer edge, which allows for the heads of the bolts on the piston, the intent being to allow the upper side of the piston to approach as closely as possible to the cylinder cover, without an absolute contact at any point.

The cylinder cover, as a whole, is an extremely strong, yet light casting. In explanation of the great depth given it at the centre, it is evident on reflection, that the piston rod acts as a lever of the first kind, in vibrating or oscillating the cylinder. The piston may be considered at all times ready to serve as a fulcrum, the crank pin is the point where the power is applied, so far as this lever-like action is concerned, and the interior of the cylinder cover the surface where the resistance is felt. When the piston is at its lowest position in the bottom of the cylinder, the strain on the cylinder cover is slight; but when the piston is at or near the highest point in its stroke, the length between the fulcrum and the bearing in the cylinder head is very short, and consequently the strain on the cylinder head is quite severe. To diminish this, and in fact, to relieve the piston as far as possible, from any side strain, and rob it, if possible, of all participation in this oscillating strain, great depth is given to the stuffing box and cylinder cover. The brass linings of the cylinder cover, and also of the gland, may be very readily replaced by others when their interiors become worn slightly oval; and by compelling the brass lining *A* to serve as one bearing, which, for perspicuity, may be termed the fulcrum, and the interior of the gland *D* has another which we may term the resistance, it will be readily seen that the greater the distance to which we can separate these bearings, and, in short, the greater the absolute depth of the cylinder cover, the less will be the strain and consequent wear.

Fig. 7 is a plan view, and Fig. 8 a vertical section of the piston. It is constructed on the same general principle as the cylinder cover, being a hollow casting strengthened by radial webs, and circular orifices, *F* being left for the removal of the core, which orifices are stopped in the manner already described. The centre of the holes is slightly tapering to receive the piston rod, which latter is retained by a very stout nut on the under side. The form is slightly conical, the increased depth in the centre being provided to compensate for the increase of strength at that part. The central boss or hub projects as represented, both above and below, and on these are accurately fitted the wrought iron rings *EE*. *DD* are sectional views of the packing rings of cast-iron, which are driven outward by their own elasticity, and also by the pressure of springs, not represented, which are inserted in the annular space between them and the body of the piston. *C* is the follower or junk-ring, an annular casting fitting into a corresponding depression turned in the upper surface of the piston, so that it stands flush with the upper surface of the piston. The ring *C* is secured to the piston by screw bolts *B*, the threads of which work in nuts previously inserted into corresponding cavities, as represented.

The piston is the agent which directly receives all the useful effect of the steam, and to which, in fact, all the other portions of the engine are subservient. There are strong reasons why every portion of a marine engine should be made as light as consistent with strength, but the piston is particularly so, from the rapidity of its reciprocating motion, and the strains induced by its momentum on the crank pin, and other parts of the mechanism. The piston of the Knoxville is of cast-iron. The rings *EE* serve as hoops to defend the boss against the violent wedging force due to the slight taper of the interior. It is common to finish the piston rod, as in this case, with a very slight shoulder at the point where it joins the piston; but this shoulder is too slight to be of much importance, and the principal force of the piston in one direction is received on the very gradually tapering interior of the bore *A*, figures 7 and 8, which tapers about ¾ inch to one foot in length, and the expansive or wedging force due to this strain is necessarily very severe, though, except when it produces an actual cracking of the piston, this strain is of no effect, or rather, is highly advantageous, by insuring a perfectly rigid union of the parts.

Fig. 9 is a vertical section, and Fig. 10 a bottom view of the upper safety valve. This valve, denoted by *A*, performs a part corresponding exactly to that quite fully described at the lower end of the cylinder, and is mounted in a similar manner on a cylindrical spur or rod *D*, projecting from the centre of the annular seat D^1. It will be observed in Fig. 10, that what we have termed the crossbar, which connects *D* with D^1, does not extend across, but only radially from the periphery to the centre. *B* is the housing which encloses the spring *H*. *G* and *F* are in effect, one casting fitted to play loosely within *B*. The stout spiral spring *H* endeavors to extend itself, acting in one direction against the head or top of the housing *B*, and in the other direction against the casting *F*, tending to urge it downwards. The rod *E* is tapped through *G*, and bears on A, and thus the whole force of the spring H is employed in holding A tightly to its seat. E plays loosely through the aperture represented in the end or top of the housing; and by applying a suitable wrench to the end presented on the outside, it may be turned so as to compress H to any degree of tension required.

Fig. 11 is a side view; Fig. 12 an end view, and Fig. 13 an edge view of the piston rod. The head B is forked as represented, the cavity C being made a little larger than the crank pin, so as to receive a brass lining. Figs. 14 and 15 show the head of the piston on a larger scale; and with the cap G in place, to confine the brasses B^1 and B^2, which embrace crank pin D. Figs. 11, 12 and 13 represent a key seat, and D^1, Fig. 14, shows the key in position. It is tapered as represented, to the extent of about three-eighths of an inch to one foot of length, and a screw thread D^3 is cut on its small end to allow of its accurate adjustment. By turning the nut D^2 this key is drawn inwards, and thus wedges the brass D^1 into tighter contact with the crank pin. The tapered part E is the seat for the piston, kept in its place by the nut F. A is the wearing part of the piston rod.

Fig. 16 shows another view of the end D^3 of the key, and also of the devices for accurately adjusting and retaining it. The object of these devices is to allow of a perfect compensation for the wear of the brasses, and to provide means for adjusting the same while the engine is running. The nut D^2 is constructed with four star-like arms, which may be struck with a hammer, to turn the nut in either direction at each revolution of the crank, even while the whole is rapidly revolving. I is a fixture bolted upon the side of the head B, by the bolts I^1, and K is a spherical ended bolt, which plays in a cavity on the interior of I, and is urged by a spring, not represented, into intimate contact with the nut D^2. There are cavities in the side of the nut D^2, as shown by the dotted lines. These cavities receive the spherical end of the bolt K, which is pressed in with sufficient force to prevent the nut D^2 from voluntarily turning itself, and yet without retaining it so firmly as to prevent its motion when the inclined sides of the shallow cavities forcibly push back the bolt K, as they are certain to do when the arms D^2 are struck with sufficient force. There are eight of these cavities in the circumference of the nut D^2, and the nut is therefore capable of adjustment to so small an extent as one eighth of a revolution. Whenever the connection of the piston rod to the crank pin is found to be to any appreciable extent loose, the nut D^2 is turned one or more of these notches, and thus the key D^1 is drawn forward so as to tighten the connection.

The cap G is secured by four bolts H, which are tapped into B to the depth shown by the dotted lines. H^1 are nuts which firmly retain it in place. G^1 G^1 represent lugs forged on the top of the cap G, and which serve to carry the pin G^2. This pin gives motion to the air-pump.

Figs. 17, 18, 19 and 20 represent respectively views of the hangers which support the rock shafts. The hanger, represented by Figs. 17 and 18, is bolted to the rectangular projection A^{14}, on the under side of the exhaust chest A^3; while the hanger Figs. 19 and 20 is bolted to the projection A^{15}, or the under side of the steam chest. In each of these drawings, A represents the bar in which the respective rock shafts are carried, and B the brass or box employed to receive the wear.

Fig. 21 is a bearing for valve stem. Figs. 23 and 24 are the bearings, cast to the cylinder, for the rock shafts of the steam and exhaust valve motion; of which A are the journals and B brasses with a screw C, and a jam nut D on its lower part, to tighten the journals.

Plate VII.

Fig. 25 is a plan, and Fig. 26 a side elevation of the condenser P, which contains the air-pump, as represented. Both these important parts are of cast-iron, the cylindrical portion of the air-pump being lined, however, with brass, in the style ordinarily adopted in American engines. The outline of the air-pump, and the thickness of the brass lining are very distinctly represented by the dotted lines in Fig. 26. The thickness of this lining is about half an inch, and its design is simply to avoid the oxidation and corrosion of the surface, which would occur with iron. The pump is denoted by A, the nozzle, by which it is connected to the hot well by A^2, the flange, by which it is bolted to the condenser by A^3, internal projections near the bottom by A^4, and the extreme bottom rim by A^5. P^1 represents the stuffing-box, by which a slip-joint connection is made with the exhaust pipe. P^2, Fig. 25, represents the gland by which the packing in this important connection is compressed. P^3 is the flange, forming the foot of the condenser, and by which it is bolted to the bed-plate. P^4 is a hand-hole to allow access to the interior. P^5 is one of several projections to support the scattering-plate. P^6 is an opening into P^7. P^7 is a channel connecting P^6 with P^8. P^8 P^8 are openings into the feed-pump. B is the opening through which the injection water is admitted, and leading up to the scattering-plate B^1. Through holes in this latter, B^1, the water is allowed to fall and condense the steam. The object of the scattering-plate is to present as large a surface of the water as possible to the steam.

Fig. 27 is a vertical section, Fig. 28 a plan view, and Fig. 29 a side elevation of the hot well, which is represented in Plate V. by Q. A^2 is the opening which is fitted to the flange A^2 on the upper portion of the air-pump. A^3 and A^4 are stout rims cast above and below the opening A^2, to receive the short screw bolts connecting the hot well to the air-pump. These parts are represented enlarged by Figs. 30 and 31. B is a simple bonnet or cover, fitted to a corresponding hand-hole. Q^1 is the casting fitted upon the top of the hot well, and which serves the purpose of a kind of air chamber, to regulate the discharge. C is a brass valve, and C^1 C^1 are delivery valves; these portions are of brass, and are represented in plan view by Fig. 32. D and D^1 represent the connection to the discharge pipe indicated by Q^1 in Plate V., and through which all the water and condensed steam discharged into the hot well from the air-pump is allowed a free escape through the side of the vessel into the sea. E are stops to prevent the valve C^1 from rising too high.

Fig. 32 represents the valve seat C, before referred to, with one valve in place. D D represent the lugs on the valve seat, which form respectively portions of the hinge to which the valve is attached, and E bars which cross the openings, and constitute it a species of grating. These valves are of prepared India rubber, and work very smoothly and noiselessly.

Fig. 33 is a plan, and Fig. 34 a vertical section of the air-pump cover. A is the gland, and B the cover. There are four ribs of moderate depth on the under side, which stiffen and strengthen this casting as represented.

Fig. 35 is a side elevation, Fig. 36 a plan, and Fig. 37 an end view of the force pump and its box. These parts are represented in Plate V. by O and O^1 respectively. The parts are principally of brass, and are represented on a sufficiently large scale to exhibit all the peculiarities. A is the pump proper, which is of the single-acting description. A^1 is the upper portion of the flange, and A^2 the ring bolted on the top, which, by the aid of suitable bolts, compress the packing.

A^3 is a pin which traverses across the axis, and forms the point to which the pump rod denoted by N in Plate V. is attached. A^4 is a stem projecting through the base, and by which motion is given to the slide valve A^5, and by which the valves on either side are brought into action or shut off from connection with the pump at pleasure. A^6 is a gland by which the packing is compressed. B is the box or base which carries all the valves of the pump. C C are two inlets through which the water is received from the hot well, or rather from the passages P^8 connecting with the channel P^7, and the passage P^6, shown in Figs. 25 and 26. I I, are handles giving motion through the shaft I^1 I^1, to the throttle valve H H, by which the influx of water is controlled. L L are glands which compress the packing in the stuffing boxes around I^1. K is a cylindrical rod jointed to B, at the point K^1 represented, and by which, through the medium of a pinching screw in a swivel on I, the position of the throttle valve H is adjusted with certainty, so as to admit exactly the quantity of water required for the supply of the boilers. D^1 D^1 are self-acting induction valves, through which the water is allowed to rise into the barrel of the pump A. D^{11} D^{11} are delivery valves, through which the water is discharged from the pump on the descent of the flange. D^2 are stops to prevent the valves from rising too high. F F are openings to which are attached the feed pipes represented by W in Plate V., through which the water is supplied to the boiler. G G are vertical polished brass rods, passing through the stuffing boxes G^2 G^2. On the lower ends of G G are secured inverted puppet valves G^8 G^8, fitting to respective seats in the horizontal partition which divides the induction from the delivery passages. G^1 is a horizontal plate maintained at a small elevation above, by the four legs represented. On G^1 is supported the stout coiled spring G^6, which exerts a constant strain on G^4. G^4 is secured by a nut so as to be adjustable on the upper extremity of G, and the effect of the whole is to retain the inverted puppet valve G^8 in its seat, with a degree of force which is adjustable by turning the nut last named. Whenever, in consequence of the partial or complete closing of the screw valves W on the boiler, more water is discharged by the force pumps through the delivery valves D^1, than can be admitted to the boiler, the pressure in the feed pipe, and consequently in the delivery passage, becomes excessive, and the inverted valve G^8 is forced downward in opposition to the tension of the spring G^6, and at each stroke allows an escape of water from the eduction or delivery into the induction passage.

Fig. 38 is a plan, and Fig. 38[1] is a side elevation of the steam pipe which is represented by B^2 in Plate V. F is the end of the trunnion. F^1 is the gland which compresses the packing around G. G is the end of the exhaust pipe. H is the broad flange, or rather rim, by which the steam pipe is secured to the trunnion pillow block. I is the pedestal which supports the weight of the exhaust pipe. C is the throttle valve spindle. C^1 is the lever by which motion is communicated thereto, and C^2 the throttle valve. E is a projection on the steam pipe, which aids in securing it to the trunnion pillow block.

Fig. 39 is an end view of what we have termed the pedestal I, and the portion by which the steam pipe is bolted to the bed plate.

Fig. 40 is a plan, Fig. 41 a side elevation, and Fig. 41[1] an end view of the exhaust pipe. C^1 C^1 are arms by which it is bolted to the cylinder, C^2 a pedestal supporting the weight, and C^3 the cylindrical end which is inserted into the oscillating trunnion C^4.

Fig. 42 is a plan view, and Fig. 43 a vertical section of the air-pump bucket. This feature of the engine is similar in general appearance to the piston, but is only about one half the diameter, and is provided with large valves opening upwards, rendering the air-pump a simple single-acting pump, for the removal of water, air, and uncondensed steam from the condenser. A is the air-pump rod, B the body of the bucket, B^1 the hub or central boss by which it is secured on A, B^2

the cavity in which the packing is compressed, *C* the junk-ring or follower, *D* the broad flat valves, *E* the guard to prevent the too great rise of the valves, and E^2 a stout nut fixed on A as represented, to maintain the connection between *A* and *B*. The guard *E* is strengthened by ribs or webs on its upper surface, and the bucket proper is strengthened, first by a very stout crossbar in the line of the hinges of the valves, and further by four radial webs B^3 B^4. *C C* represent the lugs on the interior of *C*, by which it is bolted to *B*.

Fig. 44 is a side elevation of the air-pump rod. For engines of this character, designed to work in fresh water, both the air-pump and air-pump rod may be made of iron alone. But the practice of the English, whose steamers almost invariably work in sea water, is to make both the air-pump and air-pump rod, of brass. The practice in the United States, with regard to ocean and coast steamers, is usually that adopted in this instance, making a rod of iron somewhat too small, and covering it skilfully with brass, by mounting it in a mould, and casting the less oxydizable metal around it. The diameters are given on the drawing, that of the rod finished being 4½ inches, and that of the iron within being 3¾. This method of construction makes a very strong rod, and, unless by some accident or imperfection, the brass becomes separated in some measure from the iron within, it is equally efficient and desirable to one of solid brass. The tapering portion on the left is fitted in the air-pump piston. The cylindrical portion, 5 inches in diameter, carries the guard *E E*, and nut E^1, shown in Figs. 42 and 43.

Fig. 45 is one view, and Fig. 46 another, of the air-pump cross head. *E* is the neck, which is secured on the air-pump rod by a stout key inserted in the seat *D*. C^1 is the bearing portion of the pin, and C^2, placed on each side of the pin C^1, a guide for the air-pump rod.

Fig. 47 is a section view of the bilge pumps, of which *A A* is the pump barrel made of brass, *B B* is the pump bucket, B^1 B^1 the followers, B^2 pump rod, B^3 pins to secure the rod to the bucket. *C C* are the valve chambers, C^2 C^2 valves, C^3 a branch for suction pipe, C^4 ascending pipe to carry the waste water overboard. C^5 covers, to set to the valves, C^6 C^6 guards to keep the valves from rising too much, C^7 a nozzle to keep the pump bucket in the barrel. The figure placed above this is a top view of the same.

Fig 49 is a side elevation, and Fig. 50 a plan view, of the forgings, two of which form the upper chord of the air-pump lever. The hole *a* receives the pin shown by I^1 in Plate *V*. *B*, at the other extremity, receives the pin shown by H^1, in Fig. 5, and by which it is connected by the link *H* in that figure, to the main crank pin.

Fig. 51 is a side elevation, and Fig. 52 a plan view of the lower chord of the air-pump lever, the same letters denoting similar parts to the figures last described.

Fig. 53 is a side elevation, Fig. 54 a plan view, and Fig. 55 a transverse section, of the air-pump lever complete. The pin *a* is the centre on which it turns, *B* is the centre by which it is connected to the link from which it receives its motion, and *C* the centre by which it is connected to the air-pump. *D D* are stout struts, D^1 the bolts by which the latter are screwed to the upper chords *I*, and B^2 similar bolts fastening to the lower chords *H*. *E E* are key seats by which the parts are secured in their places. *F* is the centre for the force pump rod, represented by N^1 in Plate V. F^1 F^1 are stout castings which support it, and F^2 the nuts by which it is tightened in its place. F^3 are braces which connect the upper chord to the lower, at that point, and stiffen the whole. G^1 G^1 are a series of horizontal braces employed in the upper chord, only to give the beam stiffness laterally. It acts on a principle similar to some of the most popular truss bridges. The beam is thus principally of wrought iron, and combines rigidity and lightness in a very high degree.

Fig. 56 is an enlarged view of the centre denoted by *B*. Fig. 57 is an enlarged view of the centre *C*. Fig. 58 an enlarged view of the centre *F*, and Fig. 59 shows three views of the cast-iron supports for the feed pump centre *F*, in Figs. 53 and 54. A figure not numbered, on the right of Figs. 59, shows, on a similar large scale, the beveled cast-iron supports of the centre *C* in the same drawings.

The valve motion employed on this engine, is the invention of Mr. Horatio Allen, one of the proprietors of the Novelty Iron Works, and Mr. D. G. Wells, one of the engineers connected with the establishment. The effect accomplished, as already intimated, is the rendering the point of "cut off" variable or adjustable, at the will of the engineer, while the engine is in motion, so that when, in consequence of head wind, great haste, or any other extraordinary circumstance, it is desired to use more power than is ordinarily employed, the steam may be admitted to the cylinder during a longer portion of the stroke; and when, on the other hand, in consequence of any circumstances, it is wished to economize fuel by running moderately, the admission of steam may be cut off earlier in the stroke, and the steam consequently employed more expansively and economically. This is effected by receiving two distinct motions in the valve gear. One is derived from th eccentric represented on the water wheel shaft, the other is derived from another eccentric beyond the first. Not corresponding in time with the first, by properly compounding the two motions, the first of which is employed to lift the valve, and the other to lower it, the valve is uniformly lifted and steam admitted to the cylinder at precisely the right moment; *i. e.* at, or a little before, the commencement of the stroke under all circumstances, while by properly changing the motion of the valve from one agent to the other, at any early point in the stroke, the valve commences to be lowered, and arrives at its seat and cuts off the admission of steam to the cylinder before the piston has completed its stroke, the time of such "cut off" varying according to the moment at which the motion of the valve was transferred from one agent to the other. In other words, one agency is endeavoring to lift the valve, the other to moderately close it; and the valve is allowed to cling to the first-named agent, until at a certain point in its elevation it is transferred to the other agent, and commences lowering. The time of its arriving in its seat may, by this means, be regulated very perfectly.

The importance of the expansive use of steam is too universally recognized to require much elucidation. It is well known that if the steam is allowed access to the cylinder during the whole stroke of the piston, the piston is urged with the utmost power of the fluid throughout the whole stroke, and on opening the exhaust valves, the steam escapes with great violence. If, however, the admission of steam be cut off, or stopped when the stroke of the piston is but half completed, only half the quantity of steam will be consumed in the operation, and during the remainder of the stroke, the steam previously admitted will act by expansion, and will continue to urge the piston onward, though with somewhat less force than before. This is termed working steam expansively, and is practiced more or less in all situations where steam engines are employed. In some extreme cases, as at the mammoth pumping engines in the copper mines in Cornwall, England, the steam is cut off as early as one twelfth of the stroke, that is, the steam is admitted at full pressure when the piston commences its motion, and the valve remains open till one twelfth only of the stroke is completed, after which the remaining eleven-twelths, is performed by the force of the expanding steam, and by the momentum already imparted or treasured up in the moving mass. Where, as in such instances, the mass of matter moved, consisting of immensely long timber pump rods, pump buckets, etc., is very great, expansion may be profitably carried to this extreme extent, but under ordinary circumstances in steamers, it is not well to cut off at less than about one fifth of the stroke, varying from this to one half or two-thirds. Ocean steamers generally cut off at the latter points, or, in general terms, points from one third to two-thirds of the stroke. In some, the point of cut off is variable or adjustable. The device employed on the Knoxville is simply one of the best means of accomplishing this latter result.

Leaving, for a moment, the proper order of the drawings, Fig. 88 is a plan, and Fig. 89 a side elevation of the cylinder, showing a valve motion on a larger scale than in the general view in Plate V. The parts are denoted by the same letters in both drawings, and a reference to both may be necessary to understand the whole. On the left hand side are the steam valves inclosed in the steam chests A^1 A^1. On the right hand side are the exhaust valves in the steam chests A^2 A^2. The valves are balance puppet, and are quite fully represented in Figs. 166[1] 167[1].

Plates V. and VIII.

The stem of the upper steam valve is connected to one of the toes R^9. The stem of the lower steam valve is connected to the crooked forging represented by S^{10} in Plate V., and S^{13} in Plate VIII., by which it is attached to the other toe R^8. The toes R^8 R^9, projecting in different directions, must be understood to be separate pieces, and capable each of a vertical motion, independent of the other.

The stem of the upper exhaust valve is connected to the left hand toe R^7. The stem S^{12} of the lower exhaust valve is fixed to the crooked forging S^{11}, and by this means united to the right hand toe R^7.

S^5 represents an open, triangular frame or lever, jointed to the cylinder at the apex, and to the eccentric rod at the right hand centre S^0. As the engine revolves, this triangular frame, therefore, rocks or vibrates on the upper centre or apex with each revolution. The lower centre W^1, carries a pin which is embraced by a hook on the jointed rod S^7 S^8, and thus conveys motion to the pins 6 and 7, which are fixed in arms descending from the rock shafts R^{11} R^{15}. Thus both these shafts are rocked or vibrated simultaneously by the working of the eccentric S. (See Plate V.) The triangular frame or lever S^5 is represented distinctly in its place, in Figs. 88 and 89, and represented detached in Figs. 64, 65, and 66. The casting S^6 is that by which the fixed pin S^7 is bolted to the suitable projection or lug on the cylinder A^{17}, before described. By elevating the hooks at the joint in the rod S^7 S^8, the connection between the eccentric and the valves is broken, and by lowering this joint it is again formed. This action, then, constitutes the operation of hooking and unhooking the engine. This operation is performed by the aid of the lever T, (see Plate V.,) the rod T^1, the lever T^2, the bent lever T^3 and T^4, and the curved slot in T^3, denoted by T^5. R is the starting bar, R^1 the starting lever in which it is inserted, R^2 a connection leading to the levers R^3 and R^4. R^4 works the sleeve, equivalent to the trip shaft in ordinary engines, or in other words, works the short toes R^{13}, so that either the upper or lower steam valve is open slightly when the starting bar R is much elevated or depressed. R^3 is a three-part lever, jointed to the cylinder at 2, and to a lever R^6, which is fixed to the exhaust trip toes R^9. There is an upright arm 3, on the lever R^3, and the pin 4, at its top, receives and carries the support at one extremity of the lever R^4. The portion of this lever resting on pin 4 is wedge shaped, as represented, and as the lever 3 inclines in either direction from perpendicular, the weight R^5 aids in carrying it over, or holding it in any given position. Thus the gravity of R^5 is made to counterbalance the disposition of the valves to sink into their seats, or, in other words, to aid in working the valves by partially balancing their disposition to close.

Most or all of these parts are represented on a larger scale in the accompanying details. Figs. 60 and 62 represent plan views of the connection S^7 S^8. Figs. 61 and 63 are side elevations of the same parts. Figs. 64, 65, 66 and 67, have been described as the triangular lever and its attachment. Fig. 68 is a side view of a transfer arm for cut-off, which can be seen in Plate IX.; it receives its motion by the eccentric S, placed to the right. Fig. 69 is a top view of the transfer arm for cut off. Figs. 70 and 71 are top and side elevations of a cut-off rod. Figs. 72 and 73 are respectively elevations and plans of the feed-pump rods, which are connected to A^8. (See Figs. 35 and 36.) Figs. 74 and 75 are two views of the main crank on the water-wheel shaft. Fig 75 answers equally well for the other crank, 76, but it will be observed that the hole for the crank pin in Fig. 76 is something larger than 74, and is grooved in a peculiar manner, to allow the boxes or shells enclosed, which intervene between it and the crank-pin, to work slightly, in case the shafts should be, as they almost invariably are, more or less, out of line. This allowance for working relieves the parts from the immense strain which would otherwise come on them, in consequence of the parts being inadequately fitted up, or in consequence of a straining or working of the vessel in a heavy sea. Figs. 77 and 78 are two views of the unhooking rod T^1. Figs. 79 and 80 are corresponding views of the bent lever, (represented as T^3 and T^4, in Figs. 88 and 89,) of which B is the lifter to unhook the two rods S^7 and S^8, in connection with an eccentric hook on each rod, to unite in the pin W^1, as represented in Figs. 88 and 89. A A are the journals, C a lever forged solid to its shaft, D a roller which turns on its pin D^1 and works in the slot hole T^5, Fig. 89, of the lever T^2. In Fig. 80, the lifter, in an inclined position to the left, and marked D, is the same as marked B in Fig. 79. A^1 A^1 are the bolts. Figs. 81 and 82 show two views of the unhooking handle T, and Figs. 83 and 84 two views of the hand lever R^1. Fig. 85 is the starting bar R, made to fit in a suitable socket in R^1, and Figs. 86 and 87 are two views of rod R^2, which conveys the motion of the hand lever to the trip shaft levers.

Plate IX.

Represents a grand transverse section of the ship complete. The letters of reference agree, generally with the lettering of Plate V., already described, and with the lettering of the details on Plates VI., VII., VIII. and X.

We will now attempt to define and explain the peculiarity of the valve motion. There are, it will be observed, in Plate IX. two eccentrics S and S, one on each side of the engine. The view in Plate V., and also in Fig. 89, is from the left side of Plate IX. The eccentric there seen, and the action of which is understood, is the left-hand eccentric. The right eccentric S is set considerably in advance of the other. The valve motion has been already explained to consist in lifting the valve by one movement, and transferring it, at some point in the movement, to another motion which is closing while the first is opening. This last-named eccentric, seen on the right in Plate IX., performs the closing movement. It is connected to a bent arm not represented, except in Plate IX., but which connects to the part S^{11}, in Fig. 89, and gives S^{11} and its attachments a rocking motion, independent of R^{16} and its attachments, and by which the motion of the toes S^9 S^9 are very effectually and ingeniously controlled. The toes S^9 S^9 are loose on the rock shaft. The pawls or lifters S^{10} are also loosely jointed thereto, and rest on rollers attached to R^{16}, as represented. When, for example, by the motion of the triangular frame S^8, and the connection S^7, the part R^{16}, in Fig. 89, is moved towards the left, it, by means of S^{10}, lifts the loose toe S^9, and, consequently, the toe R^8 and the valve to which it is attached. But when the point of S^{10} passes the roller, it immediately slips over the same, and quietly commences to receive the motion of the other eccentric, which, it should be distinctly remarked, is at this moment rocking in the reverse direction, and commences immediately to lower the loose toe S^9, and, consequently, to lower the valve to its seat. By turning the screw S^{11}, which is a right-and-left screw, the parts to which it is attached are turned further apart, or drawn nearer together, so that this transfer of the motion from one eccentric to the other may occur earlier or later in the stroke, at pleasure. This cut off apparatus has been the subject of litigation, and is not, at this date, introduced on new engines. A few months will end the term of the patent with which it is supposed to conflict, and will, probably, again see it in extensive demand.

Fig. 90 is a side elevation, and Fig. 91 a plan view of the pillow blocks, two of which carry the water-wheel shaft. A^0 is the shaft, A^1 the lower brass, and A^2 the upper brass. Fig. 93 is a side elevation, Fig. 94 a plan view, and Figs. 95 and 96 end elevations of the trunnion pillow block and bed plate. In Fig. 93, G is the lower brass, and G^1 the upper brass, between which the trunnions of the cylinder are supported. D is the foundation or bed plate, D^1 is the pillow block cast thereon, D^2 the cap or binder, D^3 D^3 keys, which aid in confining the inclined braces represented by F in Plates V. and IX. In Fig. 94, D^4 are projections to support the heavy exhaust pipes. D^5 D^5 are lugs, in which are secured stout transverse braces D^7. D^6 are the holding down bolts which confine the cap D^2. D^8 D^8 are upper and lower flanges of the bed plate, and D^9 D^9, etc., vertical stiffeners, extending between them. D^{10} are ribs or webs which strengthen the pillow block, and D^{11} represents a broad web on the under side, to aid in resisting vertical strains. It must be understood that the centre of this casting is, at each revolution of the engine, violently urged downwards by the pressure of the trunnions, a pressure which is resisted by the diagonal braces F at the ends, during one half the revolution. During the other half, similar strains are experienced in the opposite direction, that is, the trunnions are lifted upwards violently, while the diagonal braces F resist this pressure by holding the ends of the frame down; in other words, whenever the steam in the cylinder acts above the piston, its tendency is, by pressing against the under side of the cylinder-head, to lift the cylinder, and, consequently, the trunnion pillow block, with immense force; while with the steam below the piston, and pressing with its full power on the inside of the cylinder bottom, the tendency is downward to an equal extent.

E, Figs. 90, 91, and 92, represents one of the main pillow blocks; E^1 the cap or binder; E^2 the holding down bolts, and E^3 a flange, by which it is bolted to E^4. See, also, Plates V. and IX., which latter are transverse castings, connecting the two pillow blocks together. E^5 E^5 are the ordinary deck timbers of the vessel.

Fig. 97 is a side view of the trunnion pillow block, and the part where the steam exhausts.

Figs. 98 and 99 represents the safety valve and mountings. A is the valve, A^1 a stem by which it is guided in its seat, A^2 a rod connecting it to a pin A^3 in the lever B, which lever is hinged at A^4 to lugs rising from the cover or bonnet A^5. A^6 A^6 represent two guides cast on A^5, and which serve to support and guide the lever B. C is the brass seat of the valve, and D the casting in which the whole is supported. F represents the interior passage which connects with the boiler.

Fig. 100 is a vertical section of the injection valve, which admits cold water to the condenser. A is the valve, B is the valve stem, B^1 a stout square thread thereon, and C the valve seat. The threads B^1 act in a brass piece, firmly tapped into D, the cover or top casting. E E forms the top and bottom of the stuffing box, and F the passage through which water flows to the condenser. G is a passage communicating with the outside of the ship, and H the casting in which the whole is enclosed.

Fig. 101 is the screw-stop valve. *A* is the valve, *B* the stem, *C* the hand wheel, B^1 the thread, B^2 the nut or female screw, *D* the bottom casting, *E* the gland, E^1 the bolts, and E^2 the stuffing box. A^1 is the brass seat, *F* the passage which leads the steam from the boiler, *G* the passage which leads to the engine, and *K* a hole on which the safety-valve case, Fig. 98, is placed.

Fig. 102 is a side elevation and end view of the crank pin. *B* is the part firmly secured in the principal crank, and *A* the part loosely fitted in the other crank on the water-wheel shafts. The space between these bearings is embraced by the brasses on the head of the piston-rod.

Fig. 103 are braces to combine the bed plate with the main pillow blocks.

Fig. 104 represents one of the water-wheel shafts, the heaviest forging on the ship. *A* receives the crank, *B* is the bearing in the pillow block, *C* a broad collar which carries the eccentric, *E E* and *E* the respective rests for the stout cast iron water-wheel flanges, to which latter the arms of the water wheel are bolted, *F* is the outboard journal.

Fig. 105 is a side view, Fig. 106 a plan, and Fig. 107 an end elevation of the support for the fixed centre I^4 of the air-pump lever. It consists of a stiff casting *K* fixed in the deck beams E^6, by the bolts *N*, as represented, which latter are supported by the timbers *M M*, M^1 M^1, and the castings *O* and K^2, the last named of which rests directly upon one of the kelsons *P*.

Fig. 108 is a side elevation, 109 an edge view, 110 a plan, and 111 a horizontal section of the guides for the air-pump cross-head, a portion of the engine which cannot well be shown in either Plates V. or IX. Fig. 112 is a side view, partly in section, and Fig. 113 is a plan, of the gearing represented by U^4 in Plate V. Fig. 114 is an elevation, and 115 a plan of the stand, hand-wheel and check-wheel, shown by similar letters in Plate V. The hand-wheel *U*, by means of the rod U^3; turns the gear-wheel U^4, while the check-wheel U^1 serves to hold it firmly in any position. Fig. 116 represents the bevel gearing and small lever worked by the aid of a stand and wheels similar to those just described, by which the rod *Y*, (Plate V ,) which connects to the throttle-valve, is worked. Fig. 119 shows the ends of the rod *Y* enlarged.

Fig. 120 is a side elevation, Fig. 121 a plan view, and Fig. 122 an end elevation of the spring bearing A'', shown in Plate IX. It supports the main or water-wheel shaft A^o, at the points where it passes through the side of the ship; and care is taken to arrange it in such a manner as to allow the ship and engine to work, to a limited extent, independent of each other, by the aid of the elasticity of the materials. The great weight, and other strains to which the water-wheel shafts are subjected, is, therefore, principally received on the pillow-blocks *E* and the outboard bearing A^c. A^o is the section of the shaft. *A* is the stout shell enclosing the whole. *B* is a box, cast with shallow rectangular recesses next the shaft, as represented, which are filled with soft metal. *C* is a bar of iron standing across the front, and secured with bolts C^1, by removing which, the box *B* may be removed at pleasure. *D D* represent wrought iron keys which retain *B* in position fore and aft, and allow it to be very nicely adjusted, so as to receive a proper amount of the propelling force due to the action of the wheel upon the water.

Fig. 123 is a side elevation, 124 a plan, and Fig. 125 an end elevation of the outboard bearing, shown as A^c in Plate IX. A^o, in dotted lines, (Plate XI.,) is the small end of the shaft. *B* is the stout casting in which the whole is supported. *C C* are vertical keys, one of which is tapered, as indicated in Fig. 123. *D*, Fig. 125, is one of the bolts by which the weight is supported. *E* is a flange to aid in transferring the weight to the longitudinal timbers of the guard. *F F*, Fig. 123, are simply bolt holes, whereby it is fastened, and *G*, Fig. 125, is a key which holds the upper box in place. Figs. 126, 127, 128, and 129, show the keys independently, marked with the same letters of reference.

Fig. 130 is a representation of boiler-plates, to raise the journal of the outboard bearing. Fig. 131 is a top view, 132 an end view, and Fig. 133 a side elevation of the upper box, which is of cast iron, with soft metal linings. Fig. 134 is a top view, 135 a side elevation, and Fig. 136 a front view of the lower box for outboard bearings.

Fig. 137 shows the means adopted for tightening the packing around the piston-rod. Figs. 138, 139, 140, 141, and 142, represent independent views of the several parts. The design is to compress the packing uniformly, and without stopping the engine. *C C C C* are the bolts, provided with nuts, which carry at their bases the ratchet-wheels represented. C^1 are palls catching in the teeth of the said ratchets. C^2 are levers. *D* is a centre for the additional lever. *E* and *F* are connections, whereby a reciprocating motion, given to *E* by the aid of the hand-bar represented, is communicated to the whole. Fig. 142^1 is almost a duplicate of the drawings, Figs. 139 and 140. It represents those levers which are most distant from *E*, and which, consequently, have no provision for conveying the motion further.

Figs. 143, 144, and 145, are portions of the railing employed about the engine-room.

Fig. 146 is a plan of the upper deck of the engine-room. *B B* represent the cast-iron flooring, and *C* the circular stairs leading to the deck below. Fig. 147 is a plan of the next floor. This is the deck from which the engine is controlled. *A* is the top of the cylinder. A^2 is the top of the air-pump, and A^3 the air-pump guides. *B* is the flooring of iron. *C* the stairs. The other letters refer to the same parts as indicated by similar letters in Plates V. and IX. Figs. 148 and 149 are two additional views of the straight or lower stairs, and Figs. 150 and 151 two views of the base thereof, on a larger scale.

Figs. 152, 153, 154, 155, 156, 157 and 158 represent the valve motion, already briefly described. (See pages 45 to 48, and figures 88 and 89,) and which will now be more fully elucidated.

Fig. 152 is a side elevation, Fig. 153 a plan view, and Fig. 154 an edge view of the motion employed to actuate the exhaust valve. *B* in these figures represents the exhaust-valve shaft, (the part indicated by R^{11} in Figs. 88 and 89.) The extremity of *a B* is simply turned smaller to serve as a journal. A stout arm, *a D*, projects downward from the rock-shaft *B*, and to the extremity of this is secured a pin, *a C*, as represented. This pin receives motion from the action of the eccentric (through the rod S^6, Figs. 88 and 89) and thus gives the proper rocking motion to the shaft *B*. R^7 R^7 are two forgings, each entirely independent of the other. Each is made to embrace a fixed rectangular bar, which extends downward from the exhaust chest above. Each is, therefore, capable of a vertical motion, sliding up and down on the said bar. The one on the left in Figs. 88, 89, 152, and 153, is firmly fixed to the stem R^8 of the upper exhaust valve, the fastening being made by the aid of a nut on the latter, as represented by dotted lines in Fig. 152. The similar forging on the right side of the exhaust motion in the same figures is rigidly connected by the stout forging S^{11} to the stem of the lower exhaust valve S^{12}. The two forgings R^7 R^7 are, therefore, capable of being lifted separately, and thus of independently working their respective exhaust valves. The rock shaft *B*, which is immediately beneath the lower horizontal faces of these forgings, has two stout arms, or "toes," represented by *T T* in Figs 152, 153, and 154, and by R^{10} in Figs. 88 and 89, which toes project in opposite directions, and as it is rocked they alternately raise and lower these forgings, and consequently the valves, (Fig. 166,) which are attached thereto. It will be noticed that the upper faces of these "toes" *T T* are curved slightly, so that they raise and lower the valves by a motion which is gradual at both its commencement and its termination.

This completes the arrangement for working the exhaust valves, except the provision for working by hand. When the engine is unhooked, the rock-shafts become stationary, and the valves can only be lifted by the hand gear (shown by R, R^1, R^2, R^3, R^4, R^5, and R^6, Figs. 88 and 89.) The parts of the mechanism in immediate connection with the exhaust rock-shaft are shown very fully in Figs. 152, 153, and 154. R^6 is a lever rigidly connected, not to the rock-shaft itself, but to a loose shell or saddle resting thereon. The other extremity of said shell carries two stout toes, H H, projecting in opposite directions, similar to, but shorter than, the toes T T, by the side of which they are mounted. Their very moderate length prevents their lifting the valves to any very considerable height, but gives them increased leverage, so that it is easy for the engineer, by working the starting bar R, to rock the toes H in either direction, and thus to control the admission and escape of steam alternately, at either end of the cylinder, at pleasure.

The motion of the steam valves is more complicated, except in regard to the hand gear. Fig. 155 is a side elevation, Fig. 156 an edge view, and Fig. 157 a plan. Fig. 158 is a view of the further extremity of an internal shaft, which, passing within the rock-shaft proper, traverses across the steam passage by the side of the cylinder, and receives on the pin B, at the extremity of the bent arm, the motion of the other eccentric, which is set considerably forward of the first, and which effects the lowering of the valve. A certain portion of the valve motion is fixed on this internal shaft A, which is the rock-shaft proper, while another portion is fixed on a sleeve A^2 loosely surrounding it, or rather on an arm A^1 which extends downward from this short sleeve. The hand gear connected to the lever R^4, and working the short toes B in the manner already described for the exhaust-valve motion, is mounted on the rock-shaft, exterior to all the rest.

The forgings R^8 R^8 are independent of each other, are both mounted loosely on rectangular fixtures projecting downwards from the steam chest, and are each connected to a separate valve, being, in all respects, precisely like the corresponding forgings on the exhaust-valve motion. A^1 is the arm with its sleeve A^2 working loose on the rock-shaft A. It carries a pin S^7, and through this receives a rocking motion from the principal eccentric through the agency of the rod S^8 (Plate V.) C C are rollers carried on the widely-extended end of the arm A^1. S^9 S^9 are loose toes. They are mounted on the rock-shaft and are jointed at their extremities to the loose and peculiarly shaped lifters S^2 S^2, as represented.

The casting A^3, to the left of Fig. 156, is keyed firmly on the internal rod A, or the rock-shaft proper, and consequently receives the motion of the other eccentric. This casting, which, like all the other parts of the valve motion, is finely finished, extends forward under the rock-shaft, and supports the endless right-and-left screw D. There are two moveable pieces E, carried in a curved groove, as represented on the front face of A^3, and the position of these is controlled by the right-and-left screw, so that, by turning the latter in one direction, the moveable pieces E will be separated, while the turning in the reverse direction draws them together. There are rollers on the lower edges of these moveable blocks E, which afford a very easy and smooth means of receiving and discharging the rounded corners of the lifters S^2.

The operation of the valves by the hand gear is simultaneous with, and similar to, that of the exhaust valves. When the starting bar R (see Plate V.) is depressed, the lever R^1, and consequently the rod R^2, and the two levers R^3 and R^4 are raised. The final result of depressing the starting bar is to raise the lever R^4, and consequently to raise the right-hand toe B of the steam-valve motion (Fig. 155) and thus open the upper steam valve, and admit steam above the piston, while it also, acting through R^2, depresses the lever R^6 of the exhaust motion, raises the right-hand toe H, and consequently the lower exhaust valve, and allows the steam to escape from below the piston. Reversing the position of the starting bar R closes these valves and opens the others, and changes the motion of the piston, by allowing the steam previously admitted above it to escape into the condenser, and supplying fresh steam below.

When the engine is "hooked on," by elevating the handle T^1 and allowing the rods S^7 and S^8 to catch and receive motion from the pin W^1, (Fig. 89,) the exhaust valves are raised and lowered by the longer toes $T' T$, Fig. 152, in a manner similar to that by the hand-gear, except that they are raised higher. The steam valves also, if their motion depended entirely on that of the arm A^1, Fig. 156, would move in a similar manner, and the steam would be allowed free access to the cylinder during the whole stroke of the piston. But it follows, from the form of the lower end of the loose lifters S^2, that they are raised at the proper moment by the motion of A^2, resting meanwhile in the position of the one on the left side in Figs. 89 and 155, until it rises above the corresponding roller on the block E, when it immediately slides up on to the latter. As this portion is fixed on A^2, which latter is actuated by the other eccentric, and consequently commences to return considerably in advance of A^1, it follows that the lifter S^2, and consequently the loose toe S^9, and the corresponding steam-valve, rises until the lower end of S^2 is thus transferred from one situation to the other; but immediately on this taking place, it commences to descend. The time in which the valve reaches its seat depends on the position of the blocks E, on the face of the part A^2. If, by turning the screw D, the blocks E be brought very close together, as represented in the figure, the valves will close and cut off the admission of steam to the cylinder at a very early point in the stroke, but if, by turning D in the contrary direction, the blocks E be very widely separated, the transfer will take place at a later period, and the valve will not reach its seat till near the end of the stroke. The lifter S^2, on the right side of Figs. 89 and 155, is represented in the position it assumes after the transfer described and while the valve is closing. At or near the termination of the stroke of the piston, whatever may be the points of "cut-off," the parts invariably assume such a position that the lifter S^2 again assumes by gravity the position between E and C, proper for lifting the valve.

Figs. 166^1 and 167^1 are vertical sections of the balance puppet-valves, 166^1 being the exhaust valve, made in two parts bolted together, and 167^1 the steam valve, which latter is in one piece complete. It will be observed, that they are each equivalent to two horizontal disks rigidly connected. The seats corresponding with each disk are shown by dotted lines. The pressure of the steam on these disks tends to balance itself, but a slight surplus is allowed to press each valve down upon its seat. The space between the disks is, in every instance, in free communication with a port leading to the interior of the cylinder. Fig. 167^1 is the steam valve. The pressure of the steam on the top of the upper disk is a little greater than that on the bottom of the lower disk, and the result is, a tendency in the valve to cling tightly to its seat. Fig. 166^1 is the exhaust valve. In this valve the steam is endeavoring to escape from the space between the disks into the vacuum which exists outside. It is, therefore, politic to make the lower disk the largest, and as this renders it impossible to insert the valve complete from above, it is made in two parts as represented, the lower part being inserted through the exhaust opening from the inside of the cylinder.

Figs. 159 and 160 are different views of the loose connection of the crank-pin with one of the cranks. The design of this kind of connection is explained on page 47. It compels the rotatory motion of the crank to be communicated without requiring that perfect coincidence of position which would be necessary were the connection absolutely rigid. In short, the shafts may be out of line, to a limited extent, without any serious effect. Fig. 159 shows a section of the drag-crank in a plane parallel to the axis of the crank-pin. Fig. 160 shows the means of securing the piec which fit against the flat sides of the pin. (See Fig. 102.)

Fig. 161 is a side elevation, 162 an edge view, and 163 a plan, of one of the eccentrics, with its strap and eccentric-rod. A^0 is the hole to receive the main shaft. *B* is the eccentric of cast iron, finished with a deep groove around its periphery, to retain the strap. *C C* are two pinching or set screws which hold the eccentric in place on the shaft, and allow it to be set forward and backward at pleasure, to facilitate the economy of steam or the smooth working of the engine, as it frequently happens that thumping, or other faulty working of the engine, is found to result in practice from a very slight deviation of the principal eccentric from its proper position, a deviation which can best be found by experiment. D^1 is the strap, *D* the rod, and *E E* are braces to stiffen and strengthen *D*. *F* is the stub-end connecting to the pin S^6, Fig. 89, and *H H* are right-hand and left-hand screws, which, by the aid of a suitable coupling, connect the parts, and allow the rod to be very accurately extended, or contracted to any small extent.

Figs. 164 to 167, inclusive, are the valve stems, all differing slightly in proportions, to fit their different uses. The lower stems are in two pieces, coupled at *F*. The upper parts are adjustably secured to the flat forgings S^{10} and S^{11}, by the nuts *B B*. The lower portions pass through the stuffing-boxes in the top of their respective chests, and are secured to the valves by keys in the seats which press the valves tightly against the collars represented. *E E* are the finished ends, which pass through corresponding fixed guides, and maintain the correct position of the valves, laterally. The upper stems resemble the lower ones, except that they are in one piece.

Fig. 168 is the air-pump rod or link, connecting to the air-pump lever.

Figs. 169 and 170 are two views of one of the bilge-pump rods. *A* is a slotted piece, secured by the bolts *B B*, and which retains the pin *C* in the hook, as represented. The pin *C* is fixed in the cylinder (*a* 16, Plate IX.) *D* is the hinged connection to the plunger of the pump.

Fig. 171 is an edge, or fore-and-aft view, and Fig. 172 a side view of the ends of the link *H* (Plate V.), which connects the top of the piston-rod with the air-pump lever, and gives motion to the latter. Both ends of this link are alike. Fig. 173 and 174 are enlarged views of *F*, Fig. 161. Figs. 175 and 176 represent two views of the ends of the air-pump link, both ends alike. Figs. 177 and 178 are corresponding views of the end of the secondary eccentric-rod, where it connects to the pin *B*, in Fig. 158.

Fig. 179 is a longitudinal section of one of the boilers. Fig. 180 represents an end elevation of the same. Fig. 181 is a plan of the top of the steam chimneys, showing how the two boilers are joined so as to discharge the products of combustion into one smoke-pipe, or chimney. Similar letters of reference refer to like parts in all these figures. *A* is the grate on which the coal lies in a tolerably even stratum. *B B* are the screw-valves, and *C C* the hand-wheels, which govern them. *F* is the flange, uniting these pipes to the boilers or steam chimneys, and *G* the short pipe leading the steam from the valves. *H H* are the slip-joints which are packed like stuffing-boxes, and allow the central T shaped pipe B^1 to work slightly, as the straining of the vessel may demand. B^2 is the flange joining this pipe to another, which conveys the steam of both boilers to the engine. C^2 C^2 are the safety-valve levers of their respective boilers. C^1 C^1, Fig. 187, are flanges, connected to suitable waste-pipes, to lead off the steam blown from the safety-valves. *V* is the steam room in the boiler, or that portion of the interior which is above the water-line. V^1 V^1, etc., are the first series of flues, leading from the three furnaces to the back connection V^2. By reference to Fig. 180, it will be observed that the lower of these flues in the central furnace is as large as the upper ones, but in the two other furnaces they are made smaller, for want of room. V^3 V^3, etc., are the two tiers of return flues, leading the hot gases from V^2 to the front connection V^7; from this connection or chamber they rise through the passage represented to the base of the smoke-pipe V^4. The dimensions of all the parts are given on the drawings, and any explanation of the theory or practice connected with the construction and management of boilers is rendered impossible by the limited space at our command. We have endeavored to present all the principal points of importance connected with the engine of this magnificent steamer, making all the explanations as full and familiar as possible, in the hope to make the whole intelligible, even to those whose practice has been heretofore confined to very different styles. We will simply add, very briefly, in conclusion, a description of the process by which the cylinder-cover is removed, an operation which some have supposed, from the peculiar compactness of the framing, could not be performed without taking apart the whole framing. The piston is accessible from below by the removal of the small casting *a* 8, Fig. 1, and the piston is separable from the rod by this means, it being confined only by a nut. After effecting this separation, hoisting out the piston-rod, and turning the crank into its highest position, it is easy to remove the cylinder-cover, by turning it on edge, and the piston may as readily be removed in the same manner. The operation, therefore, of repairing or exchanging these important parts, involves little or no more difficulty than in many forms of engine, while the great stiffness and strength due to the triangular form of the framing, makes it worthy of imitation, wherever circumstances allow it.

THE PASSENGER LOCOMOTIVE

"TALISMAN."

THE locomotive is the latest, and, in some respects, the grandest development of the steam-engine Prior to the first really successful experiments in this branch of steam engineering, which took place in 1830, there were some, even among those familiar with stationary and marine engineering, who stoutly maintained, that a carriage moving itself with the boiler and all its appurtenances by steam alone, could not be made a practicable and useful construction. Now there are some 12,000 in actual use. There are many locomotives which transport loads from ten to fifty times exceeding their weight. On a trial of a new and powerful broad-gauge locomotive on the New York and Erie Railroad, in the autumn of 1855, a locomotive, weighing thirty-three tons, hauled, with the ordinary speed of freight trains, one hundred freight cars heavily laden with lumber, over all the level portions of the road. The whole mass, including the engine, tender, cars, and lumber, was computed to have weighed over 1,800 tons.

There is no branch of steam-engineering which has undergone more frequent modifications. The size has been very gradually increased since the introduction of the primitive machines of this character in Great Britain. The weight of the modern locomotive exerts so serious an influence on the track, that the distribution, size, and arrangement of the wheels, is a matter of great importance. The ability to travel rapidly depends, to a certain extent, on the size of the driving-wheels, while, on the other hand, the power to pull with great force necessarily requires wheels of small diameter. The practice in this respect, for passenger-engines, has experienced several fluctuations, the present general preference being for those of less diameter than were constructed a few years since. The location of the engines, or cylinders, the arrangement and kind of valve motion, the location and kind of feed-pumps, and other minor details, all effect, to a greater extent than might be supposed, the general design and arrangement, as also the efficiency and durability of the structure. The relative size of the furnace as compared with the other portions of the boiler, and in short, the proportions and form of the boiler itself, is also an extremely important consideration, as is also the means for promoting and regulating the draught, to incite an intense combustion at will.

The locomotive "Talisman," one of the latest constructed by the highly popular New Jersey Locomotive and Machine Company, is represented in side elevation by Plate XIII., and will be represented in detail in the several succeeding plates. The weight is principally supported on two pairs of driving-wheels, coupled together in the ordinary manner, but the forward end is carried on a swivelling truck, with smaller wheels. As is usual with American locomotives, the cylinders are placed on a level with the axles of the driving-wheels, and are outside of the track, or in general language, the engine is outside connected. The valve is a single slide, actuated by the link-motion, but with an improvement recently invented by Messrs. Uhry and Luttgens, whereby some of the principal difficulties incident to this motion are overcome.

Plate XIII. is a general side elevation. Fig. 1 is an outline, partly in section, of the same view and Fig. 2 is a plan.

Plates XIII. and XIV. A is the outside fire-box or the portion of the boiler containing the furnace. A^1 is the barrel or cylindrical portion, the latter being thickly covered with felt, or wooden lagging, to preserve it from the radiation of heat. A^2 and A^3 are domes, or elevated portions, from which the steam to supply the engines is taken by a pipe traversing the interior. A^4 is the smoke-box or stout case into which the products of combustion flow through the tubes. A^5 is the ash-pan, from which the air to support combustion is allowed to rise through the grate. B is the chimney or smoke-stack, and B^1 is the wire netting which serves to prevent the too ready escape of sparks. B^2, B^3, and B^4 are the means by which B^6 is secured. C is the cylinder, C^1 is the piston-rod stuffing-box, C^2 is the steam-chest, or the valve-chest, into which the steam is freely admitted, to be supplied to the cylinder by the action of the valve within. D is the piston-rod, and D^1 the cross-head carried in guides D^2 D^2. D^3 is the yoke or frame supporting the outer ends of the guides, and D^4 is the connecting-rod leading from the cross-head to the crank-pin on the first pair of driving-wheels. G E is the outside or parallel rod which connects the crank-pin on the first with that on the second pair of driving-wheels. E^1 E^1 are the stub-ends fitted with means for very nice adjustment, and E^2 E^2 are the brasses or boxes which are made to entirely cover and in close the ends of the crank-pins. F is the axle, F^1 the counterbalance, and F^2 the rim or tread of the hinder or secondary pair of driving-wheels. G is the axle, G^1 the balance weight, and G^2 the rim of the forward drivers. H H are the wheels, H^1 the spring, H^2 the spring-strap, and H^3 a lower brace or tie of the swivelling truck. I is the upper arm of the rock-shaft or rocker. I^1 the valve connection or valve-rod, I^2 the valve-stem, and I^0 the axis of motion, or the rock-shaft proper. J is the slide-valve, which alone performs all the operations of admitting, cutting-off, and exhausting the steam from the cylinder. K is the lever which actuates the throttle-valve L is the link, a piece of mechanism of great importance in actuating the valve; L^1 is a lever in the hand of the engineer. It controls the position of the link by the aid of the rod L^2, the bent lever or its equivalents L^3 and L^4, and a small link L^5, by which its weight is transferred to the lower extremity of the latter. M is an eccentric rod, and M^1 the eccentric. N is an additional eccentric rod, and N^1 the corresponding eccentric. The last described rod N is connected to the bottom, the first, M, is connected to the top of the link L, and being nearly opposite to each other, they impart a rocking motion to the latter, the intent or effect of which will be explained further on. O is the cam-rod, O^0 is the cam, and O^1 the cam-yoke. P is a supplementary rocker attached by its centre to the lower arm I^0 of the main rocker I. It is joined to the cam-rod O, and carries on its other extremity a pin, which is carried in a block moveable in the curved slat in the link L.

Q Q are waste cocks; Q^1 is a rod which connects the levers of Q Q; Q^2 is a slender lever hinged to Q^1; Q^3 is a light rod hinged to the upper extremity of Q^2, and by the aid of which the engineer is enabled to work the latter, and consequently to open or close the waste cocks Q Q at pleasure. When these cocks are open, there is a free communication between the interior and exterior of the cylinder C, at each end, so that any water arising from the condensation of the steam is very rapidly and violently expelled. These cocks are closed so soon as the cylinder becomes thoroughly heated. R is the frame to which is secured all the working parts of the mechanism. If all the parts of a locomotive were of an uniform temperature, the attachment of the frame would be very simple, but, as the boiler expands very considerably more than the frame, in consequence of the greatly increased heat when steam is raised, it is necessary to provide for this, by allowing the heavy mass to slip endwise within the frame at one end. This provision is made at R^1 on the side of the fire-box, the frame R being bolted firmly to the boiler at the forward end, and allowed to slide at pleasure to a certain limited extent at the point designed. The whole frame for one side of the engine is a single forging. The jaws which embrace and confine the boxes of the driving-wheels, are forged and finished separately, but each with a short piece of frame attached. To the ends of these short pieces the straight parts of the frame are subsequently welded. R^2 forms a triangular truss, bolted upon the main frame, which serves to stiffen and strengthen this portion. R^4 is a transverse beam of hard wood, covered with sheet iron, connecting the side frames, which are let partially into it as represented. R^7 is a seat on the right-hand side of the foot-board, and R^8 are uprights which support it, and also support the segment L^6, which detains the lever L^1. S is (in Plate XIII.) a single acting plunger-pump, lying horizontally, and receiving the plunger S^1, which latter is fixed to the cross-head D. S^2 is the valve-box; S^c the ingress-pipe, and S^4 a case which encloses the check-valve, through which the feed water is allowed to flow into the boiler. S^5 is the suction-pipe, through which the water is admitted to the pump. The supply contained in the tender is admitted to S^5, through a cock S^6, which cock is controlled by the engineer through the agency of the rod S^7. This cock being partly closed, the water is not allowed to pass with sufficient freedom to fill the entire capacity of the pump, and there is consequently less water forced into the boiler than would be supposed from the drawings. T^1 T^1 are merely steps to aid in mounting the engine. V is a rod to open the cock attached to the feed-pipe S^6, which is in communication with a short lever to the said rod.

W is the pilot or "cow-catcher," which has the effect to throw off any large obstruction, and prevent its getting under the wheels. It is a stout frame of wrought iron, attached to the front beam R^4 of the frame, so that it can be removed at pleasure. On the occurrence of deep snows it may be covered with boards, and made to plow open the track, throwing the snow on each side, or it may be removed altogether, and a heavy "snow-plow," supported on small wheels, provided for the purpose, may be allowed to take its place. X is a side rail of single iron bent to correspond to the contour of the driving-wheels, and which supports a light flooring, on which the engineer or fireman may travel forward on the machine to oil or examine any part. X^1 is a hand rail supported on posts X^2. The forward post contains a socket, in which is inserted a flag-staff, to give any signals which the rules of the road may require, in regard to whether a train is following or the like. Y is a stout stand bolted upon R, and which by the aid of the equalizing lever Y^1, is made to sustain or to throw upon the wheels a large proportion of the weight of the engine. [These parts are lettered S and S^1 in Plate XIV.] The very stout half eliptic springs F^5 and G^5, (Plate XIV. Fig. 1,) are supported at their centres by a connection to the boxes on the axles of the driving-wheels, and the weight of the whole hinder portion of the locomotive is transmitted to the ends of these springs by the straps and levers represented. S^3 suspends the frame R directly to the forward ends of the spring G^5, and a similar strap suspends the frame at another point to the hinder end of F^5. The other ends of the spring are connected by similar straps, not to the frame R, but the ends S^2 S^2 of the equalizing lever S^1, which is free to rock on the pin in the stand S. The effect is to ensure an equal strain on the springs F^5 and G^5 under all circumstances, and so materially soften the effect of inequalities in the track. When either driving-wheel passes a joint or other inequality, the lever S^2 rocks slightly, and causes both springs to feel the shock equally. Z Z, etc., are boiler braces which transfer the weight of the boiler to the frame R. There are four on each side, the forward, round, and obliquely placed, resting on R^4; the others, flat, and standing more or less oblique in a transverse direction, resting directly on R. The upper ends Z^1 are flattened and riveted to the shell of the boiler.

We will endeavor to explain the details of this engine as fully as consistent with the limits of our work. These volumes are intended to meet the wants of men and apprentices of very various acquirements. While with the most advanced engineers and designers it is valued chiefly as a collection of reliable detail drawings of the latest and best specimens of engines, there are many amateurs, owners of steamers or of manufactories, foreign artisans, etc., to whom the more general features of some of these structures are new, and, without explanation, quite unintelligible. In the endeavor to serve all classes,—to make our volume explain itself in the libraries and academic institutions, as well as in the workshop and draughting room—we are compelled, briefly but distinctly, to give explanations, which in the absence of these considerations may seem unnecessary and puerile.

The slide-valve J, (shown on a larger scale in Figs. 22 and 23, Plate XV,) presents the readiest means known for controlling the induction, expansion, and eduction of the steam used in the cylinder. All these operations are performed by sliding the valve-stem I^2 backward and forward. The precise motion of this rod, therefore, is a matter of great moment. The link L, (as represented in Figs. 1 and 3, Plate XIV.,) presents the simplest means known of controlling the motion of this valve. This device, sometimes known as the "Stevenson Link," is constructed in two styles, each of which has its peculiar advantages; but with either, the motion of the valve may, by a simple movement of a lever, be reduced nearly to nothing, and by a further movement of the same lever, be reversed so that the engine will revolve in the opposite direction. The link is open, as represented, and one arm of the rocker I is attached to a block (P^8, Fig. 9) fitted in the interior, which block fits tight in any point therein. When, by a movement of L^1, the link L is lowered so that the link-block P^8 stands at its upper extremity, the valve receives a throw of some five inches, admitting the steam freely so as to induce a forward motion of the engine. When the link L is raised a little, there is less motion to the valve, and this motion diminishes until the link-block stands in the centre of the link, when its motion is only $\frac{7}{8}$ of an inch, and as there is just that amount of outside lap on the valve, no steam is admitted to the cylinder during any portion of the stroke. By raising the link higher, so that the link-block stands in the lower part of the link, it is actuated in the other direction in the same manner. By proportioning the parts aright, and giving a proper curvature to the link, the motion of the valve is so controlled, that by elevating the link to various intermediate degrees, indicated by notches on the segment L^6, (Fig. 1,) the steam is admitted to the cylinder, during various portions of the stroke, at pleasure, so that by this agent is attained facilities for using the steam with various degrees of expansion. The stroke of the piston being 22 inches, the steam may be allowed to follow 20 inches, or cut off at 4 inches, as may be desired; the first being termed the lowest, and the last, the highest "grade of expansion."

The unavoidable defects of the link motion, are due mainly to the fact, that the motion of the valve is very slow. There is but a small opening of steam port on the higher grades of expansion, in consequence of which, the steam is withdrawn, and its pressure in the cylinder greatly reduced. There is also on the high grades of expansion, a too early commencement of the exhaust; and although this may be remedied by giving considerable inside lap to the valve, this modification involves a great compression before the piston, toward the termination of the stroke, an evil of some importance, especially to an engine travelling at a slow speed. The improvement introduced in this and several other of the latest engines from this establishment, is the invention of H. Uhry and H. A. Luttgens, and was patented in 1855. Its mechanical construction is simple—durable, —requires little more space than the common link motion, and is controlled in the same manner, by one reversing lever from the footboard. It can be applied to any style of engine, and consists simply in a supplementary or differential rocker, suspended on the lower arm of the main rocker. This differential rocker has for its fulcrum, the block within the slot of the link, while its lower arm is moved by a cam of regular form, thereby modifying the motion of the link in its effects on the valve; but as the leverage of the differential rocker is as one to one and two-thirds, the power to move the valve mainly devolves upon the eccentrics, except at the higher grades of expansion, when the throw of the link being reduced, the cam becomes more effective. By providing the lower arm of the differential rocker with a slot or several holes, in order to increase or decrease the effects of the cam, a valve motion may be adjusted with an accuracy and efficacy which the common link motion is entirely incapable of.

The practical objections against a regular cam motion for a locomotive valve gear—the too sudden action of the valve—is in this case removed, the resistance or power to move the valve being always more or less divided between the eccentrics and the cam drawn and its pressure on the cylinder considerably reduced, thereby rendering the otherwise economical application of the highest grades of expansion in locomotive engines almost impracticable.

A too early and slow exhaust, destroying partly the benefit of the expansion, for which the present but insufficient remedy is the inside lap, which not only increases the compression, but also tends to choke up the exhaust.

An arrangement which removes these defects must evidently be a desirable object, especially when such an arrangement is free from practical objections and difficulties in construction.

Uhry & Luttgen's improved link motion, not only obviates both the above stated defects, but also secures several other advantages.

DETAILS.

Fig. 3 shows, on the same scale as Figs. 1 and 2, the cam O^{0}, the cam yoke O^{1}, cam rod O, and auxiliary rocker P, Fig. 5, the latter being almost concealed behind the link L, Fig. 3. It shows, by a dotted circle, the position of the pin on the lower arm of the rocker I, which forms the axis of motion of the supplementary rocker. The point where the link-block stands in the link L, is indicated by the curved slot shown by dotted lines in the lower arm of the rocker. These parts are rather obscure in Plate XIII., but are distinctly and correctly shown in all the following figures, Figs. 1 to 11.

Figs. 4 and 5 are portions of transverse sections of the locomotive. The dotted line above and between them shows the relative location of the boiler, and the smaller circle on the right, the *locale* of the cylinder. The stands O^{2}, and the slight link O^{8}, which supports the cam-yoke O^{1}, is shown in Figs. 1, 2, 3, 4, 6, 7, 10 and 11. The forward eccentric M^{2}, and the backing eccentric N^{2}, are distinctly shown in Figs. 2 and 4. The collars are provided merely as a precaution against a possible loosening of the eccentrics, and a slipping sideways of the axle G^{8}.

Fig. 5 gives the clearest possible view of the peculiar novel parts, the place of the curved slot in the lower arm of the rocker being indicated by P.

Figs. 6 and 7 give enlarged views of the cam O^{0} and cam-yoke O^{1}. Cams designed in this manner, so as always to fill the space between the parallel sides of the interior of the yoke, are much used in the high pressure engineering of the interior, to give motion to the valves both of stationary and of river boat engines, as, by properly laying out the cam, almost any motion desired can be imparted to the valve or valves. Instead of a crank motion, like that given by eccentrics, hanging almost, but not quite, motionless at the ends of the throw, and then smoothly and more rapidly passing along the middle of its motion, to be checked and turned back in the same moderate manner at the other extremity of the motion, a rod, carrying a yoke, impelled by a cam, may be made to stand perfectly fixed at each end of its motion, and so leap almost instantly from one extreme position to the other. Still more, it may, by giving a suitable shape to the cam—which may be ascertained by experiment or a circle laid off in divisions—be made to throw the valve into an intermediate position, and to hold it there during a sixth or any other small portion of a revolution, and then complete the motion. The slide valves of many small stationary engines are worked by this means in such manner as, by giving considerable lap to the valve to secure tolerably well the advantages of expansion, without seriously retarding the exhaust. On engines liable to work as rapidly as locomotives, however, the violence of the very sudden jerks induced by such means proves very destructive. In combination with the eccentric motion derived from the link, however, in this device, it is believed to be quite innoxious, and is proved of great advantage, as will be shown below. With the form here adopted the motion is rapid, but not violent, from one extreme to the other of its motion.

Figs. 8 and 9 are enlarged views of the centre I^a, and the lower arm I^o, (marked P in Fig. 5,) of the principal rocker, and also of the auxiliary rocker P, jointed at its lower extremity. Fig. 9° is the other and outer extremity of the rocker shaft I^A, with a small portion of the upper arm I, which carries the valve-rod I^1, (Fig. 1.)

Fig. 10 is an enlarged view of the stand to support the cam-yoke, O^4 O^4 being legs which rest on upper surfaces of the boxes which embrace the driver axle. Fig. 10° is a section, showing the form of the inner faces adapted into corresponding V shaped vertical grooves in the boxes. O^5 (Fig. 6) shows a slight bolt which serves to confine the stand firmly in place. Fig. 11 shows the slight suspending link O^6 (Figs. 3, 4, and 6,) enlarged.

Figs. 13 and 14 are views of the piston, showing the peculiarity in the means for setting out the packing, which are omitted in the general drawings. Owing to the rapid motion of locomotive pistons, and the high pressure of the steam employed, the proper setting out of the packing rings against the interior of the cylinder is a matter of great importance. The packing, if too slightly pressed against the cylinder, allows a leakage, and if, on the other hand, it be set out too tightly, it consumes power wastefully in friction, and, what is worse, wears out the packing and the cylinder. The packing springs, as usually arranged, are set up by turning nuts on the interior of the piston, which nuts are accessible only by taking off both the end of the cylinder and the follower, and the difficulty of doing this induces the too common evil of postponing the necessity for a frequent attention to this duty, by setting it up very tightly at long intervals. The arrangement here adopted, the invention of Geo. R. Hoagland, of Port Jervis, N. Y., provides for setting out all the packing springs by turning a single screw, and that without removing the whole cylinder end.

The notation in these figures is independent of that in the other portions of the locomotive. A is the piston-rod, B the piston, C the bolts, D the follower, E a kind of compound nut secured on the end of A, F a screw tapped through E, F^1 a check-nut to confine F, G a conical wedge attached to F, and H suitable rods extending from a contact with G to the centres of the respective springs I. The lower spring I^1 is stiffer than the others, to sustain the weight of the piston. K

is the packing. The piston as represented in Fig. 13, at the extreme end of its forward stroke, and consequently very close to the cylinder rod. L is the cylinder end, and L^1 is a cap tapped in L, containing sufficient space to allow F to enter. By simply turning the cap L^1 it is removed, and then by applying the wrench to the screw F, and driving forward the cone G, the rods H are thrust outward, and the packing extended as much or little as may be desired, after which the cap L^1 may be replaced, the whole operation being performed in a very few seconds. A portion of L^1 is made hexagonal, as shown in the extra drawing annexed, so as to facilitate the removal and replacing of the same. This device has been but a short time in operation, but has so far recommended itself very heartily, and will probably come into extensive use.

Plate XV.

Fig. 14 represents a longitudinal section, showing the interior of the boiler, fire and smoke-box, the connections of steam-pipes, engine-truck, and frame. A top view of the engine-truck is fully illustrated in Plate XVI., Fig. 36. The steam-pipes C^3 and C^5, Fig. 14, are fastened on both ends to the boiler, in the manner as shown at C^4 and K^1. Drawings on a larger scale of the parts C^4, being the throttle-valve chamber, and C^5 a continuation of the steam-pipes, are clearly and fully represented in Plate XVI., Figs. 48 to 51, and will be more particularly set forth with their important arrangements, when we explain that figure. The upright pipes marked, C^8 C^8, in Fig. 14, extend into the steam-dome A^2 and A^3, and are fastened at its lower end to the horizontal steam-pipe, for the purpose of taking up the steam from the domes in a dry state, and then passing through the horizontal pipes to the throttle-valve C^4, and from C^5 into the steam-chest and cylinder. A throttle-valve rod K^1 K^1 extends from the engineer's stand, through the interior of the horizontal pipe, fastened to a lever on one end, which is illustrated in Figs. 24, 25, 26, and 27, on a larger scale, and on the other end operates the throttle-valve, either to limit the communication of steam to the steam cylinders, or to shut it off entirely, as circumstances may require. It will be seen by the capacity of the fire-box A, and the low placing of the fire-grates A^a, that wood is used in the Talisman for the production of steam. The difference of a wood and coal burner consists mainly in the dimensions of the fire-box; a wood burner requires not so much grate surface as a coal-burning locomotive. The whole space between the fire-grate A^a up to the door through which the supply of fuel is made, can be filled with wood without deadening the fire, which is on the contrary with a coal-burning engine; the fuel must be kept to a certain height to gain the full advantages by the consumption of it, the proportions and dimensions of which will be given in a treatise hereafter, that will describe the construction and its peculiarities for the building of a locomotive and other stationary engines, and will be commenced with its calculations, when we have described a series of the most valuable and important engines that have lately been built and constructed. C^6, Fig. 14, is the exhaust-pipe, connected to the side of the fire-box, in direct communication with the exhaust of the cylinder, and is fastened in the same manner as the steam-pipes C^6, which are clearly shown in Fig. 48, Plate XVI. A^0 A^0, Fig. 14, are strong braces which are secured in the square braces over the fire-box, to stiffen the upper part of it, fully to sustain the pressure of the steam, which is very great, upon its surface. The sides, front, and back part of the fire-box are strongly secured with rivets in the manner as shown at A, in drawing Fig. 14.

The steam domes A^2 and A^3 have on the top cast-iron covers A^0 and C^9, of which the latter is supplied in the centre of it with a steam whistle A^0, also with a safety-valve, more particularly shown in the longitudinal view of Plate XIII. *O O O* are copper tubes, riveted and kept with iron rings on the fire-box, and only riveted on the back part of the boiler-head.

The cylinders *C C* are firmly secured with bolts *M M M* to the sides of the smoke-box A^4; and to prevent the least vibration, strong cross-braces, above and below, to the cylinder's flanges connect them in opposite directions, to keep the cylinders in a firm position, as shown in Fig. 41, representing half a section through the fire-box, and the other half an outside view of the same. The engine-truck of a locomotive is usually placed under the smoke-box, and rests on a strong cast-iron piece *N N*, called the centre-pin, which is bolted on the bottom of the smoke-box with an extra piece of sheet-iron, that reaches the two engine frames on the two extreme ends, and is firmly secured to them; see Figs. 14, 36, and 41. The centre-pin then rests upon a cast-iron plate N^8, which has two wrought-iron braces *A A* on the bottom part, extending to the truck-frame H^8 H^8, bolted to the same; two square braces *A A*, bolted on the top of the centre plate, reach the truck-frame and are secured with bolts to it. The fore part of the locomotive then rests upon the four braces in connection with the truck-frame, the lever 6 6, and pieces of wrought-iron 8 8 8 8, which rest directly upon the journal of the truck-axles, and levers 7 7 7 7, connecting the spring H^1 with the lever 6 6, Fig. 14, causing a downward pressure upon the journals 4 4 4 4. Fig. 15 is the main connecting-rod, connected at D^1 with the crosshead which is fastened to the piston-rod, and at D^4 communicates its pressure to the centre-pins D^4 D^4, as shown, (in Fig. 2,)—these centre-pins are well fitted into the driving wheels G^2 G^2. *C*, in Figs. 15 and 16, is an oil cup; *B B B B* bolts which firmly secure the straps in its place; *a a*, two keys with a screw a^1, to keep the two keys in such a position as to prevent any unnecessary friction by the brasses upon the centre-pins, or heating and unnecessary wearing may be the cause of such a neglect. Figs. 17 and 18 is a parallel connecting-rod, combining the two driving wheels G^2 F^2, Fig. 2, by its centre-pins D^4 E^2. The only difference between this connecting-rod and that of Figs. 15 and 16, consists of an additional key to the right, for the purpose of making the two journals correspond with the distance of the two driving axles F^8 and G^8. Fig. 18 is the bottom view of it.

Figs. 19 and 21 are side views, and Fig. 20 a top view of the crosshead, of which *E* is a key to fasten the piston-rod *D*, in the crosshead D^2 D^2. The centre-pin D^1 is cast solid to the crosshead made of cast-iron, and is exactly finished on the vice. The same pin can be made separate, turned on a lathe, and bolted to the crosshead, and is done so in many instances where labor is not taken into consideration. *F F* are bolts for a plate on one side of the piston, and on the other, which is on the outside of the locomotive, fastened an attachment *G*, to carry the plunger S^1 of the pump through the whole stroke of the crosshead. S^2 is a nut for the fastening of the plunger S^1 to its attachment, which is fixed to the crosshead by the bolts *F F F*.

Fig. 22 is a side view, partly drawn in section, and Fig. 23 is a top view of slide-valve attachment, of which *J* is the valve, I^2 I^2 I^2 I^2 the valve stem, I^1 valve rod, *I* the pinion which is attached to the rocker *I*, as shown in Fig. 2. I^3 is an adjusting piece, with a screw and jam nut, I^5, for the purpose of making the valves correspond with their openings.

Figs. 24, 25, 26, and 27, are parts of the throttle-valve fixture, of which K^6 is fastened to the rod K^1, as shown in Fig. 14. K^6, of Fig. 24, is bolted to the front of the locomotive, where the engineer stands, in such a position to make it convenient for him to regulate, through the handle K^2, the admission of steam to the cylinder. The short lever K^3, in connection with the small wheel K^4 by a thread and the long lever *K*, represents a very ingenious mechanical arrangement to keep the lever *K* in a firm position for the admission of steam through the throttle-valve to the cylinder, as the engineer may find proper.

Figs. 26 and 27 show details on a larger scale of parts of the Figs. 24 and 25, of which 2 is a screw, 3 a nut, and K^4 a wheel and nut which, by turning the wheel tight, will have the effect as before described.

Fig. 28 is a side view, and 29 a top view of the eccentric straps, made of cast iron; M is the eccentric rod; 1 1 1 bolts; M^2 straps; 2 2 bolts; 3 an oil cup.

Figs. 30 and 31 represent the driving-wheel pedestal with adjusting wedges. P^3 of it is a lower cover kept firmly by two bolts P^4 P^4. The adjusting wedges P^1 P^1, are raised or lowered by the screws P P, having each two jam nuts, to bring the driving-axle boxes either to the right or left. P^2 P^2 are two bolts to hold the adjusting wedges P^1 P^1 in their proper place.

Fig. 32 is a longitudinal section view, Fig. 33 a front view, Fig. 34 a bottom view, and Fig. 35 a top view of the boxes for the driving axle. A of it is brass, C C cast iron, B a lower receptacle, of cast iron, for greasing the journals, C^1 C^1 two bolts to keep that in its place, C^2C^2 recesses for flat iron bars, marked 8 8, in Fig. 14, which rest upon the journals in the manner as previously described.

Plate XVI., Fig. 39, is a top view, and Fig. 40 a side view of the cylinder and steam-chest. C^1 C^1 flanges of the cylinder, the depth of which is shown in the side view. A^3 is the entrance for the steam into the steam-chest C^2 C^2. A is the opening when the steam enters into the same, and exhausts through the rectangular hole A^1 into A^2. B B are the two steam entrances to the cylinder and exhaust openings, whenever the steam returns from the cylinder and exhausts into the chimney of the locomotive, the arrangement of which is more particularly shown in Fig. 41, being the square section of it marked C^6. The exhaust steam, that passes by its force into the chimney, increases the draft sufficient to carry off the smoke and combustible gases without any further aid, and is therefore considered of some importance for the purposes as previously stated.

The materials for the cylinder (cast iron,) must be very compact and hard, to prevent friction and wearing between the piston-rings and cylinder; the cylinder is usually made of harder and compacter iron than the piston-rings, which wear on the inside surface of the cylinder. Piston rings can be easier renewed, and it is therefore advisable to make them of softer material—rather to wear them out than the inside surface of the other. That part of the cylinder which is covered by the steam-chest must, or ought to be, very hard, where the steam-valves move over the steam and exhaust openings, to prevent its wearing out and any unnecessary friction, which would otherwise be the case in some degrees, when such parts are not of the requisite hardness. Fig. 41 represents, half to the left, a section view, and the other, to the right, a front view of the fire-box; also, showing the engine-truck cylinders, lying in opposite directions, on the sides of the front part of the boiler. C^4 is the throttle-valve casing, conducting the steam, in two pipes C^5 C^5, to the right and left into the cylinder C C. C^2 C^2 are steam-chests. C^1 C^1 C^1 cylinder flanges, bolted to the side of the boiler, and at its lower part to the engine frames. C^6 is the exhaust-pipe; O O tubes; E^1 a door, supported at E with hinges and a pin, in opposite directions supplied with some simple mechanical contrivance to keep it shut, or to open, if circumstances may require to do so. E^2 is a small door to remove ashes which may collect on this place. A^1 is a strong wrought-iron beam, fastened in opposite directions to the engine-truck frame, marked H^5 in Fig. 36. N is a centre-pin; N^3 a plate; A braces bolted on the top of N^3 and the truck-frame. B is the chimney showing, to the right, an outside view, and to the left a section of the same. All the other parts have been heretofore explained.

Fig. 42 is drawn in a manner similar to the previous figure, showing to the right, an outside view from the back part of the locomotive, where the engineer and fireman stands; to the left, a square section view through the fire-box, representing the grate-bars *I I I I*, in section, the tubes *O O O*, braces for upper part of fire-box *M* and A°. A^{2} a section view through the steam dome, with two safety-valves A^{8} placed to the right and left, for the engineer to examine the pressure of steam in the boiler. *K* is a lever which communicates in connection with a rod to the throttle-valve, and is moved by the engineer to regulate the quantity of steam to be used in the cylinders. *N N N* are three gauge-cocks. Steam must come out of the top one when the engine is running; but when the engine is at rest, the water in the boiler is rather lower than when it is in motion, and is therefore high enough if it just reaches the middle gauge-cock. *L* is the reversing lever, supporting and turning on its centre at L° a cast-iron stand, bolted on the flooring where the engineer and fireman stand. *P* is a door, secured through hinges P^{1} P^{1} and a pin to this part of the boiler, in such a manner to open and shut it as may be required by the fireman, to make a fresh supply of fuel to the furnace, for the production of steam. *S* is a strap to support the lever S^{1}—those parts are marked *Y* and Y^{1} in the longitudinal view of the locomotive, Plate XIII. Similar letters of reference indicate the same parts which are previously described of those figures.

Figs. 43 and 44 is the feed-pump to supply the boiler with the necessary quantity of water for the evaporation of steam. The casing of the pump-valve and the two air chambers are made of composition, and highly polished. The plunger S^{7} is of a hard quality of wrought-iron, and is the part which is fastened at 3 in connection with an attachment to the crosshead D^{1}, Plate XIII., will carry the plunger through the whole stroke of 22 inches, and at each outward movement of the plunger draws the water into the pump, and on the contrary movement forces the same through the opening of the valve *A*, air-chamber S^{2}, and feed-pipe S^{8}, and to the check-valve Fig. 45, in raising the valve *B*, and enters into the boiler, either cold, or heated to a boiling degree by a pipe connected to the suction pipes and top of boiler by a small cock shown in Fig. 42, which can be set to admit the proper quantity of steam to the suction-pipe and tender to heat the water. S^{5} is the suction-pipe through which the water is drawn into the air-chamber S^{1}, of which C^{1} is a pipe extending downwards and cast to the same to allow a free space for air or steam to collect, which would otherwise injure the perfect operation of the pump in some degrees. The water enters now the pump between the openings of the valve and valve-seat *B* and B^{2}, and forced by the return of the plunger into the other compartments as heretofore described. The upper and lower air-chambers are bolted by 3 bolts, and kept, by an intermediate cylindrical piece S^{3} cast to the pump barrel *S*, at its proper distance. All the joints on this pump are ground to be air and water tight, under an equal pressure as in the boiler. S^{4} is a pipe cast to the air-chamber and extended downwards to allow a surrounding space for compressed air, which will, in many instances, prevent the bursting of pipes by the operation of the feed-pump. S^{6} is a stuffing-box which is kept by two bolts to its seat. B^{2} and A^{2} are valve-seats, *A* and *B* valves, A^{1} and B^{1} guides to prevent the valves from raising any more than is necessary to allow the quantity of water to pass through. Too much raising of the valves prevents the pump to deliver the full quantity of water, as a part of it will return into the suction-pipe if the valve does not immediately fall on its seat. Fig. 46 is a front view, and 47 a side view of the slide-yoke. $D^{3}D^{3}$ is the yoke, D^{2} the slides which are bolted with a single bolt in the yoke D^{3}, which is well braced with a rod D^{4} to the boiler, and two other braces above and below to the engine frame *R*. Fig. 48 represents a section view through the steam pipes and throttle-valve, of which *B B B B* are cast-iron rings well ground in their seat, to keep the joints steam tight. *A A* are two bolts through the centre of the pipe C^{5}, with two nuts 1 1 and 2 2 above and below to keep C^{5} and C^{4} together. The inside nuts require grinding on their seat to keep that part tight. *D D D D* are four bolts holding the pipes C^5 C^5 to the flanges of the cylinders, and kept tight in the same manner as already set forth. Fig. 49 is an outside view of the part C^4. Fig. 50 is a longitudinal section through the throttle-valve chamber, K^2 of which is the valve, K^1 a rod, and K^3 a bolt, to prevent the valve from going any further; they being placed over the centre of the openings to shut off all communication of steam with the cylinders in that position. Fig. 51 is an inside view of the cover: the valve seat and cover are cast together, and can easily be taken out and repaired if necessary.

The following are the dimensions of the locomotive Talisman:—

Diameter of cylinder, ...	17 inches.
Length of stroke, ...	22 "
Length of fire-box, ...	4 feet 5 "
Width " " "	3 " 11 "
Depth " " "	65 "
Length of flues, ...	11 "
Outside diameter of flues, . .	2 "
Space between flues, ...	⅝ "
Area of grate, ...	17 " 3
Length of boiler, including fire and smoke box, ...	18 " 1 "
Diameter of boiler, ...	3 " 11 "
Diameter of boiler near smoke-box,	3 " 9½ "
Diameter of driving-wheels, . .	5 " 6 "
Face of driving-wheels, . . .	5¾ "
Copper flues, ...	153 pieces.
Diameter of steam domes, . .	21 "
Height of steam domes, . . .	22 "
Diameter of main steam-pipe placed in the interior of the boiler, .	5½ "
Diameter of steam-pipe near cylinder, ...	4½ inches.
Diameter of plunger for feed-pumps,	2 "
Stroke of plunger, ...	22 "
Diameter of chimney, ...	13½ "
Diameter of main driving-axle, .	6¼ "
Diameter of centre-pin on the main driving-wheel, ...	4 "
Diameter of centre for the outside journal, ...	3 "
Diameter of piston-rod, . . .	2¼ "
Depth of piston, ...	5½ "
Diameter of valve-stem, . . .	1½ "

Pressure in boiler, 100 to 120 lbs.

One cord of wood (dry pine) to run 60 miles.

Length of fire-box for bituminous coal-burner, to produce the same quantity of steam, 5 feet 9 inches.

Length of fire-box for anthracite coal-burner, 6 feet, with the addition of a combustion chamber, projecting some 24 inches into the barrel of the boiler.

MANAGEMENT, CALCULATIONS,

AND DRAUGHTING.

WE propose now to devote a few pages to an elucidation of the management of steam-engines, and more particularly of marine and locomotive engines, with calculations for proportioning, and the best method of draughting the same. Stationary engines, though very numerous and important, are usually very similar to one or the other of these varieties; the small high-pressure stationaries being quite analogous to the locomotive but less complex and easier of management, while the larger condensing stationaries are very similar in their general treatment to the various classes of marine constructions. Stationary engines and boilers may in general be stouter and more cumbrous, and may occupy considerably more room than either of the varieties referred to of similar power, and in general we may assume, that parties competent to construct or manage those more compact and complex constructions, can find little difficulty with a stationary of a character in any wise analogous. We will therefore briefly treat, first on the management of locomotive and marine engines and boilers, and subsequently on the calculations and draughting of the same.

MANAGEMENT OF THE LOCOMOTIVE.

A prominent and extremely important part of an engineer's duty, while out on the road, is the observance of signals. The systems of signals vary on different roads, so that no rules can be given therefor, except that he should make it his business to study and become perfectly familiar therewith, so as to be at no loss how to act when an exigency may arise. On some roads minute printed instructions are given, providing for every possible case, and on *all* the proper signals both to be made and to be observed by the engineer are distinctly established, and their observance should, as rapidly as possible, be made a matter of habit, never for a moment to be neglected under any circumstances. Continual vigilance is expected and required on the part of an engineer while running. On some roads he is expected to stand at every moment with his hands in position for action, the left grasping the throttle and the right the reversing lever. On some roads the engineer is made responsible for any accident due to running upon a misplaced switch or turntable, and on all roads a proper regard for his own safety, as well as for that of the train behind him, demands a constant and extremely careful attention to all his multifarious duties. The hours of active service are usually few in comparison with those of many other professions, and the whole energies of the man are expected and demanded during this period.

Beyond a proper watch over the track, the engineer's chief duty is to regulate the manner of firing and the supply of water. In firing, the door must be kept open as short a time as possible, the furnace should not be crowded quite full, and the wood or coal, should never be thrown against the tubes. If an engine has a variable exhaust, open it wide in firing, to slacken the draught, and as soon as the door is closed, contract the exhaust to quicken the draught. At any rate, it is best to keep the damper wide open in firing, so as to draw some air directly through the fire, although much more will necessarily draw in at the furnace door. Every engine should have double dampers, one on the back side of the ash-pan, so as to save trouble in snows. The fire should never be put out by throwing water in the furnace, as the sudden cooling may set the tubes leaking. The damper can be shut, and the brands drawn out with tongs and extinguished outside. The fire should be very low on reaching terminal stations—enough to reach the engine-house is just right. A good fire should be on when approaching ascending grades, so as to go up with a good head of steam. The furnace may be also filled just before reaching a roadside station, the steam being partly shut off, so as to give a light draught. The engine will then have plenty of steam in starting. An engineer should be careful to slip his drivers as little as possible, as it not only strains the engine, cuts out the tires, and endangers a serious fracture of the parallel rods, but wastes steam, and consequently wastes fuel.

In regulating the water, an engine will work drier steam when the water is rather low; but great care should be taken not to allow the water to become too low, and great care is required not to be deceived by the gauge-cocks, especially when the engine is working heavily; the color of the exhaust can be generally taken as a tolerable indication of the state of the water. Care must be taken not to overpump the boiler, as well as not to let the water get down. In over-pumping, water is wasted, and all the fuel it has taken to heat it up—so also the engine is strained. The boiler should be well filled on approaching bad grades, way stations, and especially terminal stations. On long levels, and on down grades, the boiler can be supplied with one pump but a very little on. Pumping up checks the formation of steam, and care must be taken to pick the best places in the road for pumping. In standing for any length of time, the fire should not only be checked to prevent blowing off, but some steam may be turned into the tank, to heat the water, thereby saving wood.

The valves and cylinders require to be greased with melted tallow according to the character of the road. If steam is shut off in running down grade, more tallow should be put in the cylinders than would otherwise be necessary. If the engine is about to encounter a heavy pull, on a grade, the valves should be well oiled to prevent sticking. If the valve-motion is reversed in bringing up at a station, hot air will be pumped in out of the smoke-box, and more grease will be required to prevent the packing from becoming cut out. The cylinders should always be greased before going into the engine-house, so as to prevent rusting. The cylinders are of course oiled while the engine is in motion. The bearings will require oiling about every time the engine stops for wood and water.

An examination should be made of the principal bearings, also a glance at the keys, and at the nuts, and other parts of the running work, as often as the engine stops for any length of time. The bearings, especially after any repairs, or after the boxes have been taken up, should be looked after, and if found heating, should be taken out and examined, when the difficulty may discover itself in a bad fit, or rough rubbing surface.

An engine runs freer and works drier steam with the throttle only partly open—or at least not entirely open, unless the throttle-ports are very small.

The pistons require to be looked after certainly once a month, and generally oftener. Allusion was made on page 60 to the prevailing tendency to set out the packing of the piston very tightly at long intervals, instead of, as it should be, to maintain at all times only sufficient tightness to prevent the leakage of steam. Pistons working as tightly as is frequently allowed, abrade and enlarge the interior of the cylinder tenfold more rapidly than would result from proper management. Packing which is new, or with cylinders newly bored out, will require extra care. So with cast-iron rings. It can be told by the sound of the exhaust, if the pistons are blowing through.

The outside case around the smoke-pipe must be kept clean, by emptying it daily, unless the sparks are allowed to run back through a hole into the inside pipe, and to burn up in the smoke-arch. In going through stations and covered bridges, shut the damper. Avoid as much as possible the necessity of throwing green wood on a low fire just before reaching such places, as the engine is thus most apt to throw sparks. The damper should be fitted tight, and should of course be closed, or nearly closed, while waiting at stations.

Never slip the tender or car-wheels by the brakes. Slipping heats and softens the chill, thus making a flat spot.

The boiler must be kept clean, by blowing off as often as the nature of the water requires it. With hard, brackish or muddy water, the boiler will require to be often blown out, and in such cases there should be a blow-off cock at both ends, the front one being in front of the smoke-box, with a pipe back through to the bottom of the tube-sheet. The back cock should be either on the side or hind end of the furnace, so as not to throw up dirt in the bearings. The boiler should never be blown off with a fire in, or under a considerable head of steam, as it is liable to be burned in one case, and to be too suddenly cooled in the other. There should be only steam enough left to get the engine into the house.

We have on page 36, alluded to the liability of boilers to throw water instead of steam into the steam-pipe. Locomotive boilers are likely to foam in the following cases:—When the steam-room is small; when the boiler is new, or has had repairs in it which have left grease and dirt inside; where the water is bad. In the first case, the water should be run as low as it can safely be, and care taken not to slip the wheels. In the next case, use care also, and use sugar in the boiler. For bad water, use care and blow off frequently.

For scale in boilers, a handful of mahogany saw-dust will tend to loosen it, when it may be blown out after the next trip. A powder is sold which will loosen scale, and is said not to injure the boiler. Scale, by coating the tubes, destroys their conducting power, wastes heat, and is liable to burn out the tubes.

A leaky boiler may be sometimes made tight by putting a few potatoes inside.

The water spaces around the furnace need looking after. If they fill up with mud, the sheets will burn through.

The pumps require care, to be sure they are in working order. They are watched by the pet-cocks, and by the rise of water at the gauge-cocks. If they fail the valves must be taken down and examined, and the obstruction removed.

The red-lead joints require to be packed with putty of a firm, even consistency. The stuffing-boxes are best packed with hemp soaked in warm water. The sulphur in India-rubber eats off the screws of the stuffing-boxes. The packing should never be left in the stuffing-boxes until it is rotten and burnt, as it then scratches the rods in contact with it.

In taking off and putting on the cylinder-heads and steam-chest covers, the nuts must be loosened lightly all around in taking off, and tightened in the same manner in putting on. Finished bolt-heads must be driven only with a copper hammer, and nuts tightened only with a wrench.

In keying up the boxes of the connecting and parallel rods, the rods must be set to a good fit by trying them first in one place and then barring the engine half a turn along. The engine must stand on an even rail, with the boxes at the same height in both pedestals, when the parallel-rod boxes are fitted. The outer box at one end, must be a little slack so as to let one box rise or fall in the jaw without straining the rod.

When the pistons are examined, it must be seen that they are not working out of the true centre of the cylinder. When the piston is at the forward end of the cylinder, the spider or head falls, by its weight, below the proper place, unless the bottom springs are set to a little more tension than the upper springs.

Each eccentric should have its proper place marked on the axle, and the valve-rod should have its mid-throw position and its right lead marked on it. If the eccentrics slip, the effect can be heard in the exhaust, when the marks will help to get the eccentrics into their right places.

The truck and driving-wheels must be observed occasionally, to see that they are not working off the axles.

The engineer must thoroughly canvass his resources, and be prepared, as far as possible, for all the various minor accidents which are liable to occur. When a tube fails, a wooden or metallic plug, must be driven into each end of it. When a leak from this source, or any other, becomes so great that the level of the water in the boiler cannot be maintained, it will be necessary to remove the grate-bars and extinguish the fire, in order to save the other tubes and the fire-box from injury.

If any accident happens to the connections or valve-motion of one cylinder, when on the road, the rods on that side, and the parallel-rod on the opposite side, must be detached, the valve set midway, so as to cover both ports, and held there, and a trial made to get the engine in with a single cylinder. If one parallel-rod breaks, take the other off, and make the best of the way to the next station, or, if possible, to the end of the trip.

If a spring strap breaks, the spring can be held temporarily with a chain, if one is on the engine.

There have been cases of the throttle-valve becoming detached when the engine was running, so that the steam could not be shut off. In such a case, with a link motion, throw it into mid-gear and apply the brakes to stop. With V hooks the cut-off valve must be set to cover its ports, unless it works on the back of the main valve, when the steam must be immediately let down and the engine reversed if there is danger of running too far. If the engine gets off the track, but only so that the wheels just drop between the rails, it can be again run on with its own steam, by laying sticks of wood right to lead the engine on. Or, the wheels may be jacked up until the flanges are a little above the rails, and the engine then swayed over the rails by another jack-screw, braced against the wheels.

We cannot well prolong these suggestions without encroaching too much on our limited space, we must pass to the next subject.

MANAGEMENT OF THE MARINE ENGINE.

The position of chief engineer of a sea-going steamer is one of the greatest responsibility in the profession, on account of the very serious nature of the results which may accrue from a failure of the power in an emergency. There are usually a sufficient number of assistant engineers to form two watches complete, without an absolute necessity for the personal inspection of the chief, and on the assistant engineers must necessarily rest a great share of responsibility in regard to every detail. Some chief engineers make it a rule to visit the engine and fire-rooms at an unexpected moment in each watch, even in good weather and with every thing working favorably.

Steamship engineers are usually put in charge of all that directly relates to the propelling power, including the boilers and the fuel. The boilers require much attention both at sea and in port, especially if they be complex tubular constructions. The great points at sea are, the firing, the feeding, and the blowing off; the great points in port, the cleaning and repairing. If the boiler be blown off by means of blow-off cocks, the operation should be performed twice in the watch, or once every two hours. The feed should be so set that the water will rise in the course of two hours from a little below the middle to near the top of the glass gauge tube; the rule being to blow off so frequently or so much as to prevent any accumulation of scale within the boiler. *In every case in which there is an accumulation of scale in the boiler, the fault lies with the engineer*, who is either ignorant of his duty or inattentive to it.

In boilers furnished with brine pumps, reliance must not be placed upon the pumps always acting well, and once every watch some water must be drawn off from the boiler to be tested by a salt gauge, to see whether it is too salt or not. When the water has been evaporated to such an extent as to reduce its volume some three or four fold, the saltness becomes so excessive that solid salt is liable to form upon the exposed surfaces. The saltness of ordinary sea-water varies somewhat in different places, but as a general rule, there is about one pound of salt in every thirty-three of sea-water. When by boiling, the proportion of salt is increased to about $\frac{4}{33}$, the formation of a scale, consisting mainly of salt, is likely to commence. It is important therefore to blow out a portion and to supply its place by new, so often as to keep the water fresher than $\frac{4}{33}$; but on the other hand, every exchange of hot water for cold diminishes the supply of steam or increases the consumption of fuel. To ascertain the saltness of the water as accurately as possible, hydrometers or salinometers are employed on many of our best steamers. In the absence of these instruments, an engineer may make one for himself in the following manner:—Take a glass phial or eau de Cologne bottle, pour into it so much shot that it will nearly sink in sea water, and then cork it tightly. Take any convenient weight of boiling water, say 33 lbs., dissolve therein 1 lb. of salt, and then put the phial into it turned upside down, so that the shot will rest against the cork; make a mark at the point at which the water stands on the phial; this represents the saltness of sea water. Add then another pound of salt to the water, marking the point on the phial at which the water stands, and repeat the operation until 12 lbs. of salt have been added, at which point the water will have received as much salt as it can dissolve; transfer the marks upon the bottle to a paper scale, which paste on the inside of the bottle in exactly the same position as the original marks. You will then have a salt gauge which will tell the saltness of brine from the point of sea water up to the point of saturation. Reckoning sea water at 1, the water within the boiler should not exceed the saltness represented by 4, at which point the water contains $\frac{4}{33}$ds of salt. It is not probable that this rude contrivance will be often made, but the description may be of service in explaining the nature of the more elaborate and accurate instruments sold for the purpose.

In maintaining the pressure of the steam, very much depends on the skill and zeal of the firemen. One fireman will keep the steam up with a moderate consumption of coal, while another, with an increased consumption, will not prevent the pressure from declining, and yet it is hard to say wherein lies the difference of manipulation. It is a common fault, however, with lazy or ignorant firemen, to pile up the coal at the mouth of the furnaces, while the bars at the after end are nearly bare, and if there be any holes in the fire, the cold air will rush up through them and greatly diminish the efficacy of the fuel. Opening the furnace door frequently has also a pernicious effect, and should be avoided as much as possible. The greater width between the bars at the under than at the upper edge, facilitates the admission of the air, and the descent of the ashes and cinders. Bars are usually set $\frac{3}{8}$ths of an inch asunder, but this width must be diminished if the coal be very small, and can be diminished with impunity when the bars are thin.

Partially closing the damper will sometimes increase the generation of steam, and in such cases it is expedient to place a sheet-iron hanging-bridge at the end of the flues, where they enter the chimney. In the case of tubular boilers, however, this cannot be done; but a sliding perforated plate, or Venetian damper, may be so applied to the ends of the tubes, as to retain the hot air for a longer period within them, thereby increasing the efficacy of the fuel. Whatever be the steam-producing powers of the boilers, a vacuum should never be suffered to be formed within them, as it is impossible to blow off if there be a vacuum in the boilers, and the gauge cocks, moreover, will in such case cease to afford any indication of the height of the water level. If the pressure of the steam cannot be maintained, that grade of expansion must be used which will cut off the steam at such a point as will keep the pressure at the right pitch; or if there be no variable cut-off, the throttle-valve must be so far closed as to keep the steam gauge nearly up to the point answering to the load on the safety-valve. Partially closing the throttle-valve checks priming, and opening the throttle-valve or the safety-valve suddenly has a great tendency to produce it, as has already been mentioned. When the boiler primes, the speed of the engine is sure to be diminished, in consequence of the large quantity of water the air pump has then to deliver, and it will be expedient in such cases partly to shut off the injection water. In the case of the water being suddenly carried out of the boiler by priming, or of the water having been suffered to subside too far by the neglect of the feed, the best plan is to quench the fires as rapidly as possible. If from the neglect of the feed, the flues or furnaces have become red hot, on no account must cold water be thrown in by the pump, without taking care to raise the safety-valve, else the sudden pressure within the boiler thus created will be sure to make the heated places bulge down, and may perhaps burst the boiler. A plate which has bulged down may be set up again by lighting a fire against it, so as to make the plate hot, and then forcing it up with a jack-screw.

One of the greatest dangers that can occur to a boiler, is that of a safety-valve jamming or refusing to act. Every boiler ought to have a safety-valve of its own, and if this valve should jam, the steam has still a means of escape through the stop-valves into the other boilers, the safety-valves of which are not likely to be similarly affected. The existence of a dangerous pressure within the boiler is always shown by the open mercury gauges, where such are employed, the contents being blown out when the pressure of the steam becomes dangerously high.

Two or more fires are usually cleaned every watch, depending on the number of the furnaces and the quality of the coal. The fires to be cleaned are suffered to be burned down until there is very little else than clinker left upon the grate, and the whole of this clinker is then raked off, and the fire is lighted afresh. The operation of cleaning the fires is usually performed just before the termination of the watch, and the whole of the ashes are then hoisted up and thown overboard, the firemen on watch filling the ash-buckets, and the firemen about to come on watch hoisting them up and emptying them into a shoot in the wheel-house, or over the ship's side.

In the management of the engines, the first point to be looked to is that the keys are neither too slack nor too tight, and that none of the brasses are heating. In the generality of engines, the bearing most apt to heat is the crank-pin, but much depends on the proportions of the parts, which differ in different engines. The provisions for very nicely adjusting the brasses about the crank-pin of the Knoxville have been very fully described at page 41, and as the crank can be easily touched with the hand at each revolution, there would seem little excuse in such engines for allowing a serious heating at that point. But some engines are less fortified against difficulty in this respect, and in any it is probably that through some cause a heating may be commenced at such points. It is usual to lubricate the crank-pin, when it heats, with sulphur and tallow, and the same discipline is observed in the case of other bearings. This plan answers well enough when the heating is not considerable, but it will sometimes be necessary to cool the bearing by cold water applied by means of the hose communicating with the deck pump. Bearings, however, rarely heat unless they are too tightly screwed up or the supply of oil to them has been neglected, and they must be slackened and lubricated as well as cooled with water. The heating of a bearing very frequently injures the surface of the metal, and cuts away the brass very much: it should therefore be checked at the outset, and in replenishing the oil cups, which the engineer should periodically do, he ought to feel the bearings to make sure that they are not hot. He should see at the same time that none of the keys are working loose. A looseness of any of the main bearings will generally manifest its existence by a jerk in the engine, but some of the keys of the parallel motion may come out and occasion serious breakages without giving any such warning. The same remark applies to any of the main keys which are not made with a taper, such as the cross-head or cross-tail keys of side-lever engines, which may come out without giving intimation of the danger. The main keys about an engine should all be provided with screws, to prevent them from going either back or forward. Generally speaking, keys have a tendency to work further in, whereby, if the tendency be not counteracted, they will cause the bearings to heat.

The state of the vacuum will be shown by the vacuum gauge attached to the condenser, and if it be imperfect, the cause must be ascertained and the fault corrected. If the hot well be much more than blood warm, more injection water must be admitted, and if the vacuum is still imperfect, there must be some air leak, which the engineer must endeavor to discover. Very often the fault will be found to lie in the valve or cylinder cover, which must then be screwed more firmly down, or in the faucet joint of the eduction pipe, the gland of which will require to be tightened, or the leaking part puttied up. The cylinder and valve stuffing-boxes may at the same time be supplied afresh with tallow, and the door of the condenser examined, if the engine be provided with one. The joints of the parts communicating with the condenser are usually tried with a candle, the vacuum sucking in the flame if the joint be faulty.

The attendants upon engines should prepare themselves for any casualty that may arise, by considering possible cases of derangement, and deciding in what way they would act should certain accidents occur. The course to be pursued must have reference to particular engines, and no general rules can therefore be given; but every marine engineer should decide on certain measures to be pursued in the emergencies in which he may be called upon to act, and where everything may depend upon his energy and decision. If the ship springs a leak, the water may generally be kept under by injecting from the bilge, and every steam-vessel should be provided with cocks for this purpose. These cocks should not communicate with any rose within the condenser, as the water drawn from the bilge is not clean water, and a rose within the condenser would probably soon become choked up. Should there be no injection from the bilge, a great deal of water may be lifted out by partly opening the snifting valves, but should they be of such a construction as not to admit of being opened by a handle, or should they be in an inaccessible position, the cover of the foot valve, or the man-hole door of the condenser, may be slackened. If the snifting valve cannot be opened readily, the injection may be shut off, so that the engine will heat and vitiate the vacuum, and the valve will then open of its own accord during the descent of the air-pump bucket. When raised, it must be prevented from closing again by something being wedged in below it: the steam will then be condensed in the air-pump, and the water drawn through the snifting-valve will, in all ordinary leaks, soon leave the ship dry. The fire hose must on no account be used for pumping out the boilers, washing the decks, or other such purposes, but must be kept in a locker close beside the deck-pump, for the single purpose of quenching fire; and should be examined and oiled once a month. Cotton waste, if oil be spilled upon it, is liable to spontaneous combustion, especially if kept in a hot place; and such combustibles should never be kept in lockers behind or near the boilers, or in any place where there is much light wood work that would readily take fire. The coals often catch fire from spontaneous combustion, and serious damage arises therefrom; they should be quenched immediately the fire is discovered, by directing upon them a stream of water from the hose. Coal or combustibles of any kind should not be kept under the cabins, as it would be almost impossible to put out a fire among the light joiner-work of the cabins. If collision occur, and the steamer is cut down by another vessel to the water's edge, it may be necessary to blow off one wing boiler instantly, and fill up the other, so as to give the vessel a list to raise the broken part out of the water. There is no time for hesitation in such cases; the engineer must decide on the instant what he is to do, and must do it at once. In some cases of collision, the funnel is carried away and lost overboard, and such cases are among the most difficult for which a remedy can be sought. If flame come out of the chimney when the funnel is knocked away, so as to incur the risk of setting the ship on fire, the uptake of the boiler must be covered over with an iron plate, or be sufficiently covered to prevent such injury. A temporary chimney must then be made of such materials as are on board the ship. If there are bricks and clay or lime on board, a chimney may be built with them, or if there be sheet-iron plates on board, a square chimney may be constructed of them. In the absence of such materials, the awning stancheons may be set up round the chimney, and chain rove in through among them in the manner of wicker-work, so as to make an iron wicker chimney, which may then be plastered outside with wet ashes, mixed with clay, flour, or any other material that will give the ashes cohesion. War steamers should carry short spare funnels, which may easily be set up should the original funnel be shot away; and if a jet of steam be let into the chimney, a very short and small funnel will suffice for the purpose of draught.

If a crank or crank-pin breaks, the engine, if single, is of course completely disabled; but if there be two engines, the cranks must be disconnected, and the ship worked with the other engine and one wheel. If the shafts or cranks crack, the engine may nevertheless be worked with moderate pressure to bring the vessel into port; but if the crack be very bad, it will be expedient to fit strong blocks of wood, to prevent the cylinder bottom or cover from being knocked out, should the damaged part give way. The same remark is applicable where flaws are discovered in any of the main parts of the engine, whether they be malleable or cast-iron; but they must be carefully watched, so that the engines may be stopped if the crack is extending further. Should fracture occur, the first thing obviously to be done is to throw the engines out of gear, and should there be much way on the vessel, the steam should at once be thrown on the reverse side of the piston, so as to counteract the pressure on the paddle-wheel.

The following method of ascertaining the tightness of the different parts of the engine subjected to steam pressure, has been recommended to be used in every case after the engine has been fresh packed, or has been out of use for some time. After getting up steam, and while the vessel is still at her moorings, blow through, and then, after obtaining a partial vacuum in the condenser by the admission of a little water, watch the barometer to see how long the engine holds her vacuum. If the condenser gradually becomes hot, while the cylinder ports remain closed, we know that steam is passing the valves. The tightness of the piston may be proved in the same way, by admitting steam above or below it, and opening the indicator-cock on the opposite side. The injection-cock may be slightly opened for an instant, to withdraw any steam that may have collected on the opposite side of the piston, so that the passage of any steam may be the more readily perceived. The tightness of most parts of the engine may be tested in this way without moving it beyond half a stroke.

When a leakage of air into the condenser or its connections has been discovered, it may be stopped temporarily by driving in spun yarn, or gasket steeped in red lead and oil, or if the leakage be into the condenser, it is sometimes convenient to allow water to be injected through the orifice, by which means little harm is done. In several cases where, during a long voyage, the bottom of the condenser has become leaky by corrosion, (often induced by galvanic action with the copper bolts of the ship's bottom, as well as the brass foot-valve, &c.,) a water-tight tank has been constructed at sea between the side keelsons. By this means the condenser and air-pump are placed in a kind of well constantly replenished with cold water from the sea, which, forcing its way through the leaks by the pressure of the atmosphere, shares with the proper injection water the duty of condensing the steam—the injection-cock orifice being partially closed in proportion to the extent of leakage through the bottom.

When the vessel is laboring in a heavy sea, it is recommended that the supply of injection water should be diminished ; for in such a case, where the speed of the engines is subject to great and constant fluctuations, depending upon the greater or less submersion of the wheels or screw-propeller, the condenser is liable to become choked with water, thereby causing the engines to stop. The effect of working the engines with a stinted supply of condensing water is, of course, that the condensers will become hot, and the vacuum will be diminished; but this is a minor evil in comparison with endangering the machinery by subjecting it to too severe a strain.

Care must be taken, when the engines make a temporary stoppage, that the injection-cock or air-pump, does not leak, and allow the condenser to fill with water, which causes much trouble and delay in starting the engines again ; so, should this be apprehended, the sea-cock must also be closed at the same time with the injection-cock.

With regard to the adjustment of any important part, it must be borne in mind, that as spirit levels and plumb rules cannot be used on board ship, every thing must be done by straight edges and squares. Every engineer, therefore, on taking charge of a pair of engines, on their coming out of the hands of the manufacturing engineers, should see that centre lines are scored well into the framing, at a sufficient number of parts, to facilitate any future examinations as to whether the engines have altered their position in any way, as well as to facilitate his putting the engines very correctly, when so required, at half stroke, and many other operations. Athwartship lines should certainly be scored on the cylinder flanges, across the centres of the two cylinders, and on the base plates in beam engines, under the centre of the crank shafts. A fore and aft line in the centre line of each engine should also be scored along as much of the base plate as possible. It is also usual in well-constructed engines to have four horizontal points in an athwartship line on the framing, dressed off so that four points in a true line on the face of a straight edge may lie upon the whole of them, and thus prove at any time whether the engines have fallen in towards each other, or fallen away towards the sides of the vessel.

If the eye of the crank of the paddle shaft be perceived to bear hard upon the connecting rod brasses at one part of its revolution, and to separate from them at another part, the engineer may know that the centre of the paddle shaft is out of line with the centre of the intermediate shaft. To rectify this defect, place the engines on the top stroke and measure the distance accurately between the faces of the two cranks at the side of the crank pin, then put the engines on the bottom stroke and measure the distance at the same place. If the distance at the bottom be less, the outer end of the paddle shaft must be too low and require to be raised. Subtract the one distance from the other, and take one half of the remainder, and say, as the length of the crank is to the length of the paddle shaft, from the face of the crank to the centre of the outer bearing, so is this half-remainder to the amount that the outer bearing requires to be raised. Other examinations as to whether the crank and paddle shafts are true to each other, in other respects, can be made in a similar manner.

CALCULATIONS.

Before attempting to do justice to the subject of proportioning, we must explain at some length the nature of steam and of combustion.

The nature of heat is unknown. There are but two theories extant, one of which considers heat as a very thin fluid somewhat resembling a gas, but without weight, while the other considers it to consist simply in a motion of the atoms of matter, a tremulous or vibratory motion analogous to sound. Both are open to grave objections, as neither suffices to explain satisfactorily all the phenomena observed in relation to this mysterious agent. In our researches, then, into the action and nature of steam, we must be content to leave the intrinsic nature of heat or caloric, the agent which is relied on to produce all the effects observed, and by the transfers and modifications of which all the immense steam power in the world is realized; we must be content, we repeat, to leave this as one of the secrets yet locked up in the arcana of nature. Science has not yet unravelled the mystery.

But much has been done to ascertain precisely the quantities of heat required to produce certain effects, and to follow it through all the processes in which it plays a part; these we will attempt briefly to set forth.

There are two forms in which the presence of this mysterious agent is recognized;—these are sensible heat and latent heat. Sensible heat is that which can be felt by the senses, and measured by the thermometer, while latent heat is only effectual in maintaining a certain character or condition in matter. These terms, as also the ideas which they are generally made to convey, are in some confusion, and there are those who stoutly contend that the term "latent" should never be used in reference to heat. It will suffice for our purpose, however, to say, that whatever heat is palpable *as such* to the senses, or can be measured by the thermometer, is sensible heat, while that which cannot be, but which usually manifests itself in sustaining matter in some different condition from that it would otherwise possess, may be termed latent. If an open vessel of water be placed upon a fire, the temperature of the water will not rise above the boiling point, or 212° Fah., (100° Celcius or 80° Reaumur,) however long the boiling be continued, although the water must have been all the while receiving accessions of heat from the fire. All the heat received over and above that requisite to produce the temperature of 212° is expended in the formation of steam, but the steam itself does not rise above this temperature, and on account of the heat thus hiding itself, as it were, in the steam, it is called latent heat. Steam and water are intrinsically the same substance, and a thermometer immersed in each would indicate the same temperature, yet the steam is light and gaseous in its properties, while water is dense and liquid. The difference in condition is due to the latent heat which has entered, or perhaps we may say chemically combined, with the particles of water. The condition is such as to rob it, for the time being, of what we are accustomed to consider its true character, and to impart a new character to the matter, as already observed.

The quantity of latent heat in steam at the atmospheric pressure is about 1000° Fah. The sensible heat in such steam is 212° Fah. The latent heat in such steam being 1000°, the total quantity of heat therein is therefore 212° + 1000° = 1212°. If the whole quantity could be rendered sensible, it would indicate by a suitable thermometer 1212°; equal to that of red hot iron. Or, if all the heat in a given quantity of steam be transferred to 1,212 times that quantity of water, it will raise the temperature of the water 1 degree. From this it appears, that it requires about 5½ times as much heat to raise any given weight of water at the freezing point into steam, as would raise the same weight of water from the freezing to the boiling point. The freezing point is 32°, and the boiling 212° Fah., so that the amount of difference between these points is 180°, and 180° multiplied by 5½ is 990, or nearly 1000° latent heat.

One pound of steam of 212° Fah. contains also 6½ times as much heat as one pound of boiling water of the same degree, or 5½ pounds cold water at 32°, can condense one pound of steam at a temperature of 212° Fah., producing 6½ pounds of water of 212° by the addition of the one pound of condensed steam.

It must be recollected, however, in reference to condensation, when considered practically, that, in order to create a tolerable *vacuum* by the condensation of the steam, the temperature of the resulting water must be very much below this point, and that consequently a much greater quantity of water must be injected for the purpose of condensation. The temperature in the condenser, as explained in the preceding pages, should be only about 100° Fah., or at most not above 120° Fah., and to obtain this result, the quantity of injection water should be doubled, making it 11 pounds instead of 5½. This is assuming the water to be of a temperature of 32° Fah., whereas, in practice it varies from about 40° to 80° Fah., and may be assumed to average about 60° Fah., a fact which will make it necessary again to double our quantity, which is thus raised to about 22 pounds. The quantity actually required at any given temperature may be found with tolerable accuracy by the following process. Let us assume the temperature of the water at 70°, and the temperature of the condenser to be maintained at 100°, so as to give a very perfect vacuum. It is evident that each particle of injection water will be raised 30° Fah. by the heat which it absorbs or abstracts from the steam. Each pound of steam contains, as we have already found, about 1212°, (including both the latent and the sensible heat,) and as the resulting water is to be at 100°, of course, 1112° must be abstracted. It becomes, then, simply necessary to find by division how many times 30 is contained in 1112, which is 37,—showing 37 pounds of condensing water at 70°, are required for each pound of steam, if we would maintain the low temperature designated in the condenser. And by a similar calculation, the relative quantity of condensing water for any other conditions may be ascertained.

The easiest method of ascertaining the weight of steam used, is to weigh or measure the quantity of feed water supplied to the boiler. If one cubic foot of water [which weighs 62½ pounds,] is pumped into a boiler each minute, it is usual to infer that 62½ pounds of steam flow out of the boiler per minute, (although, in strict scientific accuracy a trifle should be deducted for water mingled with the steam,) and if the engine is low pressure, that an equal quantity is discharged in that space of time into the condenser. But if the volume and pressure of the steam used per minute, per second, or per stroke, be known, it is almost equally easy to ascertain the weight of the steam from the table given below. It is in the latter cases simply necessary to reduce to cubic feet the quantity of steam used, in the given time or in a single stroke, and to divide that amount by the corresponding number in the 5th column of our table,—the quotient is the quantity of water, in cubic feet or fractions of a cubic foot. And knowing that a cubic foot of fresh water weighs 1000 ounces, or just 62.5 pounds, the operation may then be conducted as above.

It is in practice very unusual and inexpedient to supply so large a quantity of condensing water, as the result of the above calculation indicates. It is impracticable to create a good vacuum with so warm condensing water as we have there presumed. The load on the air-pump, in removing so large a quantity, would be a greater hindrance than the improved vacuum would gain. It is, therefore, in such a case, better to allow the condenser to warm to 115 degrees or 118 degrees, as it frequently does with advantage in low-pressure engines, even with the water several degrees cooler than we have there assumed. In proportioning engines, pumps, etc., for a great variety of conditions, it is rare to provide for the admission into the condenser of more than thirty times as much water as is pumped into the boiler. Watt's rule was, to supply a wine pint (28.9 cubic inches) of condensing water for each cubic inch of feed; or, in other words, he injected into his condenser about twenty-nine pounds of cold water for each pound of steam which it received from the cylinder, and, of course, pumped out by his air-pump thirty pounds of warm water, together with the trifling amount of air which had chanced to gain admission with either.

In discussing the fundamental properties of steam above, we have been particular to specify steam at atmospheric pressure, in giving the figures, because in steam, at other pressures, the relative proportions of sensible and latent heat are different. Steam at a higher pressure possesses a greater degree of sensible heat, or stands at a higher temperature by the thermometer, but, it is a remarkable fact, that the quantity of latent heat diminishes in very nearly the same proportion, so nearly, in fact, that it has required very refined experience to determine that it is not precisely so. For all practical purposes, therefore, the total amount of heat, in steam at all pressures, may be assumed to be precisely the same, an amount about 1212° Fah. above zero, or 1180° above the freezing point. When, therefore, it is desired to know the amount of latent heat in steam at any pressure, it is simply necessary to ascertain its sensible heat, by the thermometer, and to deduct it from the above total sum.

Steam, at any given pressure, always stands at a certain temperature, which is termed "the temperature due to the pressure." The temperatures due to various pressures have been ascertained with considerable care, and are presented in the following Table, as are also the volumes of the steam compared to that of the water from which it is generated:—

TABLE OF THE ELASTIC FORCE, TEMPERATURE, AND VOLUME OF STEAM.

From a Temperature of 32° to 457° Fah., and from a Pressure of 0.2 *to* 900 *Inches of Mercury.*

Elastic Force in		Pressure above Atmosphere.	Temperature.	Volume.	Velocity of Escape.	Elastic Force in		Pressure above Atmosphere.	Temperature.	Volume.	Velocity of Escape.
Inches of Mercury.	Pounds per Square Inch.					Inches of Mercury.	Pounds per Square Inch.				
.200	.098		32	187407		.524	.257		60	75421	
.221	.108		35	170267		.616	.302		65	64762	
.263	.129.		40	144529		.721	.353		70	55862	
.316	.155		45	121483		.851	.417		75	47771	
.375	.184		50	103350		1.	.49		80	41031	
.443	.217		55	88388		1.17	.573		85	35393	

TABLE OF THE ELASTIC FORCE, TEMPERATURE, AND VOLUME OF STEAM.—*Continued.*

From a Temperature of 32 to 457° Fah., and from a Pressure of 0.2 to 900 Inches of Mercury.

Elastic Force in		Pressure above Atmosphere.	Temperature.	Volume.	Velocity of Escape.	Elastic Force in		Pressure above Atmosphere.	Temperature.	Volume.	Velocity of Escape.
Inches of Mercury.	Pounds per Square Inch.					Inches of Mercury.	Pounds per Square Inch.				
1.36	.666		90	30425		93.84	46	**31.3**	277.8	598	
1.56	.774		95	26686		95.88	47	**32.3**	279.2	586	
1.86	.911		100	22873		97.92	48	**33.3**	280.5	575	1690
2.04	1.		103	20958		99.96	49	**34.3**	281.9	564	
2.16	1.068		105	19693		102.	50	**35.3**	283.2	554	
2.53	1.24		110	16667		104.04	51	**36.3**	284.4	544	1720
2.92	1.431		115	14942		106.08	52	**37.3**	285.7	534	
3.33	1.632		120	13215		108.12	53	**38.3**	286.9	525	
3.79	1.857		125	11723		110.16	54	**39.3**	288.1	516	
4.34	2.129		130	10328		112.2	55	**40.3**	289.3	508	1750
5.	2.45		135	9036		114.24	56	**41.3**	290.5	500	
5.74	2.813		140	7938		116.28	57	**42.3**	291.7	492	
6.53	3.1		145	7040		118.32	58	**43.3**	292.9	484	1774
7.42	3.636		150	6243		120.36	59	**44.3**	294.2	477	
8.4	4.116		155	5559		122.4	60	**45.3**	295.6	470	
9.46	4.635		160	4976		124.44	61	**46.3**	296.9	463	
10.68	5.23		165	4443		126.48	62	**47.3**	298.1	456	
12.13	5.94		170	3943		128.52	63	**48.3**	299.2	449	
13.62	6.67		175	3538		130.56	64	**49.3**	300.3	443	
15.15	7.42		180	3206		132.6	65	**50.3**	301.3	437	
17.	8.33		185	2879		134.64	06	**51.3**	302.4	431	1816
19.	9.31		190	2595		136.68	07	**52.3**	303.4	425	
21.22	10.4		195	2342		138.72	08	**53.3**	304.4	419	
23.64	11.58		200	2118		140.76	69	**54.3**	305.4	414	
26.13	12.8		205	1932		142.8	70	**55.3**	306.4	408	
28.84	14.13		210	1763		144.84	71	**56.3**	307.4	403	
29.41	14.41		211	1730		146.88	72	**57.3**	308.4	398	
30.	14.7	**0.**	212	1700		148.92	73	**58.3**	309.3	393	1850
30.6	15.		212.8	1669		150.96	74	**59.3**	310.3	388	
31.62	15.5	**0.8**	214.5	1618		153.02	75	**60.3**	311.2	383	
32.64	16.	**1.3**	216.3	1573		155.06	76	**61.3**	312.2	379	
33.66	16.5		218.	1530		157.1	77	**62.3**	313.1	374	
34.68	17.	**2.3**	219.6	1488		159.14	78	**63.3**	314.	370	
35.7	17.5		221.2	1440		161.18	79	**64.3**	314.9	366	
36.72	18.	**3.3**	222.7	1411		163.22	80	**65.3**	315.8	362	
37.74	18.5		224.2	1377	874	165.26	81	**66.3**	316.7	358	
38.76	19.	**4.3**	225.6	1343		167.3	82	**67.3**	317.6	354	
39.78	19.5		227.1	1312		169.34	83	**68.3**	318.4	350	
40.80	20.	**5.3**	228.5	1281		171.38	84	**69.3**	319.3	346	
41.82	20.5		229.9	1253		173.42	85	**70.3**	320.1	342	
42.84	21.	**6.3**	231.2	1225		183.62	90	**75.3**	324.3	325	1004
43.86	21.5		232.5	1199		193.82	95	**80.3**	328.2	310	
44.88	22.	**7.3**	233.8	1174	1135	203.99	100	**85.3**	332.	295	
45.90	22.5		235.1	1150		214.19	105	**90.3**	335.8	282	1050
46.92	23.	**8.3**	236.3	1127		224.39	110	**95.3**	339.2	271	
46.94	23.5		237.5	1105		234.59	115	**100.3**	342.7	259	
48.96	24.	**9.3**	238.7	1084		244.79	120	**105.3**	345.8	251	1080
49.98	24.5		239.9	1064		254.99	125	**110.3**	349.1	240	
51.	25	**10.3**	241.	1044		265.19	130	**115.3**	352.1	233	
53.04	26	**11.3**	243.3	1007	1205	275.39	135	**120.3**	355.	224	2006
55.08	27	**12.3**	245.5	973		285.59	140	**125.3**	357.9	216	
57.12	28	**13.3**	247.6	941		295.79	145	**130.3**	360.6	210	
59.16	29	**14.3**	249.6	911	1407	306.	150	**135.3**	363.4	205	2020
61.2	30	**15.3**	251.6	883		316.19	155	**140.3**	366.	196	
63.24	31	**16.3**	253.6	857		326.39	160	**145.3**	368.7	193	
65.28	32	**17.3**	255.5	833		336.59	165	**150.3**	371.1	187	
67.32	33	**18.3**	257.3	810	1491	346.79	170	**155.3**	373.6	183	
69.36	34	**19.3**	259.1	788		357.	175	**160.3**	376.	178	
71.4	35	**20.3**	260.9	767		367.2	180	**165.3**	378.4	174	
73.44	36	**21.3**	262.6	748		377.1	185	**170.3**	380.6	169	2074
75.48	37	**22.3**	264.3	729	1550	387.6	190	**175.3**	382.9	166	
77.52	38	**23.3**	265.9	712		397.8	195	**180.3**	384.1	161	
79.56	39	**24.3**	267.5	695		408.	200	**185.3**	387.3	158	
81.6	40	**25.3**	269.1	679	1600	448.8	220	**205.3**	392.		2100
83.64	41	**26.3**	270.6	664		524.28	257	**242.3**	400.		2136
85.68	42	**27.3**	272.1	649		599.76	294	**279.3**	418.		2160
87.72	43	**28.3**	273.6	635		748.08	367	**352.3**	420.		2190
89.76	44	**29.3**	275.	622	1652	880.64	441	**426.3**	457.		2226
91.8	45	**30.3**	276.4	610							

It will be observed that the third column, the most prominent in the table, is headed "Pressure above atmosphere." By this is meant the apparent pressure of the steam, as indicated by a steam gauge. The first column gives the absolute pressure of the steam in inches of mercury, or the height to which the pressure would raise a column of mercury in a tube, *provided the opposing pressure of the atmosphere were removed.* The second column gives the absolute pressure in pounds per square inch, under the same circumstances. But the pressure of the external atmosphere is always equal to from 29 to 31 inches of mercury—varying with the state of the weather—and the steam-gauges employed on steam boilers to indicate the pressure do not show the *absolute* or total pressure, but only the *difference* between the pressure in the boiler and that of the external air. The pressure of the air averages about 14.7 pounds per square inch, but, for convenience, it is generally assumed at 15 pounds. We have constructed our third column by simply deducting 14.7 from the pressures shown in the second column. A column might be constructed, in a similar manner, to show the apparent pressure in inches of mercury, by simply subtracting 30 from the pressure recorded in the first column, 30 inches of mercury being the average pressure of the air. Such a column would then agree exactly with the indication of some forms of mercury gauges. The temperature in the fourth column is given in degrees of Fahrenheit's scale.

It will be observed that our column of temperature begins with 32°, which is the freezing point. It has been demonstrated, by very refined experiments, that a faint vapor, or steam, rises from ice in a perfect vacuum, but we have not deemed it worthy of examination. If a boiler, or other suitable vessel, contains water at the freezing point, 32° Fahrenheit, and if all the atmosphere be removed, the space above the surface of the water will be filled with an extremely weak fluid, precisely analogous to ordinary steam in its mechanical properties, but "cold as ice," and with a pressure of only 0.098, or about $\frac{1}{10}$ of a pound per square inch. Now, if the water be warmed to 100° Fahrenheit, or about blood heat, the pressure of the steam, if such it may be called, increases to 0.911 or nearly one pound per square inch. This, we may here repeat, is about the pressure in the interior of condensers and on the vacuum side of the piston in the best low-pressure engines. If the water be warmed to just 212° Fahrenheit, the pressure of the steam will just equal that of the external air, so that if an opening be made in the shell, neither air will rush in nor steam out. This is the point at which, in common language, steam begins to "make," because at all lower temperatures the tendency to the formation of the extremely weak steam described is resisted by the air which is usually allowed to fill the boiler and press upon the water. If the temperature be raised to 342° Fahrenheit, the absolute pressure within the boiler will be 115 pounds per square inch, but as the external air presses with a force of 15 pounds per square inch, upon the exterior of the boiler, the actual force with which the steam acts upon the shell of the boiler to burst it is 100 pounds per square inch. This 100 pounds per square inch is also the force with which such steam acts on a steam gauge, or on the piston of a high-pressure engine, and being thus, though of less scientific interest, the most *practically important* quantity in connection with the whole subject, we have made it the most prominent in the table.

Our fifth column shows the increase of volume which the water assumes in the act of changing into steam. Its usefulness has been put in requisition on page . One cubic inch of water, in changing into steam at the atmospheric pressure, as when boiling in an open kettle, becomes, as shown by our table, 1700 cubic inches, or nearly one cubic foot. The same quantity of water, in changing into steam at 100 pounds pressure, becomes only 259 cubic inches, but in changing into the extremely weak steam, denoted at the head of our table, becomes 187407 cubic inches, or about eight hogsheads of 100 gallons each. As the relative volume of the steam diminishes as the pressure increases, it is evident that a degree of temperature can be found at which no change in the volume will occur, or in which the water alone would contain all the heat necessary to manufacture it into steam, without the ordinary, and, as we are in the habit of believing, *necessary* expansion. In order to produce steam of such a character it would be necessary, of course, to allow the water to fill the whole interior of the confining vessel. No one has, as we are aware, ever experimented on the properties of such steam, but it has been generated with safety. A large cannon ball was once drilled, and a very tightly fitted screw was driven down till it pressed tightly and fairly upon a small quantity of water enclosed. The whole was then heated red hot, and kept so for some time. After allowing it to cool, the screw was removed, and the water was poured out, a very little only having been lost by the operation. There are reasons for inferring, from the experiments at gradually increasing pressures, that the temperature of "no expansion" referred to, is about 1200° Fahrenheit, or that of bright red-hot iron, and that the pressure is then about 5,000 pounds per square inch. On the principle that the sum of the latent and sensible heats are the same in steam at all pressures, it is evident that steam, or rather water, in this extraordinary condition, would contain no latent heat, but the whole of its heat would be sensible. If it should, however, chance to burst its envelope, it would at once assume the condition of ordinary steam, becoming a large volume of steam at a temperature of 212° Fah., and containing about 1000° of latent heat, as before explained.

The sixth column shows the velocity with which steam, at the given pressures, escapes through an orifice into the atmosphere, as, for example, through the safety-valve aperture of a steam-boiler. It may be of great service in estimating for some purposes, but, it must be observed, that in case the passages are long and tortuous, as in the exhaust of most varieties of engines, the velocity of exit will be retarded much beyond that denoted by the table.

In flowing into a vacuum there is but little difference in the velocity of steam. Weak steam, in consequence of its greater lightness, flows very nearly as rapidly as the highest which has yet been experimented on. The velocity with which all ordinary low-pressure steam flows into a vacuum is about 1400 feet per second.

Steam is lighter than air at the same pressure. The specific gravity of steam at atmospheric pressure is 0.488, air being 1. Whenever, therefore, air and steam are intermixed, and allowed to remain quiet, the steam gathers in the upper portion of the vessel.

Air absorbs a certain quantity of the vapor of water, without increasing its pressure. This operation is slow evaporation, as it proceeds at ordinary temperatures. The difference between slow evaporation and boiling consists in this: that whereas evaporation from the surface of the sea, etc., consists in a gradual absorption of the water by the air, the act of boiling lies in the generation of steam with such vigor that it displaces the air altogether, crowding it away to make room for itself.

The points elaborated in this table, and in the accompanying paragraphs, will do much to aid in determining the general proportions of the cylinder, feed-pumps, etc., and of the several pipes and passages required in this portion either of high or low-pressure engines, and also the proportions of the air-pumps and injection-cocks for the latter class.

We shall endeavor to give such examples below, drawn from the actual practice of our best engineers, as will serve more specifically to illustrate these points. For the present we must leave this department of the subject, and take up the subject of the generation of steam. It is evident that steam is produced by simply the addition of heat to water, and the power may be said to lie ultimately in the generation of heat. The combustion of fuel develops heat, this heat is transferred to water and produces steam, the steam urges a piston to the end of its stroke, and then escapes. In escaping it mingles with the atmosphere, if from a high pressure engine, or mingles with a large quantity of water, if a low pressure or condensing one. In either case it flows away, and the steam-boiler and engine is simply an apparatus which is actuated by the heat in its passage from the burning fuel to the aforesaid air or water. It is, in the broadest and most general view which can be taken of this operation, somewhat analogous to a water-wheel turned by the action of a reservoir or stream of water. The power originally lies in the gravity of the water. The wheel is but a machine for rendering it useful.

In the case of the water-wheel, a quantity of water is at a certain elevation, ready to descend. Its gravity impels it to descend. The wheel is so interposed, that it cannot descend without giving motion to the machine. So with the steam-engine, a quantity of heat is accumulated ready to act, and the steam-engine is so interposed, that it cannot act without giving motion thereto. But while there is only one theory with regard to the mechanical effect of gravity, there are two prominent theories with regard to the mechanical action of heat, and it is yet a disputed question among savans, how far the comparison between a steam-engine and a water-wheel will hold good. S. Carnot, of France, advocated a theory that heat gave out mechanical power by the act of diffusing itself alone; and this view is partly sustained by the undisputed fact that in every case where heat is used to generate power, whether in a steam-engine, or air-engine, or any kindred apparatus, the heat is received at a high temperature, and in a small quantity of matter, or in other words, in a concentrated condition, and is discharged at a lower temperature and in a larger mass, or at least in a larger volume of matter. In this view of the subject, the steam boiler and engine are simply machines, actuated by the diffusion of heat, very clearly comparable to the wheel which is actuated by the descent of water, or to a clock, which is actuated by the descent of a weight.

But later experimenters, and among them Holtzman and Manheim, of Germany, J. P. Joule, of England, and V. Regnault of France, have discovered, what has surprised and will continue to surprise many, that heat is not only diffused, but a certain small portion is actually *lost* in its passage through an engine. It is either absolutely annihilated or "changed into mechanical power" by the operation. If, for instance, a certain definite amount of heat be found to exist in a certain quantity of steam in a boiler, and it be allowed to blow off uselessly at the safety-valve, the same absolute amount of heat is found to exist in the steam at the close of the operation. The temperature will not be as high as before, but, as explained in former paragraphs, the latent heat will have been increased precisely to the same extent as the sensible has diminished, and the sum of both will remain the same. The steam will contain the same actual heat, though in a larger space and in a different condition. But if, instead of allowing it to blow through the safety-valve, it be worked off through an engine, and made to perform mechanical work by the operation, the actual amount of heat it contains is not as great. It will not warm as much water, will not diffuse the same heat from pipes to warm a building, nor will it melt as much ice as before. A portion of its heat has actually disappeared, beyond the reach of any scientific skill to find it. It has not been taken up by radiation from the cylinder, etc., for these obvious sources of loss have been provided against with extreme care. The conclusion is that a portion of the heat is actually changed into mechanical power. The amount of heat annihilated or changed in this manner, is so slight, that we have not alluded to it in any portion of our work, and shall not in any further investigation, allude again to the subject. It belongs among those abstract questions, proper for the philosopher rather than the engineer. Whenever the question is distinctly determined, and the precise amount of heat which disappears with the development of a given quantity of power is rigidly defined, it may be important to allow for it. At present the nearest approach to this is the estimate by Mr. Joule, that the raising of 682 pounds weight one foot high, involves the disappearance of an amount of heat sufficient to raise one pound of water one degree Fahrenheit. We may remark in passing, as a stimulus to improvement, that if, as this gentleman supposes, all the mechanical effect of any engine is due to this apparently trifling circumstance, there is yet a possibility of increasing the effect of our engines nearly a hundred-fold; as at present only about one per cent. of the heat evolved by the fire is thus disposed of in ordinary practice.

Whichever theory be true, it is very evident that the original source of the power is the heat of the furnace, and it becomes important to generate this heat as economically as possible. Fuel is a general term by which we designate any matter which is suitable to be burned for the production of heat. This will be more fully explained below.

COMBUSTION.

Combustion is nothing more than a vehement combination of the constituents of fuel with the oxygen of the atmosphere, and which only takes place at a high temperature. *Smoke* is the product of the imperfect combustion of fuel, caused either by a want of oxygen or a want of temperature, and *flame* may be defined to be aeriform or gaseous matter heated to such a degree as to be luminous, and may be produced independently of any chemical change, as is shown in the discharge of Voltaic electricity through an undecomposable gas.

The principal combustible element in all fuels is carbon; the heat necessary for steam-producing, is obtained by combining the carbon of the fuel with the oxygen of the air, forming carbonic acid gas. Carbonic acid gas consists of

Oxygen	16	Parts by weight.
Carbon	6	

Atmospheric air consists of

Oxygen	8	Parts by weight.
Nitrogen	28	

Whence, for the combustion of one pound of carbon, we require of

Oxygen 2.66

But to obtain 2.66 of oxygen from the atmospheric air, we also use nitrogen in the proportion of 28 nitrogen to 8 oxygen; whence, for converting one pound of carbon to carbonic acid we require

Oxygen	2.66
Nitrogen	9.31
Or	11.97 lbs. of atmospheric air.

From careful observations on the gases passing through the chimneys of well-constructed boilers, oxygen is found free, varying in amount from one-quarter to one-half of the quantity necessary for combustion; this is owing to the mechanical obstructions to the perfect conversion of the air, arising from leakage through the fuel.

More than the above 11.97 pounds of air should, therefore, be applied to the fire for each pound of carbon consumed. Twenty-five per cent. is found by experience to be a sufficient surplus allowance to convert the carbon.

Whence, to	11.97
add	3.03
and we have	15.00 lbs. of atmospheric air per lb. of carbon.

Air weighs .075 pounds per cubic foot, whence $\frac{15}{.075}$ or 200 cubic feet of air are necessary for the proper combustion of one pound of pure carbon.

Knowing the necessary amount of air for one pound of carbon, and also the percentage of carbon in the different kinds of fuel, it becomes a simple arithmetical operation to fix the bulk of air required for any species of coal, coke, or wood. The result of such a calculation is shown in the seventh column of the table on page 88.

In the present condition of steam engineering, the construction and management of engines, is much better understood than that of the proportioning and operating of furnaces, and the subject of combustion is one which well deserves, at our hands, a very thorough investigation. To quote from a recent contribution to an English journal by Mr. T. D. Stetson, of N. Y: "We should not be content with a mere general belief and enunciation of the undoubted fact that coal is fuel; that the carbon or solid portion of the coal is the most valuable in practice, as the combustion is now conducted; and that the hydrogen concealed therewith, and which escapes therefrom in the furnace, is practically of little value, on account of the difficulty of supplying it with just sufficient air for complete combustion. We should not be content with a general knowledge, compiled from experience and observation, that a certain construction of furnace and a certain form and proportion of grate is on the whole the best; we should, if possible, know how and why these facts are as acknowledged, not in order to let any real or supposed knowledge of this kind lead to an extravagant and reckless disregard of the well-worn rules, but to aid in a gradual and careful advance in this great department of the profession.

"However chemists may differ in the attempt to define combustion with scientific accuracy, in view of the development of heat, under all intense chemical action, it is practically tolerably correct to assume that combustion is simply a very rapid oxidation of matter. Oxygen combines very intimately with some other element, and in doing so develops heat and light.

"Oxygen exists either free or combined. The oxygen in the atmosphere is free, or comparatively so, and it leaves the nitrogen, with which it is there associated, to enter into combination with other elements whenever invited. In combining with other substances heat is developed, and when the combination is completed, there results a gas, a liquid, or a solid, which is incapable —except sometimes, by very elaborate processes—of being further oxydized or burned; and when the heat developed by the combination, or in other words the heat generated by the combustion, is dissipated and diffused into space, 'the game is played out.'

"It appears from the experiments of savans, carefully conducted under circumstances favorable for producing and sustaining an extremely intense heat, that nearly, or quite, every other simple substance in nature is capable of chemically combining with oxygen, when sufficiently heated, and that by every such combination heat is produced. In other words, every element except oxygen is fuel. Phosphorus almost ignites with the heat of the human hand or breath. Iron requires a quite intense heat to induce it to burn rapidly, though it oxydizes slowly in common air, at common temperatures, if sufficient moisture be present. Fine gold may be 'tried in the fire' with more assurance than most metals, but all simple substances will burn. The heat generated by burning a given quantity of any substance may depend mainly on the quantity of oxygen which it absorbs; but it is probable that the strength or closeness of the combination is also an element of some importance. Thus, if, as is the case, one pound of carbon, by combustion, unites with 2.65 pounds of oxygen, while a pound of hydrogen unites with 8.01 pounds of oxygen, it is inferred that the heat given out by the combustion of like quantities of the two substances is in a similar proportion. If, however, the nature of the union of the elements is stronger in one case than in the other, the quantities of heat may be affected thereby.

"The union of the elements of water may be unquestionably considered stronger than that of the elements of the first-named compound, but there seems yet to remain much doubt with regard to the quantities of heat developed by burning like quantities of these two most common fuels. M. Despretz, who probably experimented with some care, believed in the theory of Welther—*i. e.*, that the heat in any case is exactly proportional to the quantity of oxygen absorbed or combined. He proclaimed that one pound of oxygen, by combining with charcoal (carbon), warmed 5112 pounds of water one degree Fahrenheit; by combining with hydrogen it warmed 5210 pounds to the same extent; with alcohol (a chemical compound of carbon, hydrogen, and oxygen) the heat developed was sufficient to warm 4940 pounds to the same extent; with ether (a somewhat analogous compound) it warmed 5030 pounds. In other words, the heat produced by burning all these different fuels was, in each instance, proportional to the quantities of oxygen combined; or disagreed only as the numbers 29½, 29, 28½, and 28. This would make the effect of one pound of carbon equal to the evaporation from 212° of 13.5 pounds of water, and the effect of one pound of hydrogen equal to the evaporation of 41.7 pounds; so that, when, as in most of our American bituminous coals, the hydrogen and similar volatile matter is as great as 20 per cent. of the whole weight of the fuel, the heating effect which might be derived from the volatile parts may equal that from the solid; and with Liverpool coal, or with our most highly volatile varieties, the actual evaporative power of the hydrogen compounds would be some two or three times as great as that of the solid portions. But this is assuming that the heat-producing effect of hydrogen and carbon is not materially weakened by the previous weak union of the elements among themselves. It is a prominent feature in the theory of Hess, that the ultimate effect of a complete combination was a certain definite amount of heat, but that each previous partial combination developed *some* heat, which is a deduction from the amount which would otherwise be developed by the last or final union. In other words, the combustion of a compound never produces so much heat as the combustion of its several elements separately would do. M. Hess appears to have experimented mainly, or entirely, on the heat produced by the union of acids with bases, and this may be somewhat different from the effects of combustion, though the effects are probably identical. There cannot be conceived a subject which more deserves to be explored by the highest chemical talent. The figures above, from Despretz, are so entirely different from the results obtained by Grassi at a later day, and, as we may suppose, by much more carefully conducted experiments, that it is reasonable to hope there are great principles not yet struck out, which may explain all the anomalies now so perplexing. This savan's experiments appear worthy of publishing, in form for ready reference, as follows. Assuming the unit of heat to be the warming of one pound of water one degree Fahrenheit, Grassi's results indicate the following:—

	Units of Heat.		Evap., lbs. of water, from 212°.
Burning 1 lb. hydrogen produces	62,460	...	62·5
" " charcoal	14,220	...	14·2
" " alcohol	11,700	...	11·7
" " oil of turpentine	18,900	...	18·9
" " carbonic oxide	3,420	...	3·4
" " olefiant gas	16,380	...	16·4
" " marsh gas	22,500	...	22·5
" " pyroxylic spirit	10,440	...	10·4

"These figures may be considered reliable, as approximations to the truth, but serve better, perhaps, to indicate how *little* than how *much* we know of the theory of combustion. Were the heat always proportional to the quantity of oxygen simply, the hydrogen should evaporate 43.5 pounds, a result corresponding very nearly with that obtained by Despretz. The difference in the evaporation of about 18 pounds of water indicates that the union is some 40 per cent. more intense when oxygen unites with hydrogen to form water, than when it unites with carbon to form carbonic acid."

The subject of fuel is equally capable of being very closely examined with great profit.

"Substances are termed combustible or incombustible according to the readiness with which they can be burned, but the terms will not bear close examination. A combustible is generally understood to be a substance which, in burning, engenders heat enough to maintain its own combustion, and to induce the combustion of fresh and cold material of the same kind. A combustible is also generally understood to be a kind of material which will take fire with the heat of a small flame, and will continue to burn if exposed on all sides to free radiation, as is the wood in an open camp fire. Marble, oyster shells, and all the forms of carbonate of lime, are in the popular mind very far from combustible, yet when enclosed in the non-conducting walls of a suitable kiln, and properly heated by an active fire below, the heat produced by the expulsion and burning of the organic carbon in the stony masses aids very materially in the process of lime burning, as is proved by the deadness of the top of the previously glowing mass so soon as the decarbonization is completed, although the wood fire below be kept up to its full vigor. On the other hand, most of the varieties of pit coal cease to burn if the individual coals are separated and allowed to radiate their heat freely. Put to this latter test, it would appear that the great supply of fuel stored in strata in the earth, which this nation is digging out and burning at the rate of eight million tons per annum, and which Britons are digging out, burning, and exporting at the rate of sixty million tons per annum, is to be pronounced, what it was probably believed to be by all who had examined it previous to a few hundred years ago, one of the incombustible materials of our globe. Every thing is combustible. Every thing not previously saturated with oxygen, every particle of matter not already oxydized to death, may be made to yield heat by further oxydation. Limestone is simply a species of fuel which leaves an extraordinary amount of earthy product, and which generates in combustion so little heat as to be of no value *as* a fuel in the present state of the arts. The conditions under which it would be possible to burn any such material now generally considered incombustible, for the purpose of generating heat, are as fol lows:—first, the fuel or the air or both must be warmed before they are presented to each other; and second, the heat of the burning mass must be protected from the cooling effect of radiation. A portion of the heat of the gaseous products of combustion might be utilized by performing some useful effect, and another portion might be employed in the preparatory warming of the fuel, and in a similar warming of the blast. This is supposing a very extraordinary condition of affairs, and is only introduced as suggestive of the extent to which the value of bad fuel may possibly be raised by the advance of science. If the combustion of any fuel, or of any mixture of fuels, can produce any heat to spare—if it can be made to generate any more caloric than is sufficient to heat the blast of air required, it is *possible* to make it useful as a fuel. The immense quantities of bituminous shales and other combustible earths which are lying by, awaiting the exhaustion of the mines of purer coals, may give to these considerations greater importance at a future day than they now seem to deserve; though even now they possess a degree of interest in a financial point of view, which should ensure for them some consideration."

The principal fuels in use for the generation of steam, are wood and coal. Wood is a vegetable growth, possessing a great variety of characters; coal is a substance found in layers in the earth, possessing a still greater want of uniformity in its properties. Some varieties of both are capable of being partly burned or charred, an operation which changes their characters by dispelling the volatile elements, and renders them more useful for some purposes. Charred wood is charcoal; charred coal is coke. Wood is known as soft and hard; coal as anthracite and bituminous. These terms are relative only. There are varieties of wood of almost every grade of hardness; and there are coals of almost every character with regard to containing bitumen or resinous ingredients. All coal has some bituminous or gas-producing ingredients, of which hydrogen is a principal element.

Wood grows everywhere and coal is very abundant. The United States of America contain about 133,000 square miles of coal-fields, many of which contain many beds or layers, one below the other. We shall not attempt to enlarge further on the intrinsic nature of either, except to enunciate this single fact, that while the varieties of wood are distinguished by names, such as oak, pine, etc., which, (although the different specimens vary in different countries and with different soils,) are generally quite clearly descriptive of any variety in question, coal has never yet been found capable of any kindred classification, and is only known by the name of the district, or of the particular mine from which it is procured. Wood is also green or dry, according to the time it has been cut, old growth or second growth, according to the age of the tree (although little difference is recognized in the value in this respect when used for fuel), and is known as sap or heart according to its location in the vegetable organism of the tree. Coal is frequently variable in its character in the same mine, but is not capable of being designated by any such generally known terms. The size into which it has been broken constitutes almost the only general terms, except the name of the mine or district, and the general adjectives "anthracite" and "bituminous." The sizes are, lump, egg, etc., down to the finest dust, which latter is known as refuse, slack, coal-dust, etc., the names being different in different localities.

The employment of either or any of these fuels for generating steam, depends on the financial element alone. All are suitable, though some more so then others. It is particularly difficult in practice to burn coal successfully in ordinary locomotives, in consequence partly of the destruction of the boiler by the intensely concentrated heat developed, especially by anthracite. In general, anthracite coal makes an intensely hot fire, but the heat is confined almost exclusively to the spot where it lies on the grate; bituminous coal and coke make a blazing fire much less concentrated:

while wood makes a still freer blaze and is still less violent in its destructive action upon the metal. A wood fire is most easily controlled by regulating the draught of air, while anthracite is very obstinate in that respect.

The following table shows some of the most prominent qualities in the principal American woods.

Species.	Specific gravity green.	Specific gravity air dried.	Specific gravity kiln dried.	Degrees of heat which may be generated.	Percentage of Charcoal.	Quantity of heat as to volume.	Weight of one cord in lbs.	Relative value as fuel.
Hickory........				3000	44.69	25	4469	1.00
White Oak	1.07	0.71	0.66	3000	21.62	25	3821	0.81
Black Oak......				3000	23.80	25	3254	0.71
Red Oak	1.05	0.68	0.66	3000	22.43	25	3254	0.69
Beech..........	0.98	0.59	0.58	3000	32.36	25	3286	0.65
Birch	0.90	0.63	0.57	3000		25		
Maple..........	0.90	0.64	0.61	3000	27.00	25	2700	0.57
Yellow Pine				2800	24.63	28	2463	0.54
Chestnut				3000	25.25	26	2333	0.52
Pitch Pine......				2800	19.04	23	1904	0.43
White Pine.....	0.87	0.47	0.38	2800	18.68	23	1868	0.42

The following table gives the nature and value of several varieties of American coal and coke

Table showing the Weights, Evaporative Powers per Weight, and Bulk and Character of Fuels, from Report of Professor Walter R. Johnson, 1844, *who experimented at the expense of the U. S. Government.*

Designation of Fuel.	Specific gravity.	Weight per cubic foot.	Pounds of steam from water at 212° by 1 lb. of fuel.	Lbs. of steam from water at 212° by 1 cubic foot of fuel.	Weight of Clinker from 100 lbs. of coal.	Number of cubic feet required to stow a ton.
BITUMINOUS.						
Cumberland, *maximum*.....	1.313	52.92	10.7	573.3	2.13	42.3
" *minimum*	1.337	54.29	9.44	532.3	4.53	41.2
Blossburgh...............	1.324	53.05	9.72	522.6	3.40	42.2
Midlothian, *screened*	1.283	45.72	8.94	438.4	3.33	49.
" *average*........	1.294	54.04	8.29	461.6	8.82	41.4
Newcastle................	1.257	50.82	8.66	453.9	3.14	44.
Pictou....................	1.318	49.25	8.41	478.7	6.13	45.
Pittsburgh	1.252	46.81	8.20	384.1	.94	47.8
Sydney....................	1.338	47.44	7.99	386.1	2.25	47.2
Liverpool	1.262	47.88	7.84	411.2	1.88	46.7
Clover Hill...............	1.285	45.49	7.67	359.3	3.86	49.2
Cannelton, Ia.............	1.273	47.65	7.34	360.	1.64	47.
Scotch....................	1.519	51.09	6.95	369.1	5.68	43.8
ANTHRACITE.						
Peach Mountain	1.464	58.79	10.11	581.3	3.03	41.6
Forest Improvement.......	1.477	53.66	10.06	577.3	.81	41.7
Beaver Meadow, No. 5	1.554	56.19	9.88	572.9	.60	39.8
Lackawanna	1.421	48.89	9.79	493.	1.24	45.8
Beaver Meadow, No. 3.....	1.610	54.93	9.21	526.5	1.01	40.7
Lehigh	1.590	55.32	8.93	515.4	1.08	40.5
COKE.						
Natural Virginia..........	1.323	46.64	8.47	407.9	5.31	48.3
Midlothian		32.70	8.63	282.5	10.51	68.5
Cumberland..............		31.57	8.99	284.	3.55	70.9
WOOD.						
Dry Pine Wood...........		21.01	4.69	98.6		106.6

The above table exhibits the ultimate effects. As a safe estimate for practical values, a deduction (for the coals) of $\frac{14}{100}$ should be made.

On the Reading Railroad, Pennsylvania, three pounds of pine wood equal to one pound of Anthracite coal. Mr. Haswell estimates the best varieties of wood fuel to contain twenty per cent. of carbon. Walter R. Johnson found that one pound of wood, upon an average, evaporated two and one half pounds of water.

The average percentage of coke from American bituminous coal from the above table is seventy-three per cent., and the average percentage of carbon, sixty-seven and one half per cent.

The following table shows the relative properties of good coke, coal, and wood.

Name of Fuel.	Weight per cubic foot, in pounds.	Degrees of heat generated.	Percentage of Carbon in the Fuel.	Economic bulk, or cubic feet required to stow one ton.	Economic or stowage weight per cubic foot.	Cubic feet of air to evaporate one pound of water.	Equivalent economic bulk, to evaporate the same weight of water.	Weight of water evaporated per pound of fuel in ordinary practice.	Relative value as fuel, disregarding the actual cost.
Coke	63	4300	96	80	28	22.4	13	8½	100
Coal	80	4000	88	44	51	32.0	10	6	71
Wood	30	2800	20	107	21	16.0	60	2½	29

The power of fuel as usually burned depends upon the amount of carbon in it. Pure coke is almost solid carbon, hence its superior value as a heat generator.

"Fires dispose of the heat they produce by two means, each very distinct. Radiation and conduction from the brightly glowing fuel or flame, is one of these means, and in many instances, the only useful one; the heat conveyed away in the current of gases escaping from the fire to mingle with the external atmosphere, is the other. Furnaces for melting metals are examples where only the radiant heat of the combustion is made useful. The same was true of the most primitive steam boilers and their furnaces. The first boilers were simply spherical or hemispherical vessels placed over a forge fire, but all modern boilers are constructed with a view to extract much of the heat from these gases by allowing them to circulate under or through the water, and thus to convey heat thereto. There is, however, a limit to the extent to which this may be carried. It is evident that the heat of the gases must be very high on first emerging from the fire. The policy adopted in designing boilers is to hold the gases in contact with the metal of the boiler until they are cooled down to a certain extent. The water in the boiler is necessarily hot, and it is of course impracticable to abstract heat from gases below that temperature, and it is equally impolitic to attempt to utilize the heat of the gases after it has been reduced to a point ranging between 1000° and 500° Fahrenheit. The precise point to which the gases may usually be cooled to advantage depends on the facilities for promoting the draught, and on the value of space, metal, labor, &c., as compared with the cost of fuel."

There are three principal reasons why more effect is not obtained from the burning of fuels in boilers. First, the fuel is frequently not all burned; second, the products of combustion are allowed to convey away much heat as they escape; and third, a portion of the heat is allowed to escape from the boiler by radiation. These are worthy of being briefly considered separately.

When the draught is very active, as in locomotives, large quantities of coal are ejected from the stack in the form of small fragments; another portion of the fuel escapes unburned in the form of smoke. The thick clouds of sooty matter sometimes arising from fires, are unburned matter, and are valuable portions of the fuel. Smoke is solid carbon minutely divided. It is, probably, somewhat changed chemically, so as to exist in a condition analogous to the incombustible "gas carbon" which accumulates in gas retorts, but it is, nevertheless, solid and visible. Another, and usually a much larger portion of the fuel escapes unburned in the form of gas. The inflammable gases evolved by the heat are not all consumed from want of a sufficient supply of oxygen, the heat of the fire being sufficient to decompose more than can be supplied with oxygen. The thick smoke that escapes from a chimney when fresh fuel is thrown on a hot fire, is partly unconsumed gas; decomposed from the fuel, but without oxygen enough to burn, although there may have been a sufficient supply of heat. From this cause it is, that flame is seen coming from the top of a steamboat chimney which appears to be continuous from the furnace; but which, in fact, is ignited by contact with the air, having retained sufficient heat for that purpose.

All smoke-consuming furnaces, varieties much talked about in Great Britain, simply admit fresh air to the unconsumed gases above the fire; this effects their combustion whenever the gases retain the necessary amount of heat. But this only prevents the nuisance of smoke. To render the gases thus reburned useful in evaporating water, this supply of oxygen must be added while the gases are yet in the flues, and not after their escape into the chimney. To this end, James E. McConnell (England) divides the flues of his locomotives into two parts, connecting the front ends of the first part and the back ends of the second part by a space of twelve or fifteen inches, (called by him a "combustion chamber,") into which he admits any required amount of fresh air.

The heat sometimes produced by the combustion of the gases escaping from a fire is sufficient to be of very considerable commercial value. The smoke-boxes of locomotives have been made red hot by the combustion there produced, in consequence of accidental leaks which allowed fresh air to enter and unite with it.

As estimated above, it is theoretically *possible* to realize as much or more heat from the burning of the gaseous portions as of the solid portions of bituminous coal, but it is not *usual* to realize any considerable effect from the former. The heating effect of a given variety of coal is usually estimated to depend entirely on the quantity of solid carbon it contains.

But aside from the ordinary combustible gases in coal or other fuel, it is not unusual for the solid carbon itself to escape in a gaseous form without being completely burned. It is possible for carbon to combine with only half as much oxygen as is required for complete combustion, and although this combination develops heat, it is evident that the amount must be less than if it were fully burned. It is supposed by many that the heat produced by this half-burning process is exactly half that which would result from the complete combustion of the same fuel, but this matter is, as intimated above, not definitely settled. But it is well established that this half-burning effect very frequently occurs in some furnaces, and that the resulting gas (carbonic oxide) is capable of being again burned, and in doing so, of yielding a very considerable amount of heat. This latter fact alone is sufficient to prove that all the heat possible to be produced had not been developed in the first combination, and to make it desirable either to avoid this action as far as possible, or to so arrange the furnace that air is admitted to the carbonic oxide to complete its combustion before it leaves the heating surface of the boiler. Carbonic oxide forms whenever the heat of the fuel is very intense, and there is no air to unite with it. In the interior of a large fire this effect is almost invariably produced. The fuel becomes very much heated, giving it a very great affinity for oxygen, and as the air is admitted in but comparatively small quantities, so little oxygen is supplied that the fuel, failing to find a sufficient quantity for complete combustion, accepts the lesser quantity, and combining therewith, rises to escape. If it meets with fresh air before it becomes too cool to ignite, it there completes the combustion, but if not, it goes off and mingles with the atmosphere as it is. It is a law of nature that in nearly all chemical combinations the quantities combining are definitely proportioned, and it is especially so with the combustion of carbon. Oxygen and carbon never unite except in the proportions indicated. A pound of carbon either unites with 2.65 pounds of oxygen, and thus forms 3.65 pounds of carbonic acid gas, a gas utterly incombustible; or else it unites under the peculiar circumstances named with half that quantity of oxygen, and forms 2.32½ pounds of carbonic oxide. It is worthy of note that the quantity of heat developed by the combustion is, so far as has been yet ascertained, precisely the same, whether the combustion be perfected at once, or whether it be done at the two operations. We mean by this to say, that the formation of carbonic oxide generates a certain quantum of heat, and its subsequent conversion into carbonic acid develops other heat, and that the sum of these two quantities of heat is precisely equal to the heat which would have resulted from the complete change of the same amount of carbon into carbonic acid at a single operation. In firing for steam-generating purposes, therefore, it is essential to supply sufficient air, either to the fuel burning on the grate or to the gases arising therefrom. But it is almost equally important to avoid the extreme of supplying too much, as we shall see by considering the second source of loss.

The second source, as enumerated above, is the heat which escapes in the products of combustion. The fact that heat escapes in that manner has been sufficiently indicated where it was stated to be impracticable to absorb all the heat from the gases. But the subject deserves to be more amplified. If from every furnace in operation there is a current of gaseous matter flowing in more or less steady streams through the flues or tubes, and thence rising through the chimney, to be diffused in the air. As the air to form this enters cold at the ash-pit, and escapes hot, it is evident that heat is thereby taken from the boiler.

The gases, at first very hot, impart their heat to the metal of the boiler as they flow along; but they impart it less rapidly as they become cool, until they at length attain so low a degree of temperature that their heating effect is too slight to compensate for the increased difficulties involved in their longer retention. It is, consequently, impracticable to cool the gases with profit below that degree. We will illustrate this point more fully. There are a great variety of methods adopted in different styles of boilers for the utilization of this heat, but they are all reducible to, first, increasing the magnitude of the boiler, and second, increasing the heating surface by dividing the passages, or by compelling the gases to travel backwards and forwards in lieu of proceeding directly for the chimney. The first is availed of in all the examples of moderate firing, as distinguished from hard firing. When a large boiler has a small grate surface and the draught is slow, the gases remain a long time in contact with the metal, and impart their heat very thoroughly; and also when, as in a tubular boiler, the gases are compelled to travel through or among quite small tubes, so that the streams of hot gas are very small and the surface they travel over is comparatively very large; they also impart their heat very thoroughly. This last proposition is true to such extent that some locomotive boilers, although very powerfully and, consequently, it might be inferred, wastefully fired, generate steam as economically as some good double-return flue or even Cornish boilers, in which the slow combustion principle is carried out very fully. But in either method, a limit of economy may be reached. When in enlarging our boiler to generate a given quantity of steam, we produce such dimensions that the increased losses by radiation of heat from its surface equals the increased gain due to the more perfect absorption of the heat, it is evident that the utmost limit has been reached, unless the boiler can be more efficiently covered and protected. And when in multiplying the tubes and diminishing the diameter of the same, we produce such proportions that the increased friction of the surfaces against the moving current checks the draught so that the loss is as great as the gain, the limit is equally gained in this direction. In practice, the limit in each method of utilizing the heat is, owing to these considerations and the cost of the boilers and of repairs, found before the gases are reduced much below 1000° F. The loss of heat is now evidently proportional to the quantity of such gases escaping, and it becomes important to reduce the quantity of gases to the smallest possible amount. Every cubic foot of pure air which is allowed to pass uselessly though the flues is, therefore, an evil. It enters the furnace or the flue, as the case may be, in a cold state and escapes heated to nearly 1000°; this heat it conveys away and diffuses in the atmosphere. It is, therefore, a robber, and the admission of such air should be guarded against nearly as vigorously as should the escape of unburned fuel. In common practice, as before stated, from one-fourth to one-half the gaseous matter which rises through the stack is pure air, thus heated and robbing the fire of its power. Some invention which should perfectly adjust the quantity of air entering at all times might effect a very great economy. Aldrich's Hot Blast for Steam Boilers, lately patented, has a device for regulating, as well as for heating the air before its admission to the ash-pit, and in both these respects, as also by the results of the trials which have been made, it seems adapted to greatly increase the economy of steam generation.

The third source of loss enumerated is the radiation from the boiler. This may be very greatly diminished by covering the boiler with a double thickness of brickwork, and, if desired, this may again be covered with ashes to an indefinite depth, if the boilers are stationary. Marine, or locomotive boilers, cannot be thus protected, and the thin coverings of felt and wood allow a considerable quantity of heat to radiate therefrom into the atmosphere; but under any ordinary circumstances this source of loss is much less important than the other.

In order to secure conditions under which it would be possible to utilize all the heat of fuel, in making steam, it would be necessary to make an absolutely perfect protection of the boiler from the radiation of heat, and, also, to absolutely cool down the products of combustion to the same temperature, as the air is received. This latter, impossible as it may at first seem, is very nearly attained by the Aldrich arrangement above alluded to, the heat remaining in the gases after escaping being made to serve in the chimney for the heating of a stream of cold fresh air which is being drawn in to supply the ash-pit. Under these circumstances, it is of far less consequence if too much air is supplied, and it only becomes necessary to be certain to provide sufficient for perfect combustion. In that arrangement, the draft is aided and completely controlled by the action of a blower, which is driven with an automatically variable speed, as the wants of the apparatus require. This is a great step in economy, but there is still a very sensible quantity of heat lost in the escape of hot gases.

We would like to dwell longer on the extremely important subject of combustion, as applied to the generation of steam, did our space permit, but having given it a tolerably thorough canvass, we will return to the use of steam in the engine. Below is presented a series of calculations and proportions for engines and boilers, adapted each to the best present standard of their respective classes of work.

ESTIMATING POWER.

The old fashion of rating steam-engines as of a certain nominal horse-power, depending on the size of the cylinders alone, was established at a time when the pressure of the steam was almost uniformly low, and never conveyed a very definite idea of the actual work performed, even under those circumstances. Of late it has become common to include, in the estimate of power, all the conditions affecting the engine, such as the speed with which it works, the pressure of steam in the boiler, the expansion in the cylinder, etc.

With regard to how much is understood by a horse-power, there is, in this country and England, no question at all. Horses vary in their ability to endure protracted labor, and our standard may be more or less than the average of horses are able to do, but that is of little importance.

So long as the number of horses' power of an engine conveys a definite knowledge of the power, it is of little consequence what relation it sustains to the action of any particular class of animals.

A horse-power is equal to the lifting of 33000 pounds one foot high each minute. From these figures it is, of course, easy to calculate the lifting of smaller or greater weights with a greater or less degree of rapidity. Raising 550 pounds one foot per second is one horse-power, because it is precisely equivalent to the figures given above; and if a man, weighing two hundred pounds, raises his weight by climbing stairs at the rate of 82½ feet per minute, he exerts, during the time he is so working, just half of one horse-power.

One of the heaviest mechanical operations now in contemplation, is the pumping to supply the city of Brooklyn with water. Twenty million gallons of water per day are proposed to be lifted from a pump-well into the reservoir, 165 feet above: required the number of horse-power required in the engines to do this, by pumping constantly, or 24 hours per day, a gallon of water weighing about eight pounds,

$$\frac{8 \times 20{,}000{,}000 \times 165}{33{,}000 \times 60 \times 24} = 555\tfrac{5}{9} \text{ horse-power.}$$

Presuming all our readers to be familiar with the signs for multiplication and division, it is hardly necessary to explain that this operation consists in multiplying the weight of one gallon by the number of gallons, and that product by the lift, in feet, for a dividend; and then multiplying the 33000 by the number of minutes in an hour, and that product by the number of hours in a day's work, for a divisor. Dividing the first quantity by the second, we obtain 556, the number of horse-power required. To this a very considerable addition should be made in practice, to overcome the friction of the pumps, and also the friction of the water in flowing through the pipes. The calculation only refers to the power really expended in overcoming the gravity of the water.

These calculations are based, it will be observed, on the effects actually produced, and an allowance, for friction, etc., should therefore be in addition to our result. It is almost equally easy, however, in steam-engines, to base the calculations on the proportions and conditions of the engine, except that the allowance for losses, (which must, in that case, be a *subtraction* from the reresult,) can rarely be made with perfect accuracy.

The competing engines, tested at the Crystal Palace in the summer of 1858, were single engines, diameter of cylinder 12 inches, stroke of piston 3 feet, the pressure of the steam 70 pounds per square inch above the atmosphere, and the engines made 60 revolutions per minute. Assuming that the steam followed the piston full stroke, and that the full boiler pressure was maintained therein, what was the horse-power?

$$\frac{6^2 \times 3.1416 \times 70 \times 3 \times 2 \times 60}{33000} = 86.36 \text{ horse-power.}$$

In this operation half the diameter of the cylinder is squared, or multiplied by itself, and this is multiplied by the familiar constant 3.141592, or, to save figures, by 3.1416. This gives the area of the piston in square inches; this area is then multiplied by the pressure in pounds per square inch, (70,) this, again, by the stroke in feet, (3,) then this is multiplied by 2, because the piston moves in both directions to complete a revolution, and this product by the number of revolutions per minute, (60.) Dividing this product by 33000 gives the approximate horse-power, making no deduction for friction. To make this calculation exact, an allowance should first be made for the size of the piston-rod, which deducts some 4 inches from the surface of the piston when moving in one direction; then an allowance, which can never be definitely measured, should be made for the losses by friction of the engine, and, by no means least, allowances should be made for a diminution of pressure of the steam, in consequence of the difficulty it finds in flowing through the narrow passages with sufficient rapidity to follow the piston in its rapid motion, and, again, for an inability of the steam to escape instantly from the cylinder into the atmosphere when released. The packing rings of the piston and journals of engine-shafts and shaftings cause friction. It has been found in practice that the friction of an engine of this description is about 2 pounds per each square inch upon the piston, and 5 per cent. = 3.40 for additional friction, (which cannot be found by the indicator, but has been found in practice to be so,) it will thus be perceived that of 70 pounds of pressure, 5.40 are consumed by the friction of the engine without shafting. The additional friction of 3.40 pounds is found thus: deduct the 2 pounds from 70 (70−2) = 68, multiply 68 by 5 (68×5) = 340, and divided by 100 = 3.40 pounds additional friction. The 5.40 = $5\frac{40}{100}$ pounds of pressure, is supposed to work the engine with the same speed, or as many revolutions per minute as if in full working operation. Now, to find the actual, or effective horse-power of the engine, we deduct the 5.40 = $5\frac{40}{100}$ pounds of pressure per each square inch of the piston at the cylinder, for the friction of the engine, from the total pressure 70 pounds, which leaves 64.60 = $64\frac{60}{100}$ pounds per square inch to work the shafting and machinery. We have to remark, in order to find the exact amount of pressure, in pounds, to overcome the friction of the engine by an indicator, that the engine shaft must make the same revolutions in a minute as if in full working operation, with all the machinery and shafting connected. The same rule must also be observed, if the friction of the shafting should be ascertained. The actual, or effective horse-power of the engine would also be

$$\frac{6^2 \times 3.1416 \times 64.60 \times 3 \times 2 \times 60}{330000} = 79.7 \text{ horse-power.}$$

86−79.7 = 6.3, or $6\frac{3}{10}$ horse-power, would be necessary to overcome the friction of the engine. Thus, 79.7 horse-power would be to drive the shafting and machinery. As we have found the pressure per square inch upon the piston of the cylinder to drive the engine, it is now necessary to ascertain the amount of pressure upon a square inch to overcome the friction of shafting, which can be done by an indicator, and will vary according to the length of shafting, and heavy or light shafting attached to the engine. We will here remark that it will be necessary to disconnect the machinery from shafting, in order to find the amount of friction for the same, which may be from 4 to 7 pounds upon the square inch of the piston. All our engine builders, or owners of steam-engines should actually test their engines, as any deficiency in the working parts, such as piston, etc., may be discovered, and a great waste of fuel may be saved. We therefore highly recommend to them the indicators manufactured at the Novelty Iron Works of this city. A full description of it will be given below, as well as of their celebrated Manometers, Patent Engine Register or Counter.

According to Watt's rule, to designate the size of engines in measures by horse-power, steam was used at atmospheric pressure only, or 14.7 pounds, of which 4.7 pounds was supposed to be lost by imperfect condensation, and 3 pounds by the friction of the engine, leaving but 7 lbs. for effective pressure upon the piston. By the term "horse-power," therefore, as the unit of measurement for the manufacture and sale of steam-engines, was understood, engines of such specific sizes and proportions as to be able, with a uniform pressure of 7 lbs. to raise for each horse-power 33000 lbs. one foot high per minute. An increase in the amount of pressure used has since led to the distinctions of "nominal" and "effective" horse-power; the former being applied to the size, and the latter to the power of steam-engines. The effective power is found correctly only by the use of the indicator. To find the "effective power" of an engine, or that power which an engine absolutely exerts in giving motion to machinery, or in propelling a boat, it is necessary to know first what force is required to overcome its own friction. In giving motion to machinery, this power is readily told by the indicator; but in propelling a boat, the engine is seldom in motion without the paddle-wheels being in part submerged. It is therefore necessary, in that case, to estimate the power consumed by friction. By the use of the indicator, it has been ascertained that in well-made engines, when unloaded, the friction varies from one to five pounds per each inch in the area of the piston; the largest class rarely exceeding the first amount. The friction of an engine is variously estimated. In Haswell's Tables, the pressure to overcome the friction of the engine is given at one-eighth of the pressure of the steam-gauge. Brown, in his Catechism on the Steam-Engine, gives 1 pound as the friction of a locomotive when unloaded, as found by experiment, and estimates the additional friction caused by additional resistance at about .14 of that resistance. McNaught gives the friction of a 30 horse-power condensing engine at 1.75, running with a few feet of shafting; and 2.15, with cold water pumps, additional. Brown in his Treatise on the indicator, gives 1 pound as the friction of large engines, and of a small engine 1.7 pound; also, of a 25 horse-power non-condensing engine, with gearing and 40 feet of 4 inch shafting, 4.82 pounds; and 7.12 pounds at 45 and 60 revolutions respectively. Additional friction is created as resistance increases, but how much cannot be so well ascertained by the indicator. It is estimated by McNaught at five per cent. of the increased pressure upon the piston. To find the effective horse-power of an engine by an indicator, it is necessary to take two diagrams, one for the friction of the engine and the other for the gross amount of labor performed, and to deduct from the sum of the measurement of the latter that of the former, together with 5 per cent. of the remainder for additional friction created by the increased resistance. If the power to

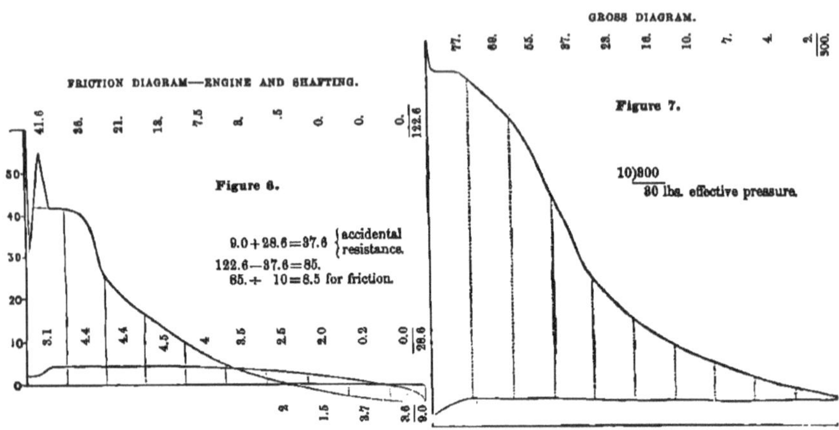

Figure 6.

Figure 7.

drive the machinery attached to an engine be the inquiry, this, obviously, is the method to be pursued—taking one diagram with the machinery in motion and another with the engine and shafting only.

Thus, in the preceding diagrams, taken by the Novelty Iron Works, of this city, from a high pressure engine (Figs. 6 and 7) used in a manufacturing establishment, it will be seen that 8.5 lbs. is the average or effective force of steam per square inch upon the piston required to move the engine with mill work and shafting, or that an average of 30 pounds is required when the machinery is in motion. If two pounds be allowed for the friction of the engine, and five per cent. = 1.4 for additional friction, it will be perceived, that of the 30 pounds of steam, 3.4 are consumed by the friction of the engine, 6.18 for the shafting, and 20.42 for driving the machinery. In other words, the gross diagram shows 30 pounds, and the friction 8.5. Of this 8.5, two pounds are to be deducted for the original friction of the engine, and also five per cent. = .32 of the 6.5 remaining. Then 8.5 – 2.32 = 6.18 for the shafting. Subtracting the amount of the friction diagram from that of the gross, we have 30 – 8.5 = 21.5 for the power to drive the machinery. But of this 21.5, five per cent. = 1.08, is for the increased friction of the engine. Therefore, we have, as before:

21.5 – 1.08 = 20.42 for machinery.
8.5 – 2.32 = 6.18 for shafting.
.2 + 1.4 = 3.4 for engine.
30.00

It is now required to find the effective horse-power of the engine, the stroke of which is 4′ feet, piston 18″ inches, and makes 30 revolutions per minute. To do this two methods are presented, varying little, if any in the result.

FIRST, BY THE USUAL FORMULA.

Rule.—Multiply the number of square inches in the area of the piston by the pressure of steam in pounds, less the amount required to overcome friction of engine; multiply this product by the number of feet the piston travels per minute, and divide by 33000, the number of pounds a horse is supposed to be able to raise through one foot of space in one minute.

```
Area of cylinder, per table,        254.47
Steam, per diagram, 30 – 3.4 =        26.6
                                   152682
                                  152682
                                  50894
Pressure on piston, in pounds,   6768.902
Speed of piston, 4 × 2 × 30           240
                                270756080
                               13537804
                   33000)1624536.480(49.22 effective, or actual horse-
                         132000                power of the engine.
                          304536
                          297000
                            75364
                            66000
                             93648
                             66000
                              7648
```

The pressure of the steam in the boiler, when the dia6ram, fig. 7, was taken, was 90 pounds, and cutting off at ¼ of the stroke. A further explanation of it will be given under the Treatise of Best Form of Diagrams.

SECOND FORMULA.

[By adding to the divisor (33000) an equivalent for the 5 per cent. additional friction, and finding the horse-power for each pound of pressure, shown in the gross diagram, over that of the friction.]

Rule.—Multiply the area of piston in inches by its speed in feet per minute, and divide the product by 34650; the quotient is the number of horse-power for each pound of steam shown in the gross diagram over that of the friction, or the sum allowed for its equivalent. The number thus obtained will serve for estimating the power of the engine, as long as the diameter of the cylinder and speed of the piston remain the same.

Example. The same diagram as before:

Area of piston 254.47
Speed per minute, 240
1017880
50894
34650)61072.80(1.76256 horse-power for each pound of steam.
34650
264228
242550
216780
207900
88800
69300
195000
173250
217500
207900
9600

1.76256
Then pressure of steam 30 lbs. – 2 = 28
1410048
352512
49.35168 or $49\frac{35}{100}$ horse-power of engine.

Before proceeding further with the calculations, proper proportions of the various kinds of steam-engines with gearings and shaftings attached to it, it is a matter of great importance to engineers and owners of steam-engines, to describe fully the advantages obtained by Stillman's Indicator and how it is to be used. We, therefore, take great pleasure in saying that the manufacturers have furnished us with every particular as to its application for high and low pressure engines, as also the diagrams and some tables, which will be illustrated hereafter. A description of Stillman's Patent Manometer Gauges and other instruments used in connection with steam engines, boilers, etc., will be set forth with illustrations.

THE STEAM-ENGINE INDICATOR.

USE OF THE INSTRUMENT.

The design of the Steam-engine Indicator is to enable those who make, use, or have the charge of steam-engines, to ascertain, at any moment desired, the condition of the parts of the engine subject to the direct action of the steam, and to what advantage the steam is applied. Its indications are automatically drawn on paper, and by the form of the diagram which it makes may be known the following particulars, viz.:

1st. Whether the valves are properly constructed and set, and if the steam-passages, or ports, are of proper size to receive and discharge the steam in time to produce the best effect.

2d. What pressure of steam there is upon the piston at every position in the cylinder, as well as its average during the stroke.

3d. What is the value of the vacuum acting upon the piston of a condensing engine in all its positions in the cylinder, and what its average.

4th. Whether the exhaust passage from the cylinder of a non-condensing engine is sufficiently large to give a free exit to the steam, and if not, what percentage of power is lost in forcibly expelling it.

5th. The actual consumption of steam in giving motion to the engine, and also what additional steam is used in giving motion to the shafting and mill work; and then what power is required to give the requisite motion to the machinery, or to any part of it, provided it offers sufficient resistance to form any measurable proportion of the power of the engine. It will therefore indicate, at any time, or under any particular condition, the actual power of the engine.

6th. In manufacturing establishments, where power is sold to tenants, it will show how much is consumed; or, in mills where extensive lines of shafting or any kind of machinery is used, the principal resistance of which is by friction of its parts, it will guide to the selection of proper oils for lubrication.

7th. It will demonstrate the degree of economy in using steam of high pressure and the use of expansion, and also the relative efficiency of different kinds of expansion apparatus.

If applied also to the air-pump of condensing engines, it will give the data by which to determine the quantity of water that would, under the circumstances, be most economically used in the condensation of the steam, or it may be advantageously applied to other pumps, when they do not work efficiently, to discover the nature or extent of defects or derangements.

The Indicator, therefore, is an invaluable appendage to the steam-engine, and where successfully applied can scarcely be too highly estimated.

DESCRIPTION OF THE INDICATOR.

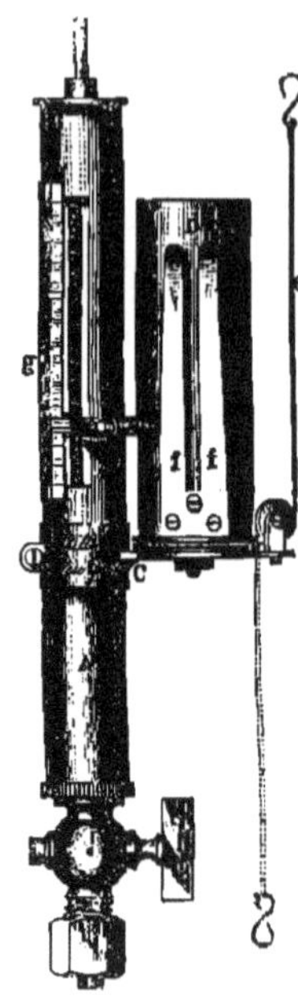

The principle and action of the Indicator are so simple, and, to most practical engineers, now so well known, that they will need but little particular description here. Some persons, however, into whose hands this treatise may fall, will perhaps be entirely unacquainted with the instrument. To such the following cut and explanation will give an idea of its construction.

A is a brass case inclosing a cylinder, into which a piston is nicely fitted. To the piston-rod a spiral spring is attached to resist the steam and vacuum when acting against it. B is a pencil attached to the piston-rod. C is an arm attached to the case, and supporting a cylinder D, which may be caused to rotate back and forth—a part of a revolution in one direction, by means of a line or cord *e*, attached to a suitable part of the engine—and in the other by means of a strong watch-spring within this cylinder D. Outside this cylinder is to be wound a paper, upon which a diagram will be made, by the combined action of the piston and paper cylinder, representing, by its area, the power exerted on one side of the piston during the whole revolution of the engine. *f f* are springs to secure the paper to the cylinder. *g* is a scale divided into parts corresponding to the pounds of pressure on the square inch. These divisions, for convenience of measuring the diagrams with a common rule, are generally made in some regular parts of an inch, as 8ths, 10ths, 12ths, 20ths, 30ths. *h* is a cock by means of and through which it is connected with the engine cylinder.

HOW TO ATTACH THE INDICATOR.

Into whatever part of the engine it may be desired to apply the Indicator, there must first be inserted a small stop-cock, with a socket to receive the one connected with the Indicator. The instrument is to be set into this in such a position, that the line attached to the paper cylinder shall lead through or over the guide pulley toward the place whence it is to receive its motion. An extension of this line should be connected with some part of the engine, the motion of which is coincident with that of the piston, and which would give the paper cylinder a motion of about three-fourths of a revolution. If the engine is of the construction denominated *beam* or *lever* engine, and is provided with a "parallel motion" the parallel bar, or a pulley on the radius shaft, furnishes the proper motion; if otherwise, the beam centre may be resorted to. In the kind denominated *square* engines, the centre of the air-pump gives it. In horizontal and vertical *direct acting* engines, it will frequently be found necessary to erect a temporary rock-shaft, or lever, connected with the cross-head. Particular care should be taken, when the power of the engine is to be estimated, that the motion communicated be perfectly coincident with that of the piston.

In nearly all forms of the steam-engine, the proper motion may be obtained by attaching a line to the cross-head, and passing it over a delicately constructed pulley, to the axis of which should be attached a smaller one, from which a line shall connect with the Indicator. The proportional sizes of the two pulleys, of course, should be as the distance travelled by the piston to the length of motion given to the paper cylinder of the Indicator. It will be necessary to attach a strong spring to the axis of these pulleys, to produce the reverse motion promptly. In an oscillating engine, it will be necessary that the Indicator, with its fixtures, should be attached to the cylinder.

As the paper cylinder cannot make more than about three-fourths of a revolution without disturbing the point of the pencil, it will be seen that the line communicating the motion must be of a definite length. It also requires to be readily connected and disconnected. To meet these demands, one end should be provided with a running loop,* and the other with a hook.

When the Indicator is to be attached to the lower end of a vertical cylinder, or to a horizontal one, it may be worked horizontally; but it is better to avoid such a position by the use of an angular attachment to the cocks.

TO TAKE A DIAGRAM.

When the Indicator and line are satisfactorily arranged, the pencil holder should be provided with a hard and well-pointed pencil. Regulate the bearing of this pencil by sliding it farther in or out of the holder; or by moving the paper cylinder forward or backward, so that it shall bear very lightly upon the paper.

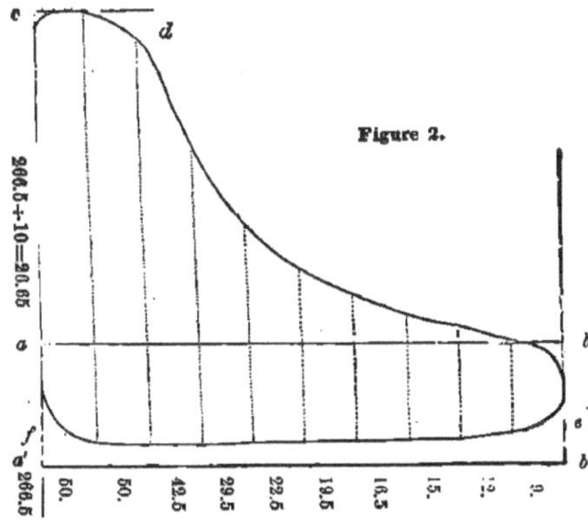

Figure 2.

Arrange the motion of the paper cylinder, by means of the running loop, so that it shall not be checked at one end by contact with its stop fixture, nor at the other by the paper springs coming in contact with the pencil, either of which occurring would alter the shape of the diagram. Lift off the movable part of the paper cylinder, and affix the paper, upon which the diagram is to be made. The most ready method of doing this is to secure one corner of the paper by one spring; then, bending it around the cylinder, enter the other between the springs, and, clasping the hand around it, slide the whole down, taking care to leave it smooth and tight. The cylinder should then be restored to its place. The cocks may now be opened, to admit the steam to expand the cylinder and piston, and to see if its motion be free. When all is satisfactory, close the cock, and bend the pencil to the paper, and it will trace the atmospheric line, represented in the diagram by the line *a b*. Then, with the hand, press the piston down until the index is at 15 below 0 on the scale of lbs.; or, more exactly, to the figure agreeing with the existing state of the barometer, and let it trace the similar line *a′ b′*, representing perfect vacuum. Let it be supposed, that the diagram to be taken be from the top of

RUNNING LOOP.

* Provide a piece of metal about an inch long, one-sixth wide, and half as thick, and divide it equally by four holes. Through this reeve the line, as here represented, and you have a very convenient running loop.

the cylinder of a condensing engine, worked by a slide steam valve, and provided also with a cut-off valve. Now, if the cocks be again opened, just as the piston begins to descend, the pencil will pass in a nearly direct line to *c*, and then, as the cylinder or paper revolves, the line *c d* will be drawn. At *d* the ingress of the steam is stopped by the cut-off valve, and the remainder of the stroke is performed by the expansive force of the steam, which produces the curved line *d b*. At *b* the exhaust-valve opens, and produces the exhaust line *b e*. The backward motion of the paper then produces the vacuum line *e f*. At *f* the exhaust-valve closes, and the remaining vapor begins to be compressed by the piston, producing the curved line *f a*, and upon the steam the effect called cushioning. Thus the pencil will have circumscribed on the paper a figure, the form of which exactly indicates the force exerted upon one side of the piston of the engine at every part of the stroke. The pencil should now be bent backward, the cock shut, and the diagram removed. The cord may be disengaged when most convenient.

For the convenience of describing these diagrams, their several parts will be designated as follows:—

The line from *a* to *b*, the atmospheric line.
" " " *a'* to *b'*, perfect vacuum line.
" " " *a* to *c*, receiving line.
" " " *c* to *d*, steam line.
" " " *d* to *b*, line or curve of expansion.
" " " *b* to *e*, exhaust line.
" " " *e* to *f*, vacuum or exhaust line.
" " " *f* to *a*, lead or cushion line.

In non-condensing engines, there being no vacuum, the line from *b* to *f*, inclusive, will be considered as the exhaust.

It should be borne in mind, that the extreme end or point of the diagram is produced at the *exact time the engine is passing its centre.* Thus it will be perceived, that the steam was in this instance admitted to the piston while the crank was on or passing one centre, and was exhausted before it reached the other, which it passed with a vacuum under it of about six pounds. It will often, and perhaps most generally, be found, that two diagrams, taken at the different ends of the cylinder, will in some respects be dissimilar. In estimating the power of the engine, it is proper that this difference should be known, and the average of the two taken. It frequently happens, that several diagrams are taken from an engine at or near the same time, but from a different place, or under some different arrangement. Therefore, to avoid mistakes, and to make it more intelligible, immediately upon its removal from the Indicator a note should be made upon the margin, naming the engine, and stating the time and circumstances under which it was taken, with the height of steam, the labor it was doing, its vacuum by mercury gauge, the height of mercury in barometer, heat of water in condenser, &c.

For the purpose of ascertaining the *condition* of the valves, piston, &c., it will, in the hand of an experienced engineer, be quite sufficient for him to see the general outline of the diagram; but if it is to be used for ascertaining the power of the engine, its average measurement is to be ascertained. This will best be done by dividing it by a series of equi-distant vertical lines, and marking in each division its average measurement as in figure 2. A division into 10 parts will, generally, be found most convenient; but for extreme accuracy, double the number may be required. The sum total of these measurements, divided by the number of them, will give the average or *effective* pressure per inch upon the piston. For all diagrams taken from non-condensing engines, and for others where only the effective pressure is sought, this manner of measurement is all that is required.

A more beautiful mode of dividing and marking the diagrams, and one which might well be pursued where the diagram is to be given or transmitted to a person who has not access to the scale of the instrument, is to draw, in different colored ink, another series of lines, parallel to the 0 line, and at distances agreeing with and numbered according to the scale of pounds on the Indicator, in the following manner.

Figure 3.

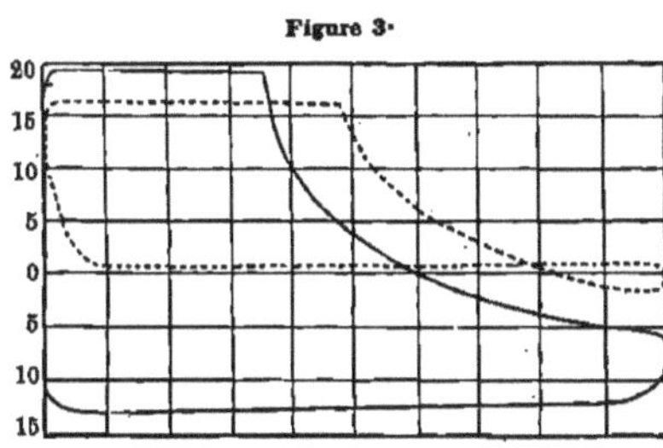

So arranged, the eye as readily comprehends their measurement as if the figures were written upon them, while their general character is still more readily observed. Some times it is desirable to find the value of the steam and vacuum pressure separately. To do this, it is necessary to measure from the 0 line upwards for the steam, and downwards for the vacuum. But in using steam expansively, it is common for the steam to expand until the line of expansion crosses the atmospheric line. In this case, the steam and expansion lines are to be measured from so long as any part of them remains above 0. When it goes below, it forms the upper boundary of the vacuum, as in the following figure.

Figure 4.

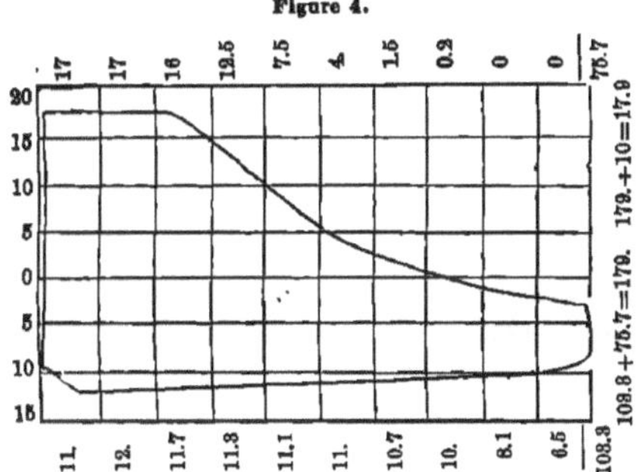

The pressure of steam is frequently, and most correctly, measured from the line of perfect vacuum. In such case the atmospheric line would be counted the 15th, and the vacuum 0. Where this mode is used, the diagram should be marked in the following manner.

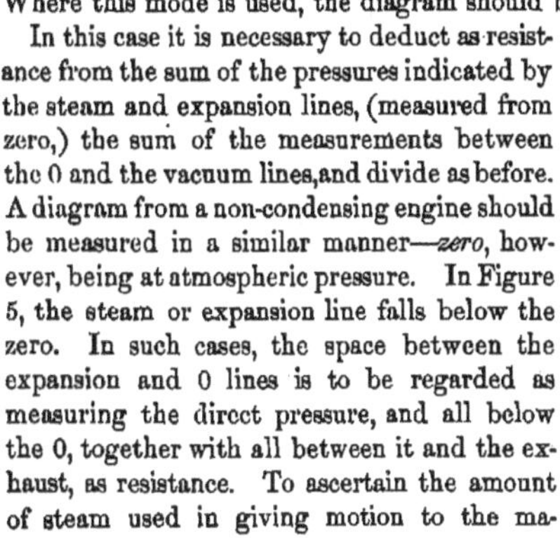

In this case it is necessary to deduct as resistance from the sum of the pressures indicated by the steam and expansion lines, (measured from zero,) the sum of the measurements between the 0 and the vacuum lines, and divide as before. A diagram from a non-condensing engine should be measured in a similar manner—*zero*, however, being at atmospheric pressure. In Figure 5, the steam or expansion line falls below the zero. In such cases, the space between the expansion and 0 lines is to be regarded as measuring the direct pressure, and all below the 0, together with all between it and the exhaust, as resistance. To ascertain the amount of steam used in giving motion to the ma-

Figure 5.

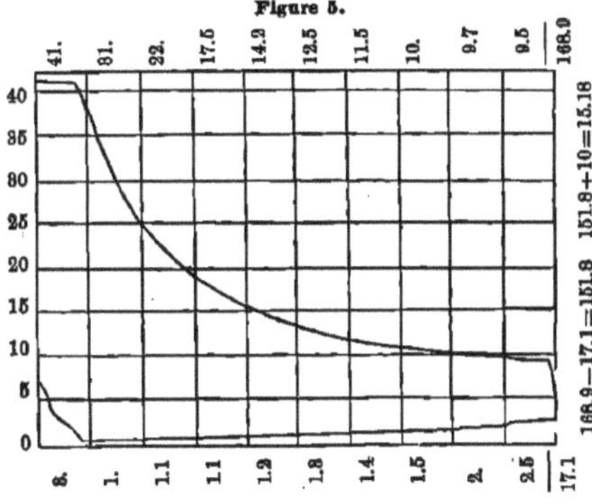

chinery, without having regard to the condition of the engine, it is sufficient to measure the space between the steam and vacuum lines, as in Figure 6.

In designating the quality of vacuum formed under the piston, there are two elements affecting it, which ought to be noticed, viz.: the weight of the atmosphere, and the temperature of the water in the condenser at the time the diagram was made. For, if the barometer stands at only 28 inches, 30″ of mercury being equivalent to 14.7 lbs., 13.7 lbs. would be a perfect vacuum; and if the water in the condenser be at a temperature of 130°, its vapor will form a resistance of 2.17 lbs.; therefore the lowest attainable vacuum would be but 13.7—2.17=11.53 lbs.; whereas, if the barometer stood at 31″, a perfect vacuum would be 15.2; and if the water was but 100°, its vapor would give a resistance of only .9, and consequently, the highest attainable vacunm would be 15.2—.9=14.3 lbs., making a difference of 2.77 lbs.*

The table showing the force with which vapor or uncondensed steam resists the ascent or descent of the piston, according to the temperature of the water discharged from the air-pump, and also the table of the variations of the weight of the atmosphere, as shown by the barometer, will make apparent the necessity of these observations.

TO KEEP THE INDICATOR IN ORDER.

After the Indicator has been used, and before putting it up, it should be taken apart and carefully cleaned and dried, to prevent injury to the springs, and to keep the dust and dirt from scratching the cylinder and piston; and, before using it, the cylinder and piston, and the axis of the paper cylinder, should be lubricated with some clean oil. If, in the use of the Indicator, the cylinder or piston should get cut or scratched, so as to interfere with the freedom of its motion, they should be delicately scraped and burnished, or ground with some nicely prepared polishing powder or tripoli.

The piston of the Indicator is, or ought to be *exactly* what it purports to be, viz.: of the area of ⅛″, ¼″, ½″, 1″, or 2″, as the case may be, and would consequently have a diameter of .4″, .56″, .8″, 1.125″, or 1.162″. To ascertain at any time whether the spring retains its original strength, it is, therefore, only necessary to turn the instrument up end down, readjust the scale to the index, and attach to the piston-rod a weight of such size as will give the spring the desired extension. A piston of ⅛″ area will be moved through each division of 1 lb., by each 2 oz. or ⅛ lb. so applied; a ¼″ by 4 oz. or ¼ lb.; a ½″ by 8 oz.; a 1″ by 1 lb.; and a 2″ by 2 lbs.

* The vacuum shown by the Indicator will generally vary from that shown by the vacuum gauge, when it is constructed with a glass tube, hermetically sealed at the top; for such gauges are designed to show the variation from a perfect vacuum, without reference to the weight of the atmosphere; but the vacuum shown by the Indicator is affected by all its variations.

BEST FORM OF DIAGRAMS.

Were all engines constructed in the best possible proportion of valves, steam passages, etc., and run at the proper speed for such proportion—and were the valves and pistons always tight, the same allowance made for clearance of piston, and no condensation of steam allowed in the cylinder—general rules might be given for the shape of diagrams from condensing and non-condensing engines, for each different pressure of steam and gradation of expansion. Thus, if steam were admitted freely from the boiler to the cylinder of a high-pressure engine during the whole stroke of the piston, the diagram should be nearly a parallelogram, the upper line of which would represent the pressure of the steam on one side of the piston, and the lower that of the exhaust steam on the other side, and the ends of it would be described while the piston of the indicator was passing from one end of the cylinder to the other. Three of the corners of such a diagram might, as a general thing, be quite well-defined angles; but the fourth—that which is produced while the piston of the indicator is passing from the steam line to the exhaust—must be more or less curved, unless the engine was going very slow, or the ports extraordinarily large, as the contents of the cylinder of the engine must be discharged during the time this line is being described.

A diagram in such a form would indicate, that the engine was receiving a power equal to the pressure of steam acting upon the piston during the whole length of the stroke. No example of such a diagram is given. The nearest approach to it is the one in the group Fig. 15, drawn in small dots. But in the engine which produced this, the steam was allowed to enter the cylinder but seven-eighths of the stroke, and only one-quarter of the usual speed was made. All the corners, it will be perceived, excepting the one where the cut-off changed the course of the steam line, are quite sharp, the exhaust line running parallel, and varying uniformly but about one pound from the pressure of the atmosphere. It is seldom that a diagram so perfect in all these respects is produced.

A diagram from a condensing engine differs from one produced by a non-condensing, in the fact that in the latter the lower line, instead of being a little above atmospheric pressure, approaches nearly to that of perfect vacuum; and, as the steam has to be condensed while the pencil is tracing the exhaust line, it is still more difficult, with such engines, to produce in the lines well-defined angles. There is also more or less of a curve produced at the termination of the vacuum line, owing either to the lead of the valves, or the compression of vapor as the valve approaches its seat. But nearly the whole of the area of the diagram lost from want of sharpness in the several parts, or their departure from right angles, except when it is the result of expansion, represents so much capacity of the engine, for which there has been an expenditure of steam, and for which there has been no other consideration exchanged than the relief to the engine from the effects of the greater percussion produced by a more instantaneous charge of the force acting upon the piston. To produce these with well-defined angles, it is evident that the pressure upon the piston of the engine must be changed from one side to the other, nearly or so far as the usual diagram would show, quite instantaneously—a usage that few large engines would be able to stand. While economy of fuel, therefore, requires well-defined angles, the stability of the engine, or economy of repairs, must direct how nearly they may be allowed to approach them. Figs. 3, 4, and 5, are examples of good diagrams in this particular.

If a diagram be taken with the paper cylinder moved by a belt from the main shaft, it will be produced in a continuous line, and show very correctly what proportion of the revolution of the engine is made while the piston is passing from the best condition of vacuum to the full pressure of steam. The following diagram was taken in this manner from one of the engines in H. B. M. model steamer *Bee*, using steam at a boiler pressure of 7 lbs., cutting off about three-sixteenths by the lap of the slide valve. It was taken by Mr. Brown, the engineer of the ship, and published in his "Treatise on the *Indicator* and *Dynamometer*," from which it has been reduced to the present scale for this work.

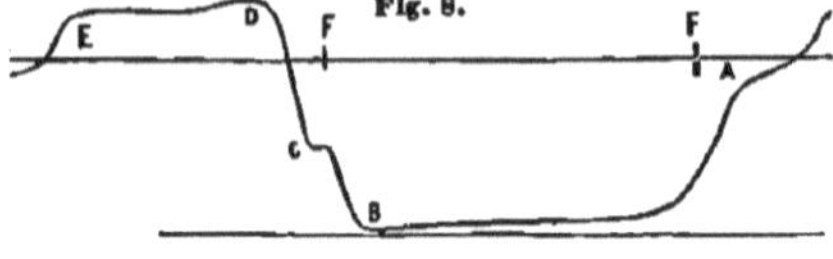

From A to B is the exhaust and vacuum lines; from B to C, the cushioning; C to D, receiving line; D to E, steam line; E to A, expansion; F F is supposed to be the ends of the stroke.

Fig. 9 is the same diagram extended more nearly to its original length, the ends joined and then folded at F F so as to represent more nearly the usual diagram, but still preserving the peculiar length and proportion of the lines. Diagrams taken in this manner expose more perfectly defective arrangements of valves; and a studious comparison of them in connection with the usual form, may often be useful in detecting defective sizes and proportions of valves, steam passages, etc., and show very correctly the proportion of time occupied in each different operation during the revolution of the engine.

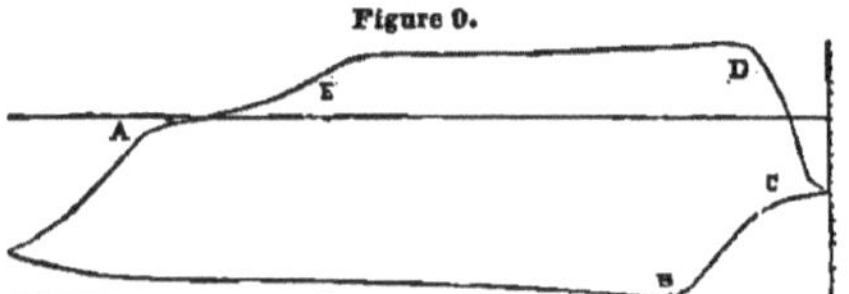

The usual form of diagram from one of the same engines, as shown at Fig. 14, represents the steam and exhaust lines as if they had been drawn nearly or quite instantaneously.

Another such diagram, taken from the engine of one of the steamers of the Pacific U. S. Mail Co., is given at Fig. 12. It is drawn, in connection with one of the ordinary form, from each end of the cylinder.

When the steam is used expansively, either in condensing or non-condensing engines, the line produced from the point where the expansion valve closes to the end of the stroke, if the valve is properly constructed, and the cylinder sufficiently protected or *jacketed*, should be nearly a hyperbolic curve, and may be thus described: Let an area be supposed equal to the longitudinal section of the cylinder, which, for the present purpose, may be 10 feet = 120 inches by 80. Let this be divided transversely into 10 sections. Now, if the steam used has an elastic force of 25 lbs. above the atmosphere, it will, with the 15 lbs. of the atmosphere, be equal to 40 lbs.; and if freely admitted to the cylinder for a distance equal to one of the above divisions, and then shut off, and the piston allowed to continue its motion to another of the divisions, the pressure of steam will be reduced one-half; and if its motion be continued to the fourth section, it will be reduced to one-fourth of its original pressure; if to the eighth, to one-eighth; and so on, reducing the pressure of steam one-half as often as the piston travels a distance equal to the whole of its previous movement. Now, if a line be drawn truly through these several points, it will represent the curve due to such a proportion of cut-off; and the area described will give the average pressure exerted by the steam during the stroke. The following figure represents such a curve, and also such other curves as would have been described had the steam been cut off at either of the other sections.

The figures at the right show to what pressure the steam would be reduced by expansion, and those at the top, its average pressure at that degree of expansion.

Figure 10.

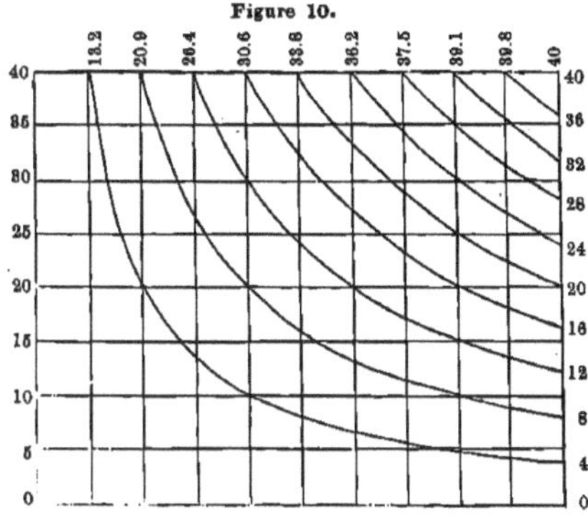

The condensation of steam by radiation of heat from the cylinder, the leakage of the steam through the piston, or valves, or the expansion of the steam in the nozzles of the cylinder, the steam-pipes, or the extra length of cylinder allowed for clearance of the piston, will vary the shape and termination of the curve. It should, therefore, always be an inquiry, upon the production of a diagram, whether this curve be theoretically correct or not. If it is not, a satisfactory reason should be found for it, in the construction of the engine, or the valves, and the piston should be thoroughly examined. A ready method to determine at what point this curve should terminate, is, to find the capacity of nozzles, pipe, &c., from the expansion-valve to the cylinder, and, adding it to the clearance, call it so much additional length of stroke. Extend the diagram in the same proportion, and then divide it into any number of parts that will divide it equally, and have one of the divisions cut it at the point where the valve closed. Take the number of those divisions on the steam line for the numerator of a fraction, and the whole number of them for the denominator, and divide with it the whole pressure upon the piston. Thus, Fig. 1 may be divided into ten equal parts, two of which measure the steam line, and therefore, the divisor is the fraction $\frac{2}{10}$, the dividend being 54, which, divided by $\frac{2}{10}$ = 10.8, for the point where the curve ought to have terminated.

A variety of diagrams, taken from different engines, and under various circumstances, selected for some peculiarity, are introduced to farther illustrate the subject—a brief explanation of which, it is hoped, with what has now been said, will enable persons using the indicator to form correct opinions of the performances of such engines as they may have occasion to examine with its aid.

Fig. 2, page 99, is from an English condensing engine, using steam at from 35 to 45 lbs. pressure above that of the atmosphere, and cutting off at $\frac{1}{3}$ of the stroke. The steam-line from *c* to *d*, would have been better had it been straight, or parallel to the 0 line. Its curve is probably produced by a contracted valve, which does not allow sufficient steam to pass to keep up the pressure, and it has, in consequence, fallen at the point of cut-off 4.5 lbs. The curve of the expansion-line is not theoretically correct, for at its termination, even had there been no condensation of steam in the cylinder, it should have fallen 4.2 lbs. farther down. This deviation has probably been produced by leakage of the cut-off valve, assisted perhaps a trifle, by the capacity of the nozzles. Nor is the vacuum-line so low as it might have been, the average vacuum being but 10.85; but its contour shows that it might have been improved by a larger supply of cold water. It is, nevertheless, a *very good* diagram, and the engine is known to have been working with very great economy.

Fig. 3 is taken from the engine of one of the Atlantic mail steamers, cutting off at a fraction above one-third, by the closing of the induction or steam-valve by the arrangement for expansion patented by Mr. Horatio Allen. The termination of the expansion-line is lower than what is due to the length of cut-off, by nearly four lbs., and it is probably about what is due to loss by radiation of heat from the cylinder. The diagram in dotted line, was produced by the engine when working without condensation. The loop at the right, was produced by the steam expanding below the atmosphere; and then, as the exhaust-valve opens, steam from the condenser returns to the cylinder to restore the equilibrium. There appears to have been, at this time, or under these circumstances, too much lead—the engine passing the centre resisted by a pressure of about 13 lbs.

Fig. 4 is taken from the upper end of the cylinder of the engine in one of the Pacific mail steamers. It was cutting off at one-fourth of the stroke, by a similar arrangement for expansion as used in Fig. 3. The expansion-line shows that the valve leaked—the termination of it due to the pressure of steam and place of cutting off, being 6.9 lbs. below the atmospheric line. The lead-line is very short, what exists being produced, probably, by cushioning, for the receiving-line commencing at 9 lbs. below 0, and curving inward, shows that the crank passed the centre, and commenced its downward stroke resisted by 9 lbs. vacuum. The curve of the receiving-line is produced by the paper cylinder starting a trifle before the steam was admitted to the piston.

Fig. 5 is from the engine of a boat designed for freighting business in the port of New York—is cutting off at about one-tenth, by the closing of the steam-valve, using the arrangement for expansion patented by R. L. Stevens, Esq. The exhaust-line shows a trifling lead, but under the circumstances, quite sufficient to produce a nearly perfect condensation of the steam by the time the engine had passed the centre—producing a very superior vacuum throughout the stroke, and an average pressure of 15.18 lbs., with a consumption of 9 lbs. steam. By comparing it with Fig. 10, it will be seen that the line of expansion terminated from 4 to 5 lbs. too high for the degree of expansion, occasioned, doubtless, by imperfection of the valve.

Figs. 6 and 7, as has been stated, are from a non-condensing, or high-pressure engine. When Fig. 6 was taken, the engine was in superior order, and expanding $\frac{5}{7}$. The curve of the expansion-line terminates at 4 lbs. below the atmosphere, or 2 lbs. above what it ought to be. At the termination of the stroke, the cylinder took steam again from the exhaust, producing the loop at the right. Fig. 7 was taken when the engine was using steam at a boiler-pressure of 90 lbs., and cutting off at $\frac{1}{3}$. A large steam-valve chamber, into which the steam had been passing, while $\frac{1}{3}$ of the stroke was made, furnished steam, in addition to what passed the governor-valve, to produce the first part of the steam-line quite parallel. The increased speed of the piston at this part of the stroke, demanded more steam to keep up this line than could pass the governor-valve, conse quently, the pressure falls off to 73 lbs. at the place of cutting off. The spires at the apex of the receiving-lines, are produced in part by the momentum of the indicator-piston, but mostly, from having a greater pressure of steam in the boiler than is required to do the work, and which is enabled momentarily to act upon the piston with nearly its full force, while the crank is passing the centre. Such spires will generally be found to indicate very nearly the pressure of steam within the boiler, and that the pressure is higher than is required to do the work. In the original of Fig. 6, this spire was continued up to 75 lbs., from whence it reacted to 31 lbs. The course of the dotted line is to be followed in measuring such diagrams.

Fig. 11 was taken by Mr. McNaught, soon after the introduction of his indicator, from an old English condensing engine, with a diameter of cylinder of 32″, and a length of stroke of 5′, working steam at atmospheric pressure only.

It is introduced to show the then existing knowledge of what was necessary to produce the best action of the engine. In this case, with a vacuum-gage attached to the condenser, exhibiting a constant vacuum of 13.5 to 14 lbs., or 27 to 28 inches, no one thought to fault its performances; but, on the application of the indicator, it was found, that the average vacuum acting upon the piston, was not over 8.63 lbs., or 17¼″. An alteration in the capacity of the eduction-valves and passages was then made, and No. 2 shows the result, it having increased the vacuum 2.81 lbs., and added to the power of the engine above one-fourth, without any change in the quantity of fuel consumed. The valves were, subsequently, more nicely adjusted, a little lead being given to them, and they were arranged to cut off at $\frac{7}{10}$. To compensate in part for this loss of power, the steam was allowed to rise 1 lb. above the pressure of the atmosphere, and as a much superior vacuum was produced, the power of the engine was quite restored, as is shown in No. 3, and with a saving in fuel of about a ton of coal per day.

Figure 11.

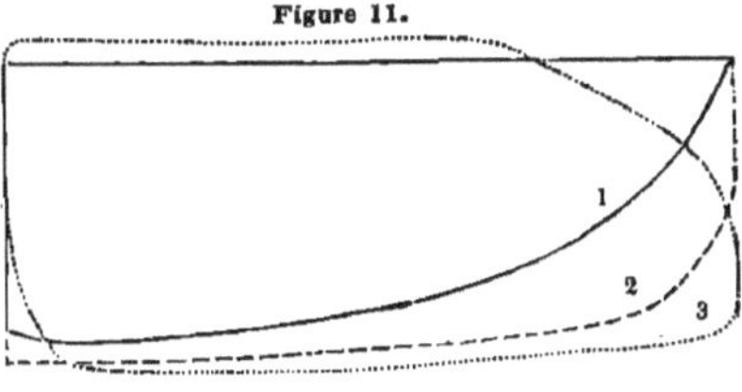

Fig 12 was taken from a steamer sailing from the port of New York, using about two atmospheres of steam, and cutting off, by the expansion arrangement of Mr. Allen, at $\frac{8}{20}$ of the stroke. When it was taken, the ship had been running a season without adjusting the valves. The diagram shows two faults in their condition—*First*, The termination of the expansion-line being two lbs. too high, shows that the valve leaked. *Second*, The lead and receiving-lines show an excessive cushioning of the steam, causing it to pass the centre against a resisting pressure of about 9 lbs., lessening the power of the engine about 5 per cent., or 25 horse power. The indicator used was out of order, the cylinder being slightly smallest at the lower end, causing the piston to bind a trifle, producing successive steps in the vacuum-line. The line below, connecting with it, was caused by pressing the piston down when it was in the position indicated. From the piston's remaining near the position after the pressure was removed, it is presumed that had there been no such restriction in its movement, it would have traced the line midway, or in the position indicated by the dotted line. Similar irregularities will be produced if the piston and cylinder of the indicator become much scratched or dirty.

Figure 12.

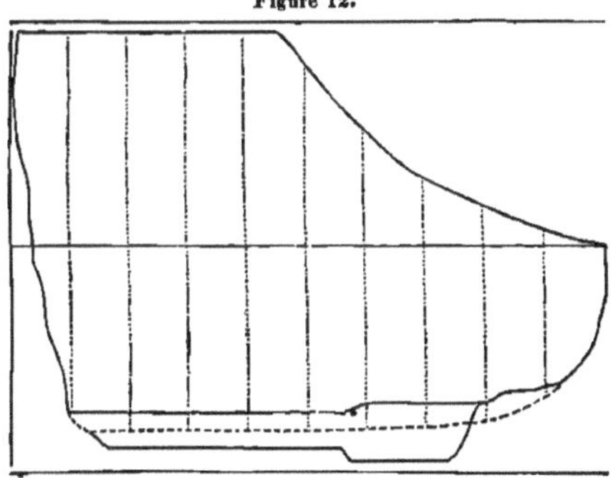

Fig. 13 is a group of three, from an engine of a similar construction and use to that making Fig. 12, the cylinder of which is 80″ in diameter, and 8′ stroke. The diagram in continuous line is from the upper end of the cylinder. Where the one from the lower end differs from it, it is drawn in dotted line. The third, in continuous dotted line, is the *Time* diagram before alluded to, (Fig. 8,) and is from the top of the cylinder. In these diagrams, the expansion-line is within one lb. of its theoretical shape. The one from the upper end of the cylinder shows a little lead to the exhaust, improving the vacuum at the beginning of the return stroke, but losing it again at its termination, by having the exhaust-valve closed too soon. By the time diagram, it may be seen, that in this case, the crank makes 18° after passing the centre, before the steam attains to its maximum pressure; and, in returning, goes from the maximum vacuum to the other centre, 27°.

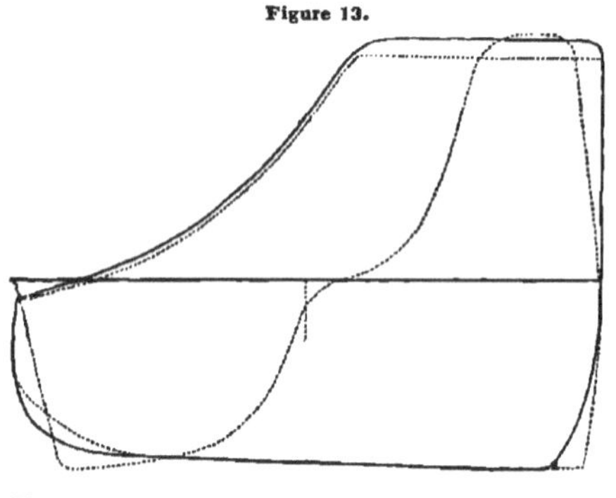
Figure 13.

Fig. 14 is a group of five, taken at different times and under varying circumstances, from the engine of the U. S. Steam Frigate Mississippi. The engines are constructed with side-levers, poppet-valves, with Stevens' arrangement for variable expansion, and with cylinders of 72″ diameter by 7′ stroke. In 1, 2, and 3, the steam was admitted freely from the boilers, until stopped by the action of the cut-off valve. The quality of the vacuum, it will be seen, is improved materially in proportion as the principle of expansion is extended. In Nos. 4 and 5, the steam was closely throttled. It is probable, that when No. 5 was taken, the steam was much higher in the boiler than when No. 4 was taken, and that the valves leaked a trifle, causing the line of expansion to terminate the same as No. 4, though receiving less initial pressure, and cutting shorter. The vacuum is also much impaired, and probably in part by the same means; but much more of it was, doubtless, to be attributed to an inadequate supply of injection water. When these two were taken, the engines were used only to keep the wheels from backing water, the vessel being under a full press of sail.

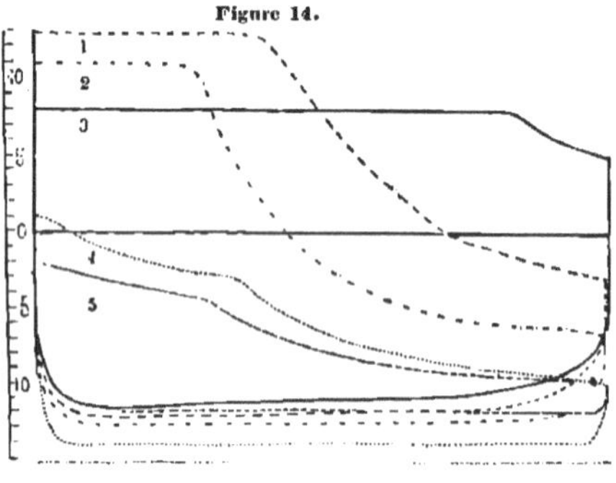

Figure 14.

Fig. 15 is a group of three diagrams, from the engines of H. B. M. model steamer Bee, having cylinders of 20″ diameter, with 2′ stroke, making 38 revolutions, and using steam at a pressure of 7 lbs above the atmosphere, the engines being constructed with the usual D valves. The first, or principal one, is a good specimen of the diagrams made by the English marine engines, as constructed and used at the present day. The one in dotted line is the continuous or time diagram of the same engine, the peculiarities of which have been already described. The small one is a diagram of the friction of the engines. It was produced by disconnecting the paddle-wheels from the Engines, and working the engines at the same speed and temperature of condenser at which it was run when the first was taken. It will be seen, that its whole area is below the vacuum-line of the gross diagram, the average pressure being 1.73 lbs.

The difference observable in the quality of the vacuum between the two diagrams, is, doubtless, to be attributed entirely to the quantity of air in the water used in condensation, and for supplying the boiler—the condition of the engines and temperature of condenser being the same when the two were taken.

Fig. 16 is a group of five diagrams from one of Stillman, Allen & Co.'s high-pressure engines, with a cylinder of 15″ by 4′, witn slide steam and cut-off valves. When No. 1, counting from the top of receiving-line, was taken, the engine was heavily loaded, and was making but fifteen revolutions, cutting off ⅘. The cut-off valve did not close tight; consequently, the expansion-line shows but a little descent until the further ingress of the steam is fully stopped by the lap of the valve. No. 2 was produced with the cut-off valve disconnected, the engine making thirty-eight revolutions, with throttle nearly closed, and steam in boiler at 70 lbs. No. 3 was produced with engine making ten revolutions, with steam at 45 lbs., throttle and governor-valves open. No. 4, the engine making twenty strokes, and No. 5 twenty-eight; and, as will be seen, with steam varying but little. In No. 5, the steam-line has a depression in the middle, owing to the cut-off valve being in a position to prevent a free flow of the steam into the cylinder, and consequently when the piston was in its greatest velocity, it fell off one or two lbs., but regained it as the velocity decreased. A slight curve in the lower part of the receiving-line indicates the lead in the valves, which was indeed very trifling, but quite as much as was desirable, and gives the fullest possible effect of the steam.

Figure 15.

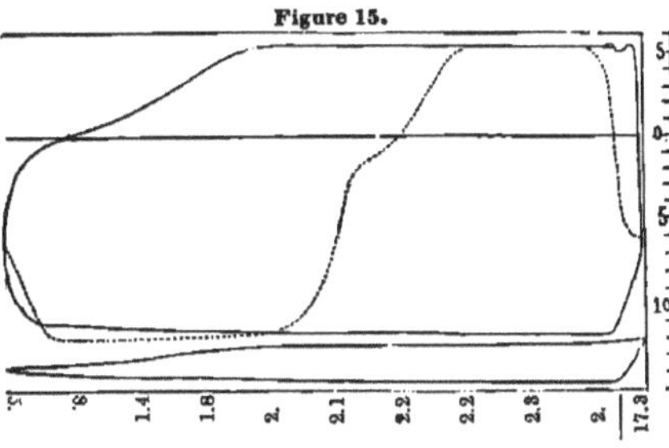

Figure 16.

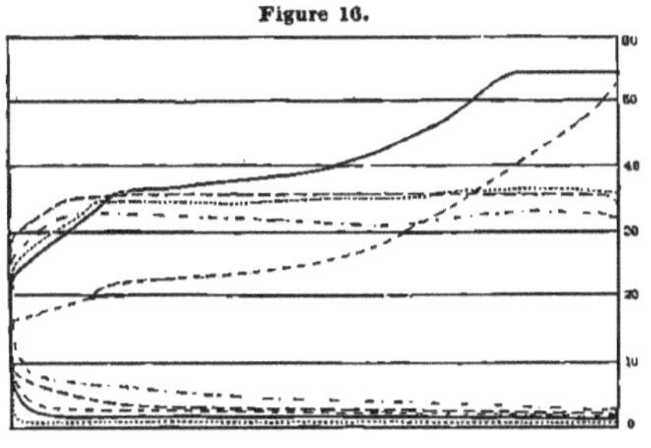

These two are also from the same engine, all the load being removed, except about 50′ of heavy shafting, and the engine making, when the one in continuous line was taken, forty-five revolutions, and when the one in dots was taken, about sixty, the steam in the boiler being about 50 lbs. A careful measurement of them will show that an average pressure of 4.82 lbs. and 7.12 lbs. was required, to produce the respective speeds of engine. But of this pressure, 1.8 lbs. for the first, and 3 lbs. for the last, were required to expel the exhaust steam from the opposite end of the cylinder, leaving but 3.02 lbs. and 4.12 lbs. for friction of the engine and shafting.

Figure 17.

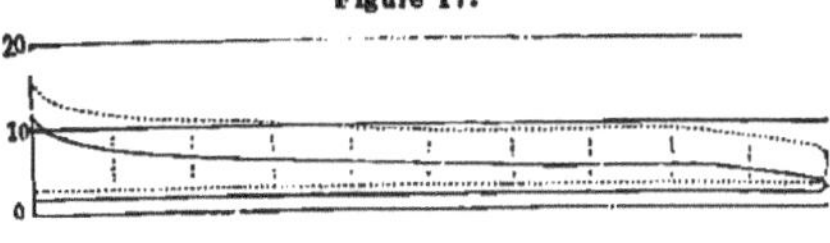

Fig. 18 is a diagram from the air-pump of a marine engine. The indicator with which it was taken, was attached to the pump near the delivering-valve. It will be observed, that the pump begins its upward stroke with a resistance of 3.5 lbs., (counting for perfect vacuum,) which gradually increases to 21.5 lbs., and that returning it, meets with a nearly uniform resistance of 3.5 lbs., which is about the same as was offered to the piston of the engine. The diagram shows the average resistance to the bucket of air-pump to be 6.32 lbs. But the air-pump is only ⅕ or ⅕ the area of that of the cylinder, and as the piston makes two strokes to one of the bucket, 6.32 must be divided by ⅕, and its quotient by 2, to get the pressure upon the piston of the engine, required to work the air-pump—which, as will be seen above, is but .63 lbs., exclusive of the friction of buckets and rods. River and stationary engines, having less height to raise the water from the air-pump, seldom require more than 5 lbs.

Figure 18.

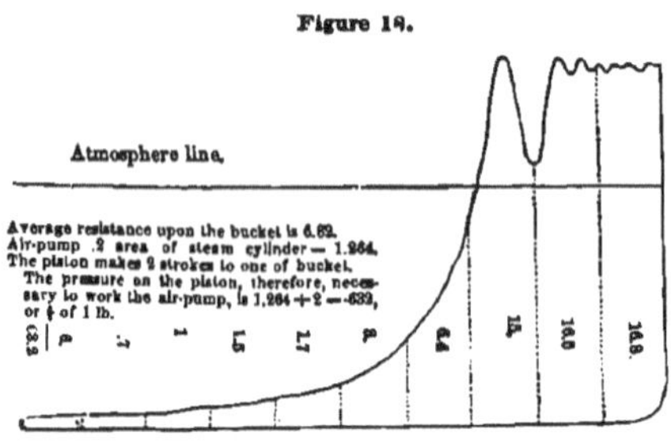

This is one of Mr. McNaught's diagrams from an engine with small ports, and illustrates the advantages in such circumstances, of cutting off and exhausting before the piston arrives at the end of the cylinder. It will be seen, if the figures are carefully measured, that, by the change of the set of the valve, so as to cause it to cut off 1/16 of the steam, and immediately exhaust, the power of the engine was increased from an average of 12.51 lbs. to 13.84, or 1.1 lbs. per square inch of the piston, thus adding 10.4 per cent. to the power of the engine, or 18.4 per cent. to the economical use of the fuel.

Figure 19.

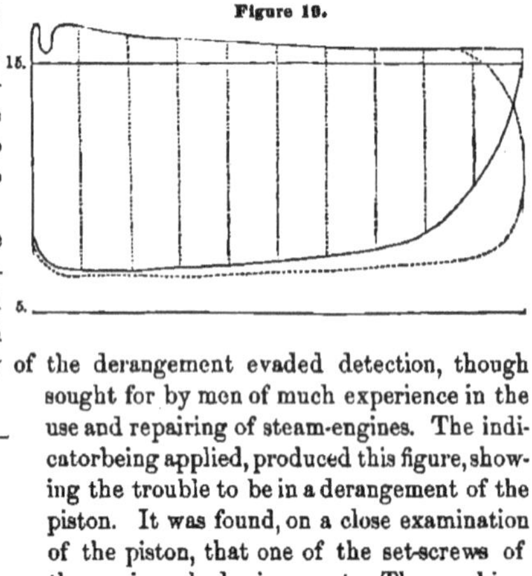

15.

5.

Fig. 20 is copies of two distinct diagrams. The one in continuous line, was from an engine laboring under a peculiar derangement, by which it all at once began to consume about twice the usual quantity of steam, while the particular locality of the derangement evaded detection, though sought for by men of much experience in the use and repairing of steam-engines. The indicatorbeing applied, produced this figure, showing the trouble to be in a derangement of the piston. It was found, on a close examination of the piston, that one of the set-screws of the springs had given out. The packing-rings were thus allowed to yield to the pressure of steam, while, by some peculiarity of construction, the screw regained its place when not under pressure, in such a manner as to have escaped attention. It will be seen that the exhaust-line falls at the end of the the stroke about one pound, and then, crossing over, it rises gradually, to about ten

Figure 20.

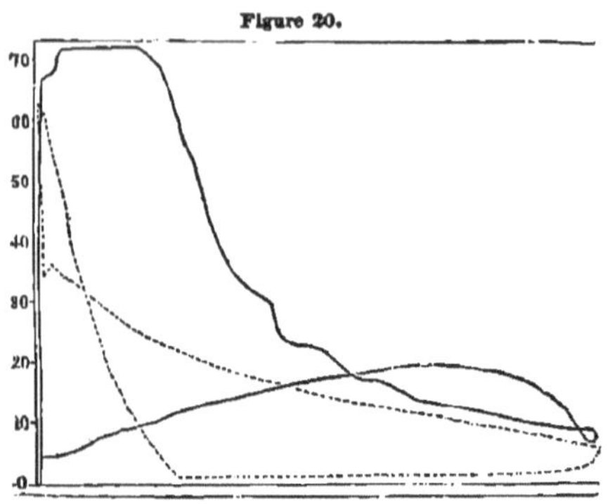

pounds above the expansion-line, and attains at two-tenths of the stroke, a resistance of about twenty lbs. above the atmosphere, and diminishes to about five at the end of the stroke. This peculiar conformation, it will be seen, could only be produced by the steam passing the piston much faster than it could find an easy exit, and consequently, at the position where the greatest pressure was acting on one side of the piston, there we find the greatest resistance on the other.

The other figure was produced by a 60-horse-power engine, in the vicinity of New York, supposed by the owner to be in fair working order, and working under a pressure of 100 lbs. per square inch. Its failure to do the required duty led to the application of the indicator. The principal defects indicated by the diagram, was a defective construction of the valve by which the exhaust-passage was closed when the stroke was but three-fourths completed, in consequence of which, the vapor then remaining in the cylinder, was confined and compressed by the piston to 64 lbs. at the end of the stroke. When this diagram was taken, the engine was running without its usual load, and close-throttled by the regulator—an average pressure of 17 lbs. of steam being admitted to the cylinder, of which an average of 7.3 was lost by the resistance.

Figure 21.

Fig. 21 is from an engine of similar construction to the one producing Figures 6 and 7. It was running close-throttled, with a full pressure of steam, and cutting off at one-fifth, but with much less than its usual load, and its motion much quickened. It illustrates the combined effects of close-throttling and expanding, the steam being admitted to the piston at about 80 lbs., and was reduced by the accelerated motion of the piston, to about 50 lbs. at the position where the expansion-valves took effect. The dotted line was produced by throttling the indicator also.

This figure represents three diagrams from successive revolutions of a 10-horse non-condensing engine, making about 60 revolutions per minute, and working expansively, the steam following about $\frac{1}{10}$ of the stroke. The great difference in the quantity of steam consumed, the burden of the engine being constantly the same, is to be attributed to an excessive motion of the governor-valve. The slide-valve also was deranged, being a little behind time, as is shown by the inclination of the receiving-line to the right; and the exhaust-opening was too narrow, as is shown by the cushioning of the steam at one end, while at the other it did not open until the piston was well advanced on the return stroke. The peculiar *hook* in the receiving-line, is to be attributed, doubtless, to the escape of the cushioned steam at the moment of time when the piston is still, and the crank passing the dead-centre. But before it all escapes, the steam-valve begins to open, being just at the instant when the crank has passed the centre,

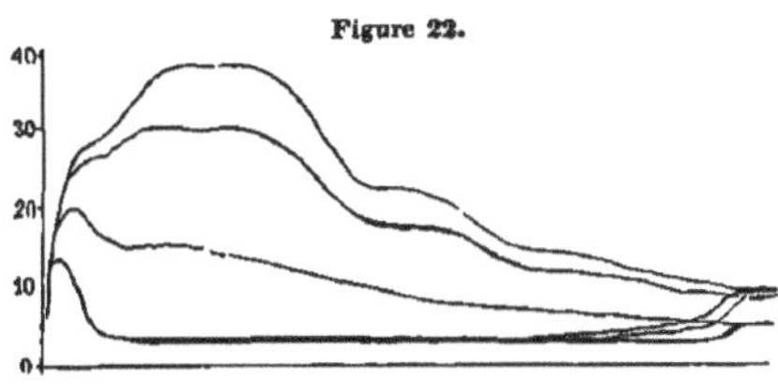

Figure 22.

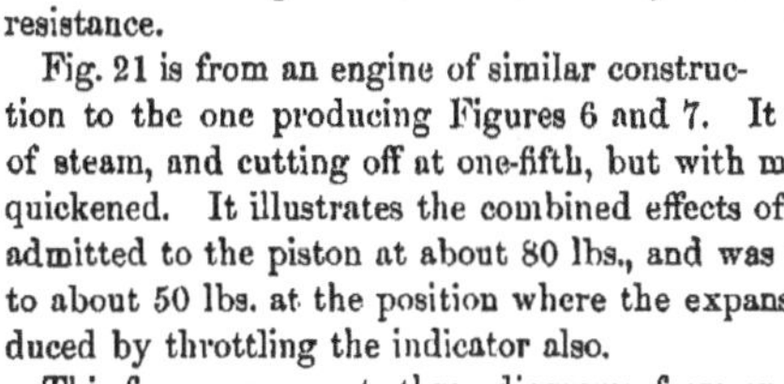

and the piston begins its forward stroke. The pencil, therefore, instantly returns on the line just made.

Similar to the foregoing, in one respect, is Fig. 23, which was produced by the same engine that produced Fig. 15. In this case, it was working as a non-condensing engine, with steam at 7 lbs. above the atmosphere. The exhaust-valve closing too soon, the steam remaining in the cylinder was cushioned or compressed to about 13 lbs., but when the steam-valve opened, the steam rushed back towards the boiler, instantly reducing the steam to working pressure. It is important to notice the distinction between the effect of a lead of a valve, and the effect called cushioning. Figs. 12, 20, 22, 23 are examples of excessive cushioning.

Figure 23.

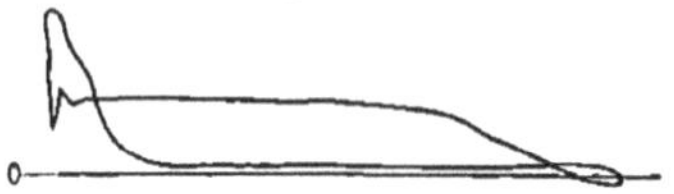

It has been a practice for many years, with English engineers, more particularly with those engaged in Cornish mining operations, to keep a record of the performances of the best engines used in pumping the mines. To carry out this purpose fully, a record is kept of the revolutions made, the load of the engine, and the coal consumed, with all other necessary particulars, and the performances are rated by what is termed the "*duty*," which means the number of pounds raised one foot high, with one bushel, or 94 lbs of coal of a specific quality. Since the publication from time to time, of the duty of these engines, as arrived at by these records, the duty has been constantly increasing, so that from the maximum attained by Smeaton, of 10,600,000 and of 27,000,000 by Bolton & Watts, it has now exceeded 100,000,000. Fig. 24 is a diagram from one of these engines, said to be doing a duty of 90,000,000, and which is of the following dimensions: Cylinder, 36″; stroke, 10′ 6″; making ten revolutions per minute; using steam at 32′, or 17″ above the atmosphere, and cutting off at 6½″ and to which 3″ were added for clearance, contents of nozzles, &c., being equal to 9½″ of steam, and expanding 13.5 times. If this diagram is examined by the rule given on page 105, it will be seen that, under ordinary circumstances, $\frac{1}{13}$ of 30 lbs. (the pressure when the cut-off valve closed,) would reduce the pressure at the end of the line of expansion to 2.22 lbs., while this terminates at about seven lbs. Whether this excess is owing to leakage of steam through the valve, or to expansion by heat in the cylinder, is not stated on the diagram from which the above was copied. The resistance from uncondensed vapor and air in the condenser, is from one to two lbs. greater than might have been expected from an engine working with that degree of expansion and economy, and 1.7 greater than is shown in Fig. 5—the diagram showing the nearest approach to this degree of expansion of any of the examples given —but in this, 3.8 lbs of steam were used, or ⅓ more than in this diagram, and consequently, subject to a proportionally increased loss of vacuum by the air carried into the condenser with the condensing water, supposing that water of the same temperature was used for condensation in both of the engines.

Figure 24.

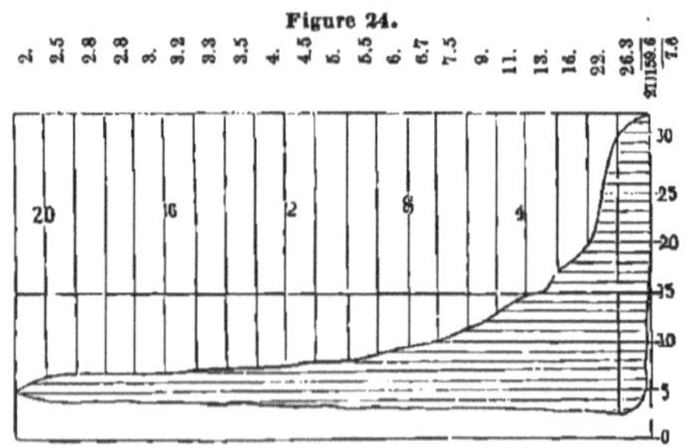

STILLMAN'S IMPROVED PATENT MANOMETER.

The value of a reliable instrument, which will indicate constantly the pressure in the boiler of a steam engine, though established in many quarters by the infallible teachings of experience, is yet, for want of experience, as a general thing, little understood. Its importance, as a preventive of explosion, and of the frightful consequences to life and limb, and ruinous pecuniary results of such disasters, is obvious on the slightest consideration; but the value of the instrument, in the economical results of its daily use, is by no means properly appreciated. The pecuniary loss arising from any considerable fluctuation of the pressure of steam, has never been properly considered by the proprietors of engines. If steam be carried too high, the surplus will escape through the safety valve, and all the fuel consumed to produce such excess is so much dead loss. On the other hand, if there be at any time too little steam, the engine will run too slow, and every lathe, loom, or other machine driven by it, will lose its speed just exactly in the same proportion, and, of course, its effective power. A loss of one revolution in ten, at once reduces the productive power of every machine driven by the engine ten per cent., and loses to the proprietor ten per cent. of the time of every workman employed to manage such machine; in short, the loss of one revolution in ten diminishes the productive capacity of the whole concern ten per cent., so long as such reduced rate continues; while the expenses of conducting the shop (rent, wages, insurance, &c.), all run on the same as if everything was in full motion. A variation to this amount is a matter of frequent occurrence, and is, indeed, unavoidable without proper instruments to enable the engineer to avoid it. A very little reflection will satisfy any one, that it must be a very small concern indeed, in which a half hour's continuance of it would not produce a result more than enough to defray the cost of a very expensive instrument to prevent it. If the engineer, to avoid this loss, keeps a surplus of steam constantly on hand, he then is constantly wasting the steam, and consequently fuel, and thus incurs another loss, which, though less alarming than the first, will yet be a serious one, and render any instrument most desirable which can prevent it.

It is evidently, therefore, of great importance to the proprietors of engines to have an instrument which should constantly indicate the pressure in the steam boilers. This would enable the engineer to keep his steam at a constant pressure, and thus avoid waste of fuel, on the one hand, and the still more serious loss of the productive power of the shop, on the other. An instrument, therefore, constantly indicating the pressure of steam, reliable in its character, and, with ordinary care, not subject to derangement, is evidently a desideratum, both to the engineer and proprietor. It would enable the one to manage his steam with ease, and the other to see whether it was so managed, and would not only protect against the disaster of an explosion, but by its every-day use promote, in a high degree, the economy and efficiency of the whole establishment.

The importance of these objects did not escape the attention of the fathers of the steam engine; and the inverted siphon, filled with mercury, and with a rod floating on its surface, under their guidance, readily answered the desired end. For steam of from two to three atmospheres, this is still the most reliable of the appliances in use at the present day; but for steam of higher elasticity, they become less reliable and convenient, in consequence of their great height, the friction of the float, and their liability to lose the mercury by its oscillation, or by excess of pressure. In addition to this source of expense, their original cost is very considerable when constructed for such use. Every method adopted to obviate these difficulties has been attended by others, to avoid which many expedients have been tried, mostly through the aid of pistons and springs. These, though theoretically good, have mechanical difficulties which prevent their use. The piston, if packed tight enough to prevent leakage of steam, is so much impeded by friction as to render it unreliable; the springs, also, should they become corroded and weakened by dampness, to which they are almost unavoidably subject, or should they lose their original elasticity, render the indications inaccurate. Other devices, had they been faultless in efficiency, would have failed of general adoption from other considerations.

The desire to avoid these disadvantages led to the introduction, to some extent, of the beautiful instrument which is the subject of this treatise—*the Manometer*—an instrument which, though attended with none of the serious causes of error attaching to the other devices, still has some others of its own.

The imperative demand for a good high-pressure steam gage, in consequence of the very general introduction of steam of great elasticity, in this country, led to the attempt to improve this instrument, which has resulted in producing a gage, occupying but a small space; attended with no *measurable* friction; having a spring not liable to be broken or diminished in its power, being adapted to any desirable pressure; costing little; its indications being constant and visible at a glance; being less liable to get out of order; and any inaccuracies it may acquire being readily detected and corrected.

According to the derivation of the word (from *manos*, rare, and *metron*, measure), the Manometer is an instrument for measuring the degree of rarification of aeriform fluids subjected to less than atmospheric pressure. The term is also, and more generally, applied to the instrument when used to indicate the density of aeriform fluids subjected to more than atmospheric pressure. The Manometer may, therefore, be defined to be an instrument for measuring the density of aeriform fluids, by means of a column of mercury in a glass tube, whether that density be greater or less than that caused by the atmospheric pressure.

Fig. 26.

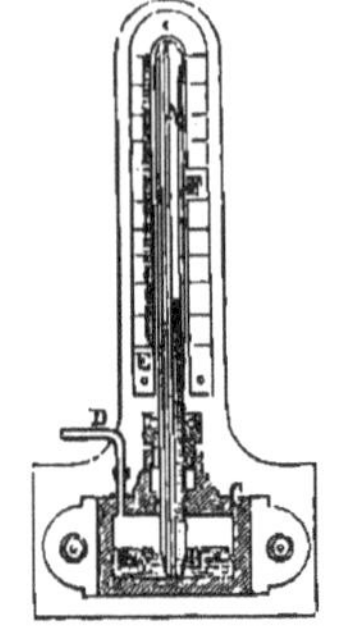

The usual construction of the Manometer, when used for its original purpose of indicating less than atmospheric pressure, is as follows: a glass tube AB, (fig. 26) sealed at one end (A), and drawn to a small aperture at the other (B), is filled with mercury, and inverted into a reservoir C, containing mercury, which reservoir is made air tight around the glass tube. By means of the pipe D, a communication is opened between the aeriform fluid, whose pressure is to be ascertained, and the interior of the reservoir (C), above the surface of the mercury. The top of the mercurial column will, at the ordinary pressure of the atmosphere, stand about 29″ or 30″ above the level of the mercury in the reservoir; but as soon as the pressure above the surface of the mercury in the reservoir becomes less than that of the atmosphere, the column of mercury will, of course, fall in the tube just in proportion to the diminution of the pressure. By the scale EF, which is attached to the glass tube (AB), and marked with inches, or other proper division, the degree of pressure will at all times be constantly indicated.

When used for measuring pressure above that of the atmosphere, the instrument (as usually constructed) is, in all respects, the same as just described, except that the tube AB is not filled with mercury, but inverted while full of atmospheric air into the reservoir (C), and the scale (EF) is differently marked. When the pressure on the surface of the mercury in the reservoir (C) is that of the atmosphere, the mercury will rise in the tube nearly to the level of that surface (but slightly lower, owing to the resistance of the air in the glass tube). As soon, however, as the pressure, communicated through the pipe D, exceeds that of the atmosphere, the mercury will be forced up into the tube, and the enclosed air condensed, until its elastic resistance is just equal to the pressure. The height of the mercurial column will, of course, vary with any variation of pressure, and thereby indicate the degree of pressure at every moment by means of the scale (EF), which is divided, according to Mariotte's law, into atmospheres, pounds, or the like.

The high degree of pressure to which the last described form of Manometer may be subjected without error from friction or loss of mercury; the permanent elasticity, and the everywhere existing and exactly-defined qualities of the material of resistance (atmospheric air or other fluids of the same nature); its comparatively small dimensions, and convenient form, make it a very desirable instrument for measuring the pressure of steam. As usually constructed, however, it has defects, which have prevented its general use as a steam-gage. Among these defects were: the coating and consequent opacity of the glass tube, by the deposition of an oxyde of mercury when acted on by the enclosed atmospheric air; the expansion and partial loss of air from within the tube whenever any partial vaccum was produced in the boiler, and so allowing the mercury to rise higher in the tube with the same pressure; its oscillation, especially when there is a varying pressure, as in engines working expansively; the almost constant tendency of the condensed steam to insinuate itself between the mercury and the glass, and to find its way into the tube above the mercury; and the great inequality in the divisions of the scale, arising from the peculiarities of the law that governs the volume of aeriform fluids under pressure.

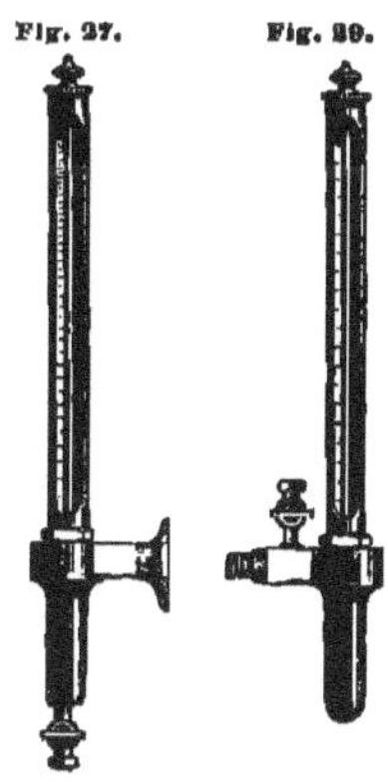
Fig. 27. Fig. 29.

The improvements by which these defects have been remedied, at the same time rendering it more serviceable for determining pressures less than that of the atmosphere, have recently been made the subject of a patent.

The New Patent Manometer is represented in the subjoined cuts:

Fig. 27 is the usual form of the Patent Manometer for showing a pressure up to eight atmospheres.

Fig. 28 represents the form of one for showing a pressure up to twenty atmospheres.

Fig. 29 is the form used for showing less than one atmosphere. The arrangement of the glass tube is quite similar in all the forms usually given to the instrument.

Fig. 30 is a longitudinal section, through the centre of the glass tube, in which A is the tube; B is an iron piece in which the tube is firmly secured by means of the stuffing-box G. It is screwed at one end, to receive the brass case C; and in the middle, to confine it in the reservoir of mercury into which the lower end of the tube is to be immersed. D D are scales divided into atmospheres, pounds, or inches of pressure, as desired. E E are blocks to secure the scales in their proper places. F is a gland which protects the lower end of the tube, and compresses the packing in the stuffing-box G. H is a cap or plug, loosely screwed into the gland to facilitate the operation of charging the tube, and also by admitting the mercury into the tube only through the interstices of the screw, prevent its oscillation, and at the same time allow the orifice to be made the full size of the tube whenever it may be necessary to clean the tube.

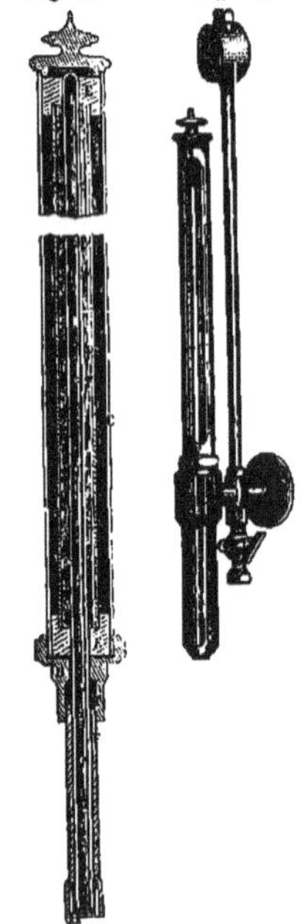
Fig. 30. Fig. 28.

In Fig. 27 the reservoir for mercury is a deep cell, with an iron tube communicating from the cock at the bottom to the middle of the chamber above the surface of the mercury. In Fig. 28 it is divided, the glass tube being inserted into a cell of greater depth, while the reservoir of mercury is in the bulb, to which a sufficient elevation is given to compress the gas within the tube to two or three times the density of the atmosphere, according to the density of the steam of which it is to serve as the gage. In this, as in the other form, an iron tube communicates the pressure from the cock below to the surface of the mercury in the bulb above. The subdivisions of the scale are by this means much more uniform and distinct than when used at atmospheric pressure only.

In all cases, the mercury should be seen above the junction of the tube with the tube holder, so as to indicate the initial pressure, or 0. In Fig. 27 it is brought up by partially exhausting the tube at the time it is erected. In Fig. 28 it is forced up by the superincumbent weight of the mercury in the bulb. The oxydation of the mercury within the tube is prevented in the latter form of the instrument by charging the tube with nitrogen or hydrogen gas; but in the former, on account of the difficulty of preventing the admixture of atmospheric air, while exhausting a portion of the contents of the tube, for the purpose above referred to, atmospheric air only is used, and a drop or two of naphtha, or other fluid answering the end, is introduced within the tube, on the surface of the mercury, to prevent the oxydation.

When designed to show a pressure less than atmospheric, but not less than that shown by two inches of mercury, the tube is to be perfectly filled with mercury, and inverted in the reservoir, and the pressure will be determined by the number of inches sustained above the level of the mercury in the reservoir below; but if it is to be used for a pressure less than the weight of two inches of mercury—that being the distance from the lowest visible part of the glass tube to the surface of the mercury in the reservoir—it will be necessary to use the bulb shown in Fig. 28, but with such an elevation only as will bring the surface of the mercury in it to a height equal to the lowest visible part of the glass tube; or, it may be done equally well by using the form shown in Fig. 28, if a scale is properly made for the purpose, and the bulb elevated so as to compress the air so high in the tube as to allow the mercury to have sufficient fall without going out of sight, when the pressure of the atmosphere is removed from the surface of the mercury in the bulb above.

It will be seen that either of these arrangements would resist the tendency of such partial vacuum as is generally formed in steam-boilers, when they are allowed to cool down, from disturbing the quantity of air within the tube of the Manometer.

If the initial quantity of air or gas in the tube be deranged by a change of temperature, or by any other cause, it becomes necessary to know the extent of the variation occasioned thereby. To ascertain this (if inexpedient to correct it at once), a simple arrangement is adopted, viz.: first, to remove the pressure by closing the stop-cock and opening the small waste-cock between it and the reservoir—this will allow the mercury to fall to a place in which it will be at equilibrium with the atmosphere; second, to note the point to which it descends. The variation from the original place of 0 will be, in addition to the lbs. shown on the scale-plate, such part of the whole as the variation from 0 bears to the whole length of the tube above 0. To determine this proportion, a series of decimals is placed on the scale at fixed distances, and the one of these nearest to where the base of the column of air within the tube rests, is to be used as a multiplier by which the pressure of steam indicated on the scale, less the lbs. of variation shown, is to be multiplied. Their product will be the true pressure. Thus, for example, if the mercury in the tube falls until the base of the column of air rests at the decimal .96, which would be near to the place due to 1 lb. pressure, and if, on opening the communication to the boiler again, it should rise to 130 lbs., from this apparent pressure of 130 lbs., 1 lb. is to be deducted, and the remainder multiplied by .96, thus giving as the true pressure 123.8 lbs., showing a variation of 6.2 lbs. The indication, in all cases, is from the top of the spirits.

HOW TO ERECT THE MANOMETER AS A STEAM-GAGE.

FORM SHOWN IN FIG. 27.

Select a place for its erection where it will be exposed to a good light, and where, with due respect to light and convenience, it may be as little as possible liable to dust, to accidental blows, or to a high or variable temperature; and at such a height, if possible, that the figures on the scale-plate shall be nearly or quite on a level with the eye. It should be secured to its support with sufficient firmness to withstand the use of the wrench in screwing down and unscrewing the tube-holder. Two 1½ inch No. 16 wood-screws, well driven through the flange into good pine wood, will generally be found quite sufficient for this purpose. A connection should be established between the instrument and the boiler by means of a small pipe, in the construction of which the following particulars should be carefully observed:—1st. When the gage is one or more stories above or below the boiler, provision should be made to prevent the pipe from filling, as it is liable to do, with the condensed steam, because every two feet of water thus vertically suspended, will affect the gage about one pound—and its aggregate will make it show materially, more or less, as the case may be, than the pressure of the steam. Therefore, if the gage is *above* the boiler, and the pipe can descend all the way into it, let the pipe be of at least ½ inch internal diameter, which will generally be found adequate to the purpose; but if its direction be horizontal any portion of the way, or if the instrument be below the boiler, provision must be made to have the water escape. Or, if preferable, a circulating pipe can be used, having one end connected with the upper part of the boiler, or steam-pipe, and the other with the lower part, the instrument connecting with it by a branch. Such an arrangement will suit for any height above the boiler. But if the gage be upon the same story, its aggregate will not be likely materially to disturb the calculation—and a pipe of ⅛ inch will, as a general thing, be quite sufficient—and no provision need be made for the condensed steam.

It is not advisable to have the instrument always subject to the heat of the steam, as this is liable seriously to injure the joints, and cause them to leak. For this reason it is improper to attach it directly to the boiler, steam-pipe, or cylinder. An upward direction should, likewise, be given to the pipe, for one or two feet, at least, near the gage, in order to retain within the reservoir the water condensed there, which, being a poor conductor of heat, will at all times preserve the instrument at a proper temperature.

The pipe for a boiler using salt water, should be of copper, but ordinary iron gas pipe will be found quite sufficient where fresh water is used, and is generally the most convenient article for the purpose.

When the reservoir is secured in its place, and the pipe connected with the boiler, it should be allowed, before charging the tube, to take the heat to which it will be subjected when in use—after this, the tube-holder should be unscrewed, taken out, and inverted. The screw-cap (represented at H, in the section, fig. 30), is next to be removed, and the tube filled with mercury to the inverted arrow-heads [≺ —— ≻] adding on top of the mercury about four drops of sulphuric acid, or such other lubricating agent as shall be preferred—then replace the cap. Put the remainder of the mercury furnished for the gage into the reservoir, and then restore the tube to its place in the fountain. If the charge has been properly effected, the mercury in the tube will fall to the place of 0—(which, in order to make the scale as long as possible, is frequently omitted to be marked on the scale-plate, its position being understood as at a distance below 1 equal to the space between 1 and 2)—or at the lowest visible point of the tube. If it should not fall thus far, more air should be introduced. This is to be effected by inclining the tube towards a horizontal position, when, in consequence of the previous rarefaction of the air within the tube, a further supply will be drawn in from the atmosphere. Care, however, must be taken that too much is not admitted, lest the mercury fall too low—for which the only remedy would be to invert the tube-holder again, and re-charge the tube as at first. A single bubble of air, as it passes through or up the tube, will generally be found sufficient to lower the mercury ⅛ of an inch.

If accurate measurement of the steam is necessary, great care must be taken in this part of the erection of the Manometer; for, if either more or less air be in the tube than was designed, a given amount of pressure will not compress it into a like space; and, as the volume of air within the tube is affected by temperature, there is obvious necessity of attention to the direction, that the temperature, when the gage is erected, shall approach as closely as possible to the temperature at which it is to work.

In charging the tube, it is well to introduce the mercury through paper wound into the shape of a tunnel, and to be particular not to fill the space between the gland and the outer surface of the tube; and care should be taken that the cap over the lower end of the tube be not too tightly screwed, nor yet left so loose as to give too much oscillation to the mercury.

In re-adjusting the instrument, the screw-cap may serve as a dipper, by adding a splint-handle, to take mercury from the reservoir, if necessary.

FORM SHOWN IN FIG. 28.

In this form of Manometer, the tubes are considerably longer, and the mercury is forced up, by an external column of mercury, to a place susceptible of observation at zero. This compression gives the air twice the density of that used in the other form, by which means the divisions on the scale-plates are much more regular.

As this arrangement does not favor the use of the usual lubrication of the tubes, they are filled with nitrogen or hydrogen gas, which, as before stated, entirely obviates the necessity of it.

To keep this gas from mixing with atmospheric air, after it is in the tubes, and before it is secured by the mercury, the open end of the tube is covered with an India rubber bag, filled also with gas.

When the reservoir is secured, and connections made, a small quantity of mercury—sufficient to cover the end of the tube when in its place—is to be put into the reservoir. The bag is now to be removed from the tube, while the tube is kept in an upright position, and the tube put instantly into its place in the reservoir, and a good joint made with cord moistened with tallow and white lead, between it and the tube-holder. The remainder of the mercury furnished is now to be put into the upper reservoir, which will, if all is properly done, cause the mercury in the tube to rise to the 0 on the scale-plate.

If it varies from that position, it will affect the indication of the pressure, and is to be corrected by the rule on page 117.

WHEN USED AS A VACUUM GAGE.

The same general directions should be followed in regard to location, &c., as have been given for erection when to be used for steam, excepting what is said of temperature; but additional care should be taken to have the cocks and joints of the pipe quite tight.

It will be remembered, that the tube is to be quite filled with mercury, and the accuracy of the instrument will depend upon the extent to which the perfect exclusion of the air has been effected. Simply to *pour* it full then will not suffice—for during this operation minute bubbles of air will attach themselves to the tube. These must be removed by a wiper, or piston of cotton thread or silk, on the end of a small rod of wood, or covered wire. The method usually pursued is, to provide a pine rod, $\frac{1}{8}''$ square, and two feet long, and, after notching or ragging the end of it, wind on it some cotton wick, to the full diameter of the tube. Then, putting a sufficient quantity of mercury into the tube to fill it for three or four inches, introduce the wiper, working it downwards through the mercury to the bottom. This action will have removed the air from the wiper, and better prepared it to expel the air from the tube on its passage upward. A few movements downward and upward will get it under, and effectually disengage such particles of air as yet adhere to the tube. With a rotary movement, the wiper should now be carefully withdrawn, and during its withdrawal the tube should be kept constantly full of mercury, so that, on the arrival of the wiper at the top, the tube shall be thoroughly occupied with mercury. The gland surrounding the tube should also be perfectly filled. More effectually to secure the exclusion of air from the interior of the gland, the surface is amalgamized so as to unite with the mercury. The cap is now to be restored to its place, taking care that the air enclosed by inverting it into the mercury be entirely expelled by opening the little vent at the end of it, after which it is to be closed again by the small screw in the bottom.

If this has been well done, the gage may be again erected, without loss of mercury, unless considerably agitated. The residue of the mercury furnished is now to be put into the reservoir, which will suffice with what is in the tube, to fill it to the point indicated for a perfect vacuum. If the quantity of mercury should become impaired, and it become necessary to determine the quantity to be supplied, it is to be done in the following manner:—There is in the reservoir an overflow at such a position below thirty inches as shall have been found equal to the contents of the tube. This overflow is stopped with a screw in the front of the reservoir. In charging the gage, therefore, a quantity, judged sufficient, is to be put into the reservoir, the overflow screw removed, and the tube, prepared as before directed, restored to its place; when the excess of mercury, if any existed, has flowed out, the screw is to be replaced, and the whole made tight, as directed for the steam gage.

It is supposed that the foregoing instructions have been amply sufficient to enable any person to erect and advantageously use the Manometer for a vacuum or a steam gage; nevertheless, as derangements have occurred, and are still possible, some general directions are subjoined to guard against or correct them.

OF ITS DERANGEMENTS.

The derangements most likely to occur when used as a steam gage, are, the fouling of the tube, and the alteration or change in the volume of air within it.

The fouling of the tube may proceed from the oxydation of the mercury, or the spirits or substance used for lubrication, or from impurity of the mercury. Whenever the lubricating agent is of a nature to imbibe the oxygen from the air, as in the case of most of the oils, naptha, kreosote, &c., the tube would necessarily become foul. This should be cleaned with the wiper before described. If wire be used for this purpose, unless it is thoroughly coated with cotton to the extent of its insertion into the tube, or if particles of grit adhere to the stick employed, it will greatly endanger the tube ; for, a scratch on the inner surface of a glass tube, although imperceptible to the eye, will most likely make it split within a few days, or perhaps hours. A *fixed* coat of oxyd will best be removed from the tube by wetting the wiper with nitric acid. The volume of air within the tube is sometimes impaired by the absorption of its oxygen without producing a coating upon the glass. This will be perceived by the instrument indicating a greater, pressure than is known to exist, or by the fact, if the cock be shut off and the pressure removed from the gage while the volume of air is so impaired, that the mercury will not fall to the place marked 0, on the scale. In this case, the correction should be effected without changing the air already in the tube, which has been deprived of its oxygen. This is easily accomplished in instruments which have the chamber, or reservoir of mercury, at the lower end of the tube, by lifting the tube-holder out of the reservoir, and carefully inclining it toward a horizontal position—when, in consequence of the rarefied state of the air already there, an additional supply will be admitted from the atmosphere. Care should be taken, however, that too much be not admitted, for then it would be necessary to fill it anew. To guard against a danger incident to the restoration of the tube to its place in the reservoir, viz.—the admission into it of water from the reservoir through the capillary influence of the screw, it is necessary carefully to remove the water from the reservoir with a sponge previous to replacing the tube.

If this absorption occur in instruments designed for steam of very high elasticity (fig. 28), a similar course must be pursued ; but, in consequence of the descent of the mercury into the part of the tube concealed from view by the reservoir and surrounding case, the process will be attended with greater difficulty. Re-filling, in such cases, may be the only remedy. In this case, in consequence of the mercury being in the elevated chamber above, when the tube is taken out of the reservoir, the mercury will overflow, and an iron, glass, or earthen vessel, should be at hand to catch it as it flows out.

Should a tube be accidentally broken, its place must be supplied by one of the same length and taper as that for which the scale was made; and, when setting it, care should be taken to have its sealed end even with the top of the brass case, and to have it carefully packed, both at the top and bottom of the joint, with small rings, or *gaskets* of thread, with putty made stiff with red lead, placing in the middle of the joint a ring or thread or two of India rubber. If the instrument is continually subject to the full heat of the steam, the packing is very liable to crack, and allow the mercury to leak out through or around the bottom of the brass case. If this occur, the first gasket, and the putty immediately under it, should be removed, and fresh putty applied. Glass tubes to fit the scale can always be supplied by the manufacturer at a trifling cost, if an order for them is accompanied with the letter and number of the instrument; but if it be from a gage of the form shown in Fig. 28, it will probably be best to return it to the manufacturer to have it replaced. If the gas escapes from the tube, it will, as a general thing, be best to lubricate the tube, as in the other form. If but two or three drops of the spirits be used, and they are allowed to settle down to the extreme end of the tube, it may be erected and immersed in the mercury before it could get into contact with the metal gland and cap at the bottom.

Water has sometimes filled the tube instead of mercury. This may be expected, if there is not a sufficient supply of mercury in the reservoir. It cannot, with these instruments, occur otherwise, unless the tube be broken, or the gland encasing its lower end leak.

The pipe connecting the gage with the boiler, or the passage between the cock and boiler, has sometimes filled with clay, lime, or sediment of some kind in the water. The best remedy for this is to suddenly remove or put on the pressure, which will soon wash it away, unless compactly filled. In clayey water, formerly, when no provision was made for the sudden removal or application of the pressure, this was a source of much annoyance.

When used for a vacuum gage, it may be deranged by accumulation of air or moisture in the tube. It cannot indicate correctly except by the total exclusion of both. The latter will be detected by its appearance on the surface of the glass, but the former can only be known by exposing the upper end of the tube when subject to atmospheric pressure. Its presence must be the result of neglect in filling, or of leakage around the packing of the glass. In the former case, it will be necessary to invert the tube, when it will rise to the surface, and may be expelled by additional mercury, or screwing down the cap. In the latter, the leak would probably be stopped by white lead, tallow, or oil, applied to the tube where it enters the tube-holder.

LUBRICATION FOR THE TUBE.

For lubricating the tube many articles have been used, with various success, no one of which, however, has uniformly given perfect satisfaction. Some of them under the pressure to which they are subject, absorb oxygen from the air, and change color, or thicken and adhere to the tube, and in such cases, diminish the volume of the air—others are decomposed, or mix with the impurities in the mercury, and coat the glass—while others attack the glass, and corrode its surface.

What is desired is a lubricating agent that will, under the circumstances in which it is used, prevent the oxydation or decomposition of the mercury, that will, at the same time, remain itself unchanged. For the safety of the tube, it should also not congeal in cold water.

Some one of the following will be found in every case to give a good result:

1st. Hydrochloric or muriatic acid. This gives the glass a bright and clear appearance, and remains clear upon the surface of the mercury; it sometimes forms a sort of film, which, after a week or two, settles down out of sight; but it most generally absorbs the air to a serious extent—in some instances, as much as one-fourth of the volume, increasing the apparent pressure from 75 lbs. to 100.

2d. Sulphuric acid, chemically pure. The crude or common acid becomes black, but does not adhere to the tube, and does not very seriously impair the volume of the air. The chemically pure acid answers a very good purpose.

Watch oil, in most cases, gives satisfaction; but it absorbs air slowly for a while, and becomes white—and sometimes under a very high pressure, the thicker portion of it adheres to the tube.

It will be found advisable to be careful, in using any of the acids mentioned, to keep them from touching the iron gland and screw-cap, lest they be corroded, and iron rust become mixed with the spirits in the tube.

PATENT ENGINE REGISTER, OR COUNTER.

A description of this instrument is given in a condensed extract from a Report of the *Committee on Science and the Arts* of the Franklin Institute, made 1849.

Extract from the "Report on Mr. Paul Stillman's Engine Register and Manometer Steam and Vacuum Gage," *by the Committee on Science and the Arts of the Franklin Institute.*

After an explanation of the Manometer Gages, the insertion of which is rendered unnecessary by the preceding pages, the Committee describe the Register as follows:

"This instrument is designed for application to marine steam engines, and is, in outward appearance, similar to the marginal sketch. It consists of a circular cast-iron box, faced with a dial, in which are cut, side by side, six (or more as may be required) slots, through which may be seen the numbers representing the revolutions of the engine; this is denominated the "counter," or "register." By an attachment to any suitable part of the engine, a vibratory motion is communicated to an arm attached to a central horizontal shaft, placed parallel to the dial, and within the cast-iron box—to the ends of which is also fixed a frame carrying a small shaft parallel to the former, on which six palls, or arms, are attached, side by side, and at a certain distance apart, in such a way that the right-hand pall may fall without the others, but cannot rise without carrying all the rest.

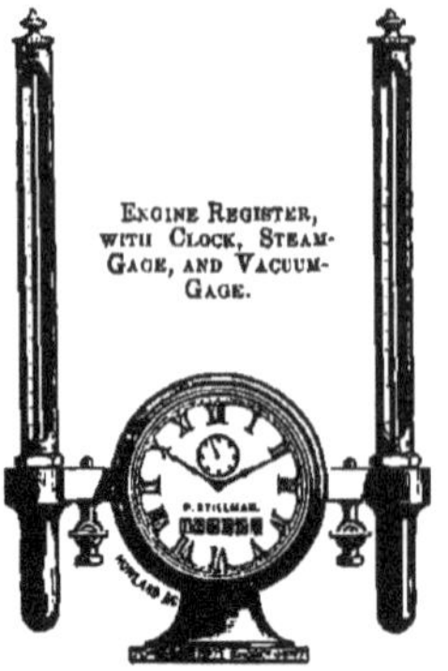

Engine Register, with Clock, Steam-Gage, and Vacuum-Gage.

This frame-work, with the pall-shaft, &c., is made, by the motion of the arm attached to the engine, to describe an arc of 36°, or to move through one-tenth of a circle.

The ends of the palls respectively rest on, and slide over, six cylinders placed side by side on the central shaft, all of which are free to move in the same direction, and independently of each other, and are arranged in the following manner:

For the sake of clearness, we shall number them 1, 2, 3, 4, &c., beginning with the right-hand one.

On the right-hand edge of each cylinder are cut 10 slots, and on the left-hand, which overlaps the edge of the next, only one slot, these slots being of such a size as will admit the end of one of the palls; then, on the back motion of the frame-work, &c., the pall is carried back till it drops in, when the forward motion carries with it the cylinder so locked.

In the central spaces (between the laps) in each cylinder, and opposite to one of the slots in the dial face, the numbers 1, 2, 3, &c., to 0, are engraved at equal distances round the circumference. The palls are placed one over each of the slots, so that the pall can fall into the inner cylinder only when the slot in the outer one comes directly under it; and, as this occurs only once in a whole revolution, and as the motion of the palls is only through one-tenth of a circle, it follows that cylinder No. 2 can only be moved through one-tenth of its circumference, after cylinder No. 1 has moved a whole revolution, or ten times that space, and so on. Thus, if the figures on No. 1 represent units, those on No. 2 will be tens, on No. 3 hundreds, &c.; and, extending the same principle, No. 1 must move round one hundred thousand times to produce one revolution of No. 6. It will be observed, that every revolution of the engine must insure one-tenth of cylinder No. 1 to move round, inasmuch as the ten slots in its right-hand edge are not covered by any other cylinder, as is the case with the rest.

The cylinders being free to move in the direction of their motion, or *forward*, they may be adjusted at any time to their starting point, without deranging any of the palls, or even opening the case.

The Committee judge the following to be the advantages of this arrangement:

1. The compactness and symmetry.
2. The ease with which the result may be read. And,
3. The facility of adjustment.

The two latter being important considerations in an apparatus of this kind.

Other uses are obvious to which this principle may be applied; as, for example, to gas meters, or any other instrument requiring a register, although in the former case the increased power required to turn the cylinders when the higher grades of figures come in play, *might* prove an objection. Of this, the Committee are not capable of judging, for the model submitted to them is intended for a marine engine, and, consequently, is of sufficient strength to withstand the rough usage it might meet with, and, consequently, requires greater effort to move it.

The arrangement is also, probably, of less expense than the old form of counter.

In view of all which points, the Committee are of opinion, that Mr. Stillman is entitled to the First Premium awarded at the Exhibition, where it was placed by him in October last.

Persons giving orders will please state—1st. Whether a pedestal is required; and, if so, 2d. Whether the motion can be most conveniently given through the back, side, or pedestal. 3d. If without pedestal, whether the back of the case goes against the engine-frame, or whether a bracket, or flange, is required. 4th. Where it is attached to the engine-frame, whether the motion cannot be communicated directly through the frame into the case.

Be careful to have the instrument so connected with the engine as to secure the requisite motion without fail.

STILLMAN'S IMPROVED GLASS WATER-GAGE.

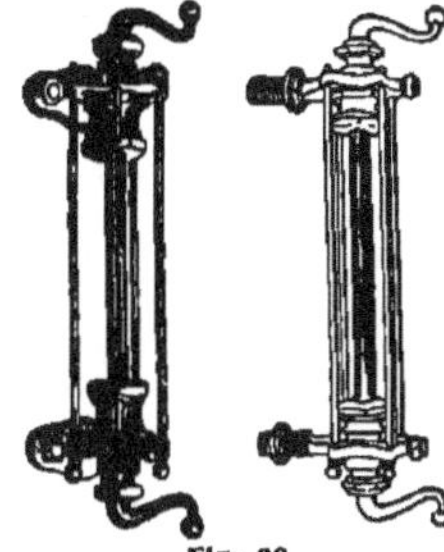

Fig. 32.

No method of determining the height of water in steam-boilers can be more reliable than a well-made glass water-gage. It has been universally used with the English low-pressure engines, with full confidence, from their first introduction to the present day. Their use, however, has been confined to steam of low temperature, in consequence of the trouble of making the glass bear the increased temperature, and of keeping it tight with the higher pressure. The improvements introduced in this arrangement of the instrument have not only obviated those difficulties, but have given it most ample protection from danger by accidental blows, and, at the same time, so simplified its construction as to give it, even in point of economy, claims to a very general adoption. All of the ocean steamers constructed at the Novelty Iron Works, and some others, are now provided with these gages.

DIRECTIONS FOR PUTTING UP.

The lower cock must be attached to the boiler on a line with the top of the flues, or the lowest water line; the upper one, at such distance above as will give the necessary allowance for the introduction of the glass tube, which should be of such length as will allow for the proper fluctuation of the water, say from 10 to 20 inches. They should be placed as nearly on a line with each other as possible. Two ½ or ⅝ inch tap-bolts in each will be sufficient for each part. When securely attached, let the gland of the packing-boxes be slacked up, and if there be no water at the upper cock, take out the key and pass the tube through the barrel into its place, taking care that the glass at either end does not rest against the cock. If the packing be of good material—such as is sold with the gages—and a metal ring be between it and the gland, and of proper size for the tube, a gentle screwing of the gland by the hand will be sufficient to make them quite tight against any required pressure. The guards should then be put in their places, to protect the tube from accident, and the levers of the cocks turned in the direction shown in the cut. If subsequently there should be occasion to replace the tube, and there be steam or water in the boiler, to prevent the key of the cock from being taken out, one of the stuffing-boxes must be taken off to put in the tube.

When a glass water-gage is in use and in good order, a little fluctuation of the water will be constantly observed. If it is not seen, the lever of the upper cock should be turned to the opposite side, to see, by blowing it through, that the passage way in the cock is not choked.

If a tube gets broken when the boiler is in use, both of the cocks should be closed by turning the levers opposite to the direction shown in the cut.

☞ The indications of the glass water-gage are those of solid water, however much the water in the boiler foams.

☞ The only indication as to which is to be used for the lower cock, is by the direction given to the pipe for blowing through.

STILLMAN'S MARINE HYDROMETER, SALINOMETER, OR SALT GAGE.

Fig. 33.

Ocean steamers, using sea water in the boilers, require frequent change of the water to prevent incrustation, or deposit of salt and earthy matter upon the flues and within the legs of boilers. This exchange should be made with regularity and care, lest, on one hand, the object sought should not be attained, or, on the other, a waste of heat should be occasioned in discharging hot water too freely from the boiler, the place of which is to be supplied with cold. It is, however, better, as a general thing, to err on the side of a too liberal use of the blow-off cock, for the loss of heat would, probably, be less from this cause than it would be if the boiler were allowed to become incrusted with a non-conducting substance. At what exact degree of saturation incrustation begins, may not be stated. It appears to vary considerably, in different circumstances, and, at present, the wisest course appears to be to feel the way to a place of safety. It is necessary, therefore, to have some test by which the saltness of the water may be known; and having this, to adopt such a system of blowing off as will keep the water uniformly at the degree fixed upon. The degree of saltness of the water, is ascertained by the Hydrometer. It is made of either glass or metal, and marked to show the amount of salt held in solution, or other quality affecting specific gravity.

This Hydrometer is made of metal, and of unusual size and strength, to withstand the hard usage to which such instruments are subject on ship-board. The gradations of it are designed to show the number of ounces of salt contained in one United States standard gallon (8.338 avoirdupois pounds) of water, which is believed to be the best yet devised for this purpose, for the following reasons: 1st. It is perfectly intelligible to every one likely to require them. 2nd. It is just as applicable to the water of one locality as another. 3rd. It is more easily noted, or *logged*, than most others. 4th. It is in just such proportion that a change of temperature of 20° makes a variation equal to one division or ounce of salt; so that a trial is not strictly confined to a particular temperature, for the reason that the necessary correction for temperature is so simple, that it is scarcely liable to be a source of mistake. It is graded from fresh water to saturation at a temperature of 70°: and on the reverse side is a scale containing the divisions most used, adapted to a temperature of 190°, to which small quantites of water, drawn from the boiler, will fall in about five minutes. For general use, this temperature will be found very serviceable; but, for exact measurement, 70° is much preferable. A standard glass Hydrometer should accompany them, to correct them in case of being bruised or broken.

The water of the North Atlantic contains in 1,000 parts 42.6 parts, or $\frac{1}{23}$ of salt, equal by this Hydrometer to $5\frac{1}{3}$ oz., and by the Hydrometer of Gray & Keen, Liverpool, (imperial standard, and generally used in English steamers,) 7 oz. In some of the steamers using this Hydrometer 6 oz. has been adopted as to the extent to which the saltness of the water is allowed to go; others go as high as 10 or 12 oz.

Sea water is often spoken of as containing $\frac{1}{33}$ by weight of salt and earthy matter. But it varies considerably in different places, as will be seen in the following

TABLE

Of the amount of saline contents of sea water from different localities.

Baltic Sea	6.60 = $\frac{1}{152}$	Mediterranean	39.40 = $\frac{1}{25}$
Black Sea	21.60 = $\frac{1}{46}$	Atlantic at Equator	39.42 = $\frac{1}{25}$
Arctic Sea	28.30 = $\frac{1}{35}$	South Atlantic	41.20 = $\frac{1}{24}$
Irish Sea	33.76 = $\frac{1}{30}$	North Atlantic	42.60 = $\frac{1}{23}$
British Channel	35.50 = $\frac{1}{28}$	Dead Sea	385.00 = $\frac{11}{28}$

Hoping that the foregoing description of the Steam Engine Indicator, with the diagrams of the working of steam engines, and the Manometer showing pressure on boiler, will be a valuable assistance to builders and owners of steam engines, to see the defects, and to calculate the actual power of their engines, we must now leave this subject, and proceed with our calculations, explaining, first, the difference of actual and nominal horse-power:

ACTUAL AND NOMINAL HORSE-POWER.

Having, heretofore, stated that the actual horse-power is a force capable of raising 33,000 lb. one foot high in the minute, or a dynamical effort to evaporate a cubic foot of water per hour, what will be the value of a nominal horse-power? *Ans.*—A nominal horse-power is considered by some to raise 52,000 lbs., by others 60,000 lbs., and by others 66,000 lbs., one foot high in the minute, by each nominal horse-power; and, therefore, no comparison can be made between the performances of different engines, unless the power actually exerted be first discovered, which can be done by the Indicator.

How do you find the nominal horse-power of a high pressure steam engine? According to "Bourne on Steam Engines," the nominal horse-power of a high pressure engine has never been defined; but it should obviously hold the same relation to the actual power as that which obtains in the case of condensing engines, so that an engine of a given nominal power may be capable of performing the same work, whether high pressure or condensing. This relation is maintained in the following rule, which expresses the nominal horse-power of high pressure engines: Multiply the square of the diameter of the cylinder in inches, by the pressure on the piston in pounds per square inch, and by the speed of the piston in feet per minute, and divide the product by 120,000; the quotient is the power of the engine in nominal horse-power.

Colburn, in his "Practical Series on Locomotive Engines Relating to Stationary Engines," says: "The standard of a horse-power serves as a standard of comparison, and its utility as a unit of reference is not impaired, whether it represents the actual power of one horse or three, so long as the standard is universal." But as the work of a locomotive is all of one character, it becomes an object to know the actual power of an engine in drawing freight or passengers, in preference to referring it to any doubtful standard. For the assistance of such as may have occasion to estimate the horse-power of a stationary, or even a locomotive engine, we give the usual rule. It is as follows: Multiply the area of the piston, the pressure of steam per square inch, the number of revolutions per minute, and the length of stroke, together, divide the product by 33,000, and take $\frac{7}{10}$ of the quotient. It is a received law in mechanical science, that the effect of a machine is to be estimated from its weight, or elemental power, multiplied into the space through which the power acts. Our readers will detect that in the above rule we have directions to employ but one-half the speed of the piston to get the power of the engine. For instance, a 16-inch cylinder engine is usually rated as a 50-horse engine, but if, in calculating its power, we employ the actual speed of the piston in feet per minute, we shall find our engine to have 100 horse-power.

If a horse raises 150 lbs. through 220 feet in a minute, or, through the application of wheels and axles, levers, &c., it raises 33,000 lbs. one foot high in a minute, then, what is usually termed a one-horse engine will raise 66,000 lbs., through the same distance, and in the same time.

Example: What would be the nominal horse-power of the steam engines tested at the Crystal Palace. Diameter of cylinder 12 inches, stroke of piston 3 feet, pressure of steam 70 lbs. per square inch above the atmosphere, and the engine-shaft makes 60 revolutions per minute:

Bourne's Rule:

```
              12 diameter of cylinder.
              12
              ---
             144
              70 pressure of steam in pounds upon a square inch.
           -----
           10080
             360 feet in a minute.
         -------
          604800
          30240
         -------
120000 ) 3628800 ( 30 nominal horse power.
          360000
         -------
           24800
```

Colburn's Rule:

```
Area of 12 inches diameter = 113.04 square inches.
                                70    pressure of steam per square inch.
                              ----
                              7910
                               180    half the speed in a minute.
                            ------
                            632800
                            7910
                          --------
                33000 ) 1423800 (  43
                         132000    0.7
                        -------   ----
                         103800   30.1 nominal horse-power.
                          99000
                        -------
                           4800
```

Both rules show also very little difference.

To ascertain the nominal power of a high pressure steam engine of high speed, working about four times the usual speed, the following rule is adapted, according to "Bourne on Steam Engines": Multiply the square of the diameter of the cylinder by the pressure on the piston per square inch, less a pound and a half, and by the cube root of the stroke in feet, and divide the product by 235; the quotient is the power of the high speed engine in nominal horse-power.

Nominal horse-power of a condensing engine, working at about four times the ordinary speed, may be ascertained by the following rule: Multiply the square of the diameter of the cylinder in inches by the cube root of the stroke in feet, and divide the product by 12; the quotient is the power of the high speed condensing engine in nominal horse-power.

If we talk of an engine of 200 horse-power, it is not meant that the impelling efficacy is equal to 200 horses, each lifting 33,000 lbs. one foot high in a minute. Such was the case in Watt's engines; but the capacity of cylinder answerable to a horse-power has been increased by most engineers since his time, and the pressure on the piston has been increased also, so that what is now called a 200 horse-power engine, exerts, almost in every case, a greater power than was exerted in Watt's time, and a horse-power has become a mere conventional unit for expressing a certain size of cylinder, without reference to the power exerted.

The following table, we obtained from Mr. Wm. Burden, a well-known steam-engine builder, of 102 Front Street, Brooklyn, giving the nominal horse-power of his horizontal high-pressure steam engines, stating diameter of cylinder, and stroke of piston, at which amounts of horse-power his engines are sold, although the actual power may be about twice the same as of a nominal horse-power:

3 Horse Engine.		$4\frac{1}{2}$ inch Bore.		10 inch. Stroke.	
4	do.	5	do.	12	do.
5	do.	6	do.	12	do.
6	do.	5	do.	15	do.
7	do.	6	do.	15	do.
8	do.	7	do.	15	do.
9	do.	8	do.	15	do.
8	do.	6	do.	20	do.
9	do.	7	do.	20	do.
10	do.	8	do.	20	do.
12	do.	9	do.	20	do.
10	do.	8	do.	24	do.
12	do.	9	do.	24	do.
14	do.	10	do.	24	do.
16	do.	11	do.	24	do.
20	do.	11	do.	30	do.
25	do.	12	do.	30	do.
30	do.	13	do.	30	do.
35	do.	14	do.	30	do.
30	do.	12	do:	36	do.
35	do.	13	do.	36	do.
40	do.	14	do.	36	do.
45	do.	15	do.	36	do.
50	do.	15	do.	42	do.
60	do.	16	do.	42	do.
70	do.	16	do.	48	do.
100	do.	20	do.	48	do.
125	do.	20	do.	60	do.

We illustrate below another table from Messrs. H. & F. Blandy, an extensive portable steam-engine builder for saw-mills, of Zanesville, Ohio; engine attached to the boiler with one cylinder:

3	Horse-power Engine.	4	inch. Diam.	8	inch. Stroke.	Price, $300
4	do.	4½	do.	9	do.	400
6	do.	5	do.	10	do.	600
8	do.	5½	do.	12	do.	800
12	do.	6	do.	12	do.	1000
15	do.	7	do.	14	do.	1200
18	do.	8	do.	14	do.	1350
20	do.	8½	do.	16	do.	1400
22	do.	9	do.	14	do.	1600
25	do.	9½	do.	16	do.	1750
30	do.	10	do.	16	do.	2000
35	do.	11	do.	18	do.	2200

The boilers for each size cylinder, containing 15 square feet heating surface, or over that, to each horse-power.

DESCRIPTION OF BLANDY'S 15 HORSE-POWER PORTABLE ENGINE AND BOILER.

Boiler, 11 feet 3 inches long, 40 inches diameter; fire-box, 34 inches diameter, 54 inches long, with open grate surface; 56 two-inch, lap welded, safe-end flues, 6 feet 11 inches long; fire-box, entirely surrounded by water, 225 square feet of heating surface, breeching and smoke stack; heads of boiler ⅜ inch thick; fire-box, $\frac{5}{16}$ inch thick; cylinder, 7 inches diameter, and 14 inches stroke; pistons, spring metallic packing; iron band wheel, 5 feet diameter (solid), 12 inches, face turned; main shaft, wrought iron, 3½ inches diameter; weight of the 15 horse-power engine, from 5500 to 6000 lbs.

SIZE OF CIRCULAR SAW MILL.

52 inches diameter of saw, 24 feet carriage, and 48 feet ways; complete, with head blocks, price, $450.

LIST OF PORTABLE STEAM ENGINES WITH BOILER

OF THE NEWARK MACHINE WORKS (OHIO).

Horse-Power.	Diameter of Cylinder.	Length of Stroke.	Diameter of Fly-Wheel.	Face of Fly-Wheel.	Fire Surface of Boiler.	Weight.	Price.
	Inches.	*Inches.*	*Inches.*	*Inches.*	*Square feet.*	*Pounds.*	
3	4	8	24	4	45	1600	$300
5	5	12	36	5	75	3000	550
6	6	12	42	7	90	3500	650
8	7	14	54	8	120	4000	825
10	8	16	60	10	150	5200	1000
12	9	16	60	10	180	6000	1200
15	9½	18	72	12	225	6800	1400
20	10	20	72	12	300	9200	1600

CIRCULAR SAW MILLS

WITH 24 FEET OF CARRIAGE, 48 FEET OF WAYS, AND 2 HEAD BLOCKS, COMPLETE FOR USE. (BELTING EXCEPTED.)

Single Mills.		Double Mills.
No. 1, with 36 inch Saw, $260	No. 5, with 52 inch Saw, $390	With 42 and 24 inch Saw, $375
" 2, " 42 " " 285	" 6, " 54 " " 410	" 48 " 26 " " 425
" 3, " 48 " " 340	" 7, " 56 " " 440	" 50 " 28 " " 440
" 4, " 50 " " 350	" 8, " 60 " " 485	" 52 " 30 " " 480
		" 54 " 30 " " 500

NOMINAL HORSE-POWER FOR CONDENSING ENGINES.

The rule most commonly used in the United States for denoting the nominal horse-power of all condensing engines, is that first laid down by Watt, as follows :

Multiply the square of the diameter of the cylinder, in inches, by the cube root of the stroke, in feet, and divide the product by 47 ; the quotient is the number of nominal horse-power of the engine. This rule supposes a uniform effective pressure upon the piston, of 7 lbs. per square inch. Mr. Bourne, in his treatise on steam engines, in commenting on this point, says : " Mr. Watt estimated the effective pressure upon the piston of his 4-horse-power engines at 6.8 lbs. per square inch, and the pressure increased slightly with the power, and became 6.94 lbs. per square inch in engines of 100 horse-power ; but it appears to be more convenient to take a uniform pressure of 7 lbs. for all powers. Small engines, indeed, are somewhat less effective in proportion than large ones, but the difference can be made up by slightly increasing the pressure in the boiler ; and small boilers will bear such an increase without inconvenience."

The universal employment of the non-condensing system, with a very greatly increased boiler pressure for all small engines, has rendered these remarks very inapplicable to such at the present day ; but, for condensing engines of all sizes, it may be considered an excellent standard rule. In the form of a formula it appears as follows :

$$\frac{d^2 \times \sqrt[3]{S}}{47} = \text{nominal H. P.}$$

d means diameter of cylinder, in inches, and S stroke of piston, in feet.

The question of nominal horse-power is one of interest mainly to the builder and purchaser of engines, as a conventional mode of designating and comparing sizes. It is of little consequence in practical science ; and the use of the term will, probably, be soon discarded in American engineering. In Great Britain, every steam vessel is rated at a certain nominal horse-power ; in the United States, on the contrary, the dimensions of the cylinder are almost universally expressed in feet and inches, the nominal power being considered of little moment, and the actual power being well understood to depend on the effective pressure in the cylinder and the speed of the piston, both of which elements may be varied at will within certain limits.

EXPANSION OF STEAM.

RULE FOR CALCULATING THE ACTUAL HORSE-POWER OF HIGH PRESSURE STATIONARY STEAM ENGINES USING STEAM EXPANSIVELY.

The meaning of working steam expansively is, in adjusting the valves so that the steam is shut off from the cylinder before the end of the stroke, whereby the residue of the stroke is left to be completed by the expanding steam.

The benefits of using steam by expansion are: It accomplishes an important saving of steam, or, what is the same thing, of fuel; but it diminishes the power of the engine, while increasing the power of the steam. A larger engine will be required to do the same work, but the work will be done with a smaller consumption of fuel. If, for example, the steam be shut off when only half the stroke is completed, there will be more than half the power exerted; for, although the pressure of the steam decreases after the supply entering from the boiler is shut off, yet it imparts, during its expansion, *some* power, and that power, it is clear, is obtained without any expenditure of steam or fuel whatever.

The rule for ascertaining the amount of benefit derivable from expansion is: Divide the length of the stroke by the length of the space into which the steam is admitted; find in the table on the next page, the hyperbolic logarithm of the number nearest to that of the quotient, to which add one. The sum is the ratio of gain.

Example.—Suppose the steam to enter the cylinder at the pressure of 50 lbs. per square inch, and to be cut off at one-fourth of the length of the stroke; what is the mean pressure, the stroke being ten feet?

$$10 \div 2.5 = 4 \text{ Hyp. log. of } 4 = 1.38629 + 1 = 2.38629.$$

$$\text{Then, as } 4 : 2.38629 :: 50 = \frac{2.38629 \times 50}{4} = 29.82862 \text{ lbs.}$$

If a given quantity of steam, the power of which, working at full pressure, is represented by 1, be so admitted into a cylinder of a certain size, that its ingress is concluded when one-half the stroke has been performed, its efficacy will be raised by expansion to 1.69; if the admission of the steam be stopped at one-third of the stroke, the efficacy will be 2.10; at one-fourth, 2.39; at one-fifth, 2.61; at one-sixth, 2.79; at one-seventh 2.95; at one eighth, 3.08. The expansion, however, cannot be carried, beneficially, so far as one eighth on all engines, unless the pressure of the steam in the boiler be very considerable, on account of the inconvenient size of cylinder, or speed of piston, which would require to be adopted. Also, the friction of the engine, and resistance of vapor would become relatively greater with a smaller urging force.

TABLE

TO FIND THE MEAN PRESSURE BY HYPERBOLIC LOGARITHMS.

No.	Logarithm.	No.	Logarithm.	No.	Logarithm.	No.	Logarithm.	No.	Logarithm.
1.25	.22314	3.5	1.25276	5.75	1.74919	8.	2.07944	15.	2.70805
1.5	.40546	3.75	1.32175	6.	1.79175	8.5	2.14006	16.	2.77258
1.75	.55961	4.	1.38629	6.25	1.83258	9.	2.19722	17.	2.83321
2.	.69314	4.25	1.44691	6.5	1.87180	9.5	2.25129	18.	2.89037
2.25	.81093	4.5	1.50507	6.75	1.90954	10.	2.30258	19.	2.94443
2.5	.91629	4.75	1.55814	7.	1.94591	11.	2.39789	20.	2.99573
2.75	1.01160	5.	1.60943	7.25	1.98100	12.	2.48490	21.	3.04452
3.	1.09861	5.25	1.65822	7.5	2.01490	13.	2.56494	22.	3.09104
3.25	1.17865	5.5	1.70474	7.75	2.04769	14.	2.63905		

The engines tested at the Crystal Palace, of which we have given calculations, on page 92, the steam following the whole stroke of the piston, we obtained 86.36 horse-power, without deducting friction of the working parts.

Now, what would be the amount of horse-power if the steam is cut off at half stroke?

Diameter of cylinder, 12 inches; stroke of piston, 3 feet; the pressure of the steam 70 lbs. per square inch, above the atmosphere, and the engines made 60 revolutions per minute.

The average, or mean pressure, of steam upon a square inch of the piston will be by expansion:

$$3 \div 1.5 = 2 . \text{Hyp. log. of } 2 = 0.69314 + 1 = 1.69314.$$

$$\text{Then, as } \frac{1.69314 \times 70}{2} = 59.2599 = 59\tfrac{1}{4} \text{ lbs., mean pressure.}$$

Multiply the area of the piston, in inches, by 59 lbs., and this product by speed of piston, in feet, per minute, and divide by 33000; the result will be:

$$\frac{6^2 \times 3.1416 \times 59 \times 3 \times 2 \times 60}{33000} = 72.7 \text{ horse-power.}$$

Deduct for friction of engine about . . . . . 6.3 "

Leaving . . . . 66.4 for driving machinery and shafting.

Cutting off steam at one-fourth of the stroke:

$$3 \div \tfrac{3}{4} = 4. \text{ Hyp. log. of } 4 = 1.38629 + 1 = 2.38629.$$

$$\text{Then, } \frac{2.38629 \times 70}{4} = 41.76, \text{ average pressure upon a square inch upon the piston.}$$

$$\frac{6^2 \times 3.1416 \times 41.76 \times 3 \times 2 \times 60}{33000} = 51.5 \text{ horse-power.}$$

Deduct for friction of engine, calculated on page 93, . . 6.3 horse-power.

45.2 horse-power.

```
                 6
                 6
                36
            3.1416
               216
               36
             144
             36
           108
           113.0976
              41.76
           6785856
          7916832
         1130976
        4523904
        4722.955776
                  3
       14168.867328
                  2
       28337.734656
                 60
33000 ) 1700264.079360 ( 51.5 horse-power.
        165000
         50264
         33000
         172640
         165000
          76407
```

It is herewith shown, that the same engine will produce 79.7 actual horse-power, if the steam in the cylinder follows the whole stroke of piston; 66.4 horse-power, if the admission of steam into the cylinder is shut off at half stroke; 45.2 horse-power, if the admission of steam is shut off at one-fourth of the stroke, using also only one-fourth the quantity of steam; while it requires four times the same quantity to produce the 79.7 horse-power.

The mean or average pressure of steam on a piston can also be found in the following manner:

Rule. — Divide the length of the stroke, added to the clearance in the cylinder at one end, by the length of the stroke at which the steam is cut off, added to the clearance (if great accuracy is required, the space between the cylinder and the steam valve must be added to the clearance), and the quotient will express the relative expansion it undergoes.

Find, in the following table, in the column of expansion, a number corresponding to this; take out the multiplier opposite to it, and multiply it into the full pressure of the steam per square inch as it enters the cylinder.

TABLE SHOWING THE MEAN PRESSURE OF STEAM.

Expansion.	Multiplier.	Expansion.	Multiplier.	Expansion.	Multiplier.	Expansion.	Multiplier.	Expansion.	Multiplier.	Expansion.	Multiplier.
1.0	1.000	2.2	.813	3.4	.654	4.6	.549	5.8	.479	7.	.430
1.1	.995	2.3	.797	3.5	.644	4.7	.542	5.9	.474	7.1	.427
1.2	.985	2.4	.781	3.6	.634	4.8	.535	6.	.470	7.2	.423
1.3	.971	2.5	.766	3.7	.624	4.9	.528	6.1	.466	7.3	.420
1.4	.955	2.6	.752	3.8	.615	5.	.522	6.2	.462	7.4	.417
1.5	.937	2.7	.738	3.9	.605	5.1	.516	6.3	.458	7.5	.414
1.6	.919	2.8	.725	4.	.597	5.2	.510	6.4	.454	7.6	.411
1.7	.900	2.9	.712	4.1	.588	5.3	.504	6.5	.450	7.7	.408
1.8	.882	3.	.700	4.2	.580	5.4	.499	6.6	.446	7.8	.405
1.9	.864	3.1	.688	4.3	.572	5.5	.494	6.7	.442	7.9	.402
2.	.847	3.2	.676	4.4	.564	5.6	.489	6.8	.438	8.	.399
2.1	.830	3.3	.665	4.5	.556	5.7	.484	6.9	.434		

Example.—Suppose the steam to enter the cylinder at a pressure of 70 lbs. per square inch, and to be cut off at one-half the length of the stroke of the piston, the stroke being three feet.

3 feet = 36 inches + 0.5 for clearance = 36.5.
Stroke $\frac{1}{2}$ = 18 inches + 0.5 " = 18.5.

Then 36.5 ÷ 18.5 = 1.97, the relative expansion which falls between 1.9 and 2. Referring to the table, the multiplier for 1.9 is 0.864; and the difference between that and the multiplier for 2 is 0.017. Hence, multiplying 0.017 by .07, and subtracting the product 0.011, the remainder, 0.853, is the multiplier for 1.97. Therefore, 0.853 × 70 = 59.710 lbs. per square inch, the mean effective pressure on the piston.

FORCE OR FEED PUMPS.

To find the quantity of water to feed the steam-boiler by the feed pumps at each revolution in a steam-engine, the capacity of the cylinder, pressure of steam, and the number of revolutions being given, $\frac{C \times 2r}{V}$, when C represents cubic feet of steam expanded in the cylinder, r the number of revolutions, and V the volume of the steam at the pressure in the cylinder.

Example.—What quantity of water will an engine require per revolution, the diameter of cylinder being 12 inches, the stroke of the piston 3 feet, and the pressure 70 lbs. per square inch, steam cut off at one-half of the stroke of piston. Area of cylinder $6^2 \times 3.14 = 113.04$ inches; one-half of stroke $\frac{36}{2} = 18$ inches.

Then, $113.04 \times 18 \div 1728 = 1.17$ cubic feet.

Volume of steam at the above pressure, per table, pages 77, 78 = 408.

Hence, $\frac{1.17 \times 2}{408} = 0.0057$ cubic feet.

Or, $0.0057 \times 1728 = 9.8496$ cubic inches water, per revolution.

This refers to the expenditure of steam alone; in practice, however, a large quantity of water (differing in different cases) is carried into the cylinder in mechanical combination with the steam. Allowance must, therefore, be made for any losses by waste, clearance of piston, capacity of steam-port, and condensation of steam in passing from steam-boiler to the engine.

A feed-pump, with about three times the capacity of 9.8496 × 3 = 29.5488 cubic inches, will answer, when a single pump is used.

It is preferable to apply two feed-pumps, one of the pumps to contain half the capacity of $\frac{29.5}{2} = 14.75 = 14\frac{75}{100}$ cubic inches to supply the water to the boiler and keep the other pump in reserve.

The constant supply of water of such a capacity as is required to supply the steam-boiler, to evaporate the same amount of steam as is used by the steam-engine, calculating for waste, is a decided advantage, and should be recommended by builders to owners of engines; the additional cost of another pump should not be considered. Its cost will soon be balanced by the gain of fuel. Steam-pumps are decidedly the best for such purposes, as they are worked direct by the steam of the boiler, and can be so regulated, either to give a constant supply of feed water to the boiler, or the speed of pumps can be increased to supply a larger amount of water, as the engineer may find necessary.

CONDENSING STEAM ENGINES.

To calculate the actual horse-power of a condensing engine, it differs somewhat from that of a high pressure engine, on account of the air-pump. The following rule is to be observed:

I. Find the mean effective pressure, in pounds, upon the cylinder piston.

II. Find the pressure upon the piston, in pounds, necessary to overcome the friction of the engine, and pressure upon cylinder piston to overcome the friction of the air-pump and its gearing, and deduct these two amounts from the whole mean effective pressure, in pounds, upon cylinder piston; then multiply the velocity of cylinder piston, in feet, per minute, with the remainder.

III. Find the atmospheric pressure, in pounds, upon the air-pump bucket, and multiply the same with the velocity of air-pump bucket, in feet, per minute; the amount deduct from the product of (II.), and divide the remaining sum by 33000: the quotient will be the amount of actual horse-power of the engine.

Let S represent velocity of cylinder piston, in feet, per minute; n, velocity of air-pump bucket, in feet, per minute; *P, mean effective pressure upon cylinder piston, in pounds; m, pressure upon cylinder piston necessary to overcome the friction of the air-pump and its gearing; *b, the pounds pressure upon the air-pump bucket; f, the pounds pressure upon the piston necessary to overcome the friction of the engine.

The value of m is about 2 lbs. per square inch, that of b, 9.5 lbs., and f, at a fair estimate, is one-fifth of the pressure, per steam gauge.

$$\text{Then, } \frac{P - f + m \times S - nb}{33000} = \text{horse-power.}$$

Example.—The diameter of the cylinder of the low-pressure engine of the steam-boat *Francis Skiddy*, illustrated on Plate XVII, is 70 inches; the stroke of the piston, 14 feet; the revolution of main-shaft, 16 per minute; the diameter of the air-pump, 46 inches; and 4 ft. 4 in. stroke; the pressure of the steam, 33 lbs. upon a square inch. Steam cut off at 9 ft. from commencement of the stroke of the piston.

Then, $S = 14 \times 2 \times 16 = 448$ feet.

$n = 4$ feet 4 inches $= 4.33 \times 2 \times 16 = 138.56$.

P = the mean effective pressure of steam in cylinder, will be,

$14 \div 9 = 1.5$ Hyp. log of 1.5 is (see table on page 131) $= 0.40546 + 1 = 1.40546$.

$$\text{Then, as } \frac{1.40546 \times 33}{1.5} = 30.92 \text{ lbs. mean pressure.}$$

0.937
33.
2811
2811
30.921 lbs.

Or, find, in the table on page 132, the multiplier for 1.5, which is $= 0.937$, and multiply the same with 33 lbs., pressure of steam; the product will be $0.937 \times 33 = 30.92$ lbs. mean pressure.

Then multiply the mean pressure of steam, which is 30.92 lbs., with area of piston, in square inches.

* These values are best obtained by an Indicator; when none is used, refer to rule and table on pages 131, 132.

30.92 × 3848.4 = 118992.5 lbs., mean effective pressure upon cylinder piston.

m = 3848 × 2 = 7696.8 lbs.

b = area of 46 is, 1661.9 × 9.5 = 15788. lbs.

f = 30.92 lbs. = 0.2 × 3848.4 = 23798.5 lbs.

$$\frac{118992.5 - 23798.5 + 7696.8 \times 448 - 138.56 \times 15788}{33000} = 1121.5 = 1121\tfrac{1}{2}, \text{ horse-power.}$$

```
 23798.5          118992.5                       15788
  7696.8           31495.3                       138.56
 -------          --------                      -------
 31495.3           87497.2                        94728
                       448                       78940
                  --------                      126304
                   6999776                      47364
                  3499888                      15788
                 3499888                        ----------
                 ----------                     2187585.28
                 39198745.6
                  2187585.2
             -----------------
             33000 ) 37011160.4 ( 1121.5, actual horse power.
                     33000
                     -----
                      40111
                      33000
                      -----
                       71116
                       66000
                       -----
                        51160
                        33000
                        -----
                        181604
                        165000
                        ------
                         16604
```

The foregoing calculation can be simplified in the following manner: Find mean pressure upon cylinder piston, in pounds, and multiply the same with velocity of piston, in feet, per minute, and with 0.695, and divide the product, as usual, by 33000; the quotient will be the actual horse-power.

$$\frac{118992.5 \times 448 \times 0.695}{33000} = 1122.4, \text{ actual horse-power.}$$

The multiplier, 0.695, is for resistance of engine for all working parts, and can be used for all condensing engines from 100 up to 2000 horse-power.

According to Tredgold, the resistance for a Watt's low-pressure engine is calculated in the following manner:

The pressure of steam in the boiler being	1.000
there will be lost (not including the counter pressure of the condensed steam),	
1. The force for the velocity of steam in cylinder	0.007
2. Loss through the condensation of steam in cylinder and in the pipes .	0.016
3. Friction of piston and loss of steam through packing	0.125
4. Force for the steam to pass through the openings and pipes . .	0.007
5. Force to move the different valves and friction of shafting . .	0.063
6. Loss through the early cutting off of steam	0.100
7. Power to move the air-pumps	0.050
Remains	0.632
Counter pressure of condensed steam	.063
Multiplier in full for resistance	0.695

The calculations heretofore given of the engines of the *Francis Skiddy* are, as the boat is running now at a slow speed, starting from New York at six o'clock in the evening, and arriving next morning at five o'clock, making the voyage to Albany in 11 hours, a distance of 146 miles without stopping. When she made her trips as a day-boat, she started in the morning, at seven o'clock, and arrived at Albany at twenty-four minutes past two o'clock in the afternoon, making seven landings—losing one hour time to do it; made, also, the voyage in six hours and twenty-four minutes, with a pressure of steam of 45 lbs. upon a square inch, and making $21\frac{1}{2}$ revolutions with main-shaft, cutting off steam at 7 ft. 6 in. of the stroke of piston. Mean pressure of steam is, 39.1 lbs.

$$\text{Then, } \frac{39.1 \times 3848.4 \times 602 \times 0.695}{33000} = 1907, \text{ actual horse-power.}$$

At the speed producing the 1907 actual horse-power by her engine, the hull of the *Francis Skiddy* was 38 feet wide, and draft of water, unloaded, 7 feet; when loaded, 7 feet 6 inches, running with an average speed of $22\frac{8}{10}$ miles per hour. But, as she is running now, the hull of the boat is enlarged 14 feet, making the hull 52 feet wide, draft of water, unloaded, 5 feet 4 inches, when loaded, 6 feet 4 inches, running with an average speed of $13\frac{27}{100}$ miles per hour with $1121\frac{1}{2}$ actual horse-power by her engine.

SIZE OF BOILERS.

The *Francis Skiddy* has four cylindrical flue boilers, and four chimneys.

Extreme length of steam-boilers	28 feet 1 inch.
Diameter of shell	8 feet.
Two separate fire-grates for each boiler.	
Length of fire-grate	6 feet 6 inches.
Width of each fire-grate	3 feet 11 inches.
Height of fire-box from top of grate-bar	2 feet 7 inches.
Length of combustion chamber	20 inches.
Lower flues for the whole boiler	6 pieces at 11 in. diameter.
" " "	2 pieces at 15 in. "
" " "	2 pieces at 10 in. "
Length of lower flues	17 feet 2 inches.
Upper flues for the whole boiler	5 pieces at 16 in. diameter.
Length of upper flues	21 feet 6 inches.
Width of smoke chamber, foremost end	24 inches.
Width of smoke chamber, below chimney	42 inches.
Diameter of steam chamber	66 inches.
Height of steam chamber	10 feet.
Diameter of chimney	42 inches.
Water space between outside shell and shell of fire-box .	4 inches.

FIRE AND GRATE SURFACE.

Whole amount of fire surface for the four boilers .	5132 square feet.
Whole amount of grate surface, do. . . .	208 square feet.
Area of lower flues for one boiler . . .	1079 square inches.
Area of upper flues for one boiler . . .	1004 square inches.
Area of chimney for one boiler	1384 square inches.

SCREW PROPULSION.

THE principle on which the propulsion of vessels depend is very simple; but the mathematical investigation of any system, involves such a number of important elements, that it becomes one of the most severe and complex problems ever presented to the mechanical engineer. The fundamental principle in acting upon the water by mechanism, whether artificial or natural, is, to urge the water backward, or, in other words, to endeavor to impel a quantity of the water in one direction, and thereby to obtain from the inertia of that water a force, sometimes termed reaction, which impels the vessel in the opposite direction. This principle is alike availed of in the screw-propeller, the paddle-wheel, and the oar; and although, perhaps, less plainly apparent, it lies at the foundation of the movements made in swimming, whether by fishes, animals, or birds. In each instance, the water is acted on by a portion of the mechanism which tends to move it in a direction the reverse of that in which the object itself is moving. The water may, or may not, at the same time receive a sidewise motion. If it does, that part of its motion is of no effect in propelling, but it consumes the power of the mechanism uselessly.

Three points more will suffice for the consideration of the general subject of propulsion. The first is, the effect of the *quantity* of water acted upon. We have seen, that the power, or energy, expended on the propelling apparatus produces two effects, one of which is, the urging of a quantity of water backward, and the other, the urging of the vessel, or equivalent object, forward. Power is expended in producing both these effects; but it does not necessarily follow, that the amounts are, in all cases, or in any case, equal. If the propelling blade, whether of the screw, the paddle, or the oar, moves backward through the water with the same velocity as the vessel moves forward, then the power is equally distributed between the two effects. The useless effect of driving the water backward, and producing a slight current in the sea, to be gradually extinguished by the friction of the adjacent particles long after the vessel has passed, has absorbed precisely the same amount of steam power, or other dynamic agency, as the driving of the vessel usefully on her way has done in the same period. If, in a steam-vessel working under such conditions, the engine has actually worked with a power of 400 horses, only 200 has been finally spent in the direct work of urging the vessel forward. Fifty per cent., or one-half of the whole power, has been spent in the backward motion of the paddles, or screw, through the water. The term employed in engineering, for this loss, is "slip." The paddle-wheels, or the screw, are said to "slip" backwards through the water. There is always a sensible loss from this cause, but it rarely reaches the extent of fifty per cent., except in the case of tug-boats struggling with a large ship, or other load, in tow. The average slip in sharp vessels, with well-proportioned propelling means, is only from twelve to twenty-five per cent. of the whole power. The currents of water set in motion backward by the wheels of a paddle-steamer are very sensible; but they are sluggish compared to the active movement of the vessel in the other direction. This difference is due, in a great degree, to the sharpness of the vessel, as compared with the form of the paddles; but, allowing the form of the hull and paddles to remain unchanged, it is directly affected by the quantity of water seized and acted on by the paddle-wheels. The force urging the vessel onward is derived entirely from the hold of the paddles upon the water, consequently the total, or gross amount of force, measured in pounds, is equal on both the vessel and the water. If a constant force, equal to three tons, is required to be applied to a rope, in order to impel a vessel with a certain speed, it requires precisely an equivalent amount of force when applied by the engines, through the medium of the paddles or the screw, and the amount of "slip," or yielding of the water, under such a force, depends on the quantity of water grasped. If the paddles were indefinitely large, other things being equal, the "slip" might be diminished to nothing; or, if, on the other hand, the paddle-wheels be made very narrow, and their "dip," or immersion, in the water be very slight, or, if a screw be made disproportionately small, the loss by "slip" increases.

In the language of Tredgold, we may say, that "the area of the surfaces in action, multiplied by the resistance, must be equal to the area of the vessel, multiplied by the vessel's resistance when reduced to the same direction." The latter being constant, with any given model and speed, it follows, that any diminution in the area of the propelling blades must be compensated for by the attainment of an increased resistance of the water on the unit of surface, and this increased resistance can only be attained by giving the blade a more rapid backward motion through the water, or, in other words, a greater "slip."

The second point to be considered is, the agitation or churning of the water, generally designated, "the loss by oblique action." Each float of a paddle-wheel, especially when the wheel is deeply immersed, plunges into the water in an inclined position, impelling the particles of water downward toward the bottom. Thence, it sweeps around and acts directly backward for an instant, after which it emerges with a lifting action impelling the water upward, and allowing it to escape in a violently agitated condition, by falling, as it were, from its edge, or edges. This plunging and lifting consumes power, and as uselessly as does the backward "slip" of the wheel. It happens, that the paddles of a wheel cannot well be increased in size without increasing the loss from this cause, nor diminished, without increasing the loss by "slip," so that the generally accepted form and proportion of the paddle-wheel is the result of a species of compromise between these two conflicting elements.

Screw propulsion, although it may, at first view, seem peculiarly open to this objection, because the screw surfaces always act obliquely upon the water, is not, in fact, much affected by this element; but, as an offset, there appears, in this form of mechanical action, an effect peculiar thereto, and which is the third and last which we have proposed to consider. This is, the friction of the acting blades upon the water, usually denominated "the friction of the screw." The great area of the screw glides smoothly along in its helical, or cork-screw-like course, making no dashing and churning like the paddle-wheel, and thus almost entirely free from the loss by oblique action; but its pressure against the water on its rear surface, and to a less extent a pressure of the fluid on its front side, induces a great resistance even in so apparently frictionless a medium as water.

To recapitulate briefly, it appears, that all propulsion, properly so called, is performed by the urging back of water; that the power wasted in thus moving water depends on the quantity acted on, the power wasted becoming less as the quantity of water acted on is increased; that there is in the paddle-wheel a loss by oblique action, or by the plunge and lift of the paddles; and that there is in the use of the screw an offset thereto, in the friction of the screw itself upon the water. There is, therefore, no very great fundamental advantage or disadvantage, in theory, possessed by the screw, which is not offset by, or identical with, some feature of the paddle-wheel. The differences are mainly practical. The paddle-wheels are cumbrous and unsightly, and often exceedingly inconvenient: they do not allow of the vessel being careened over by a press of canvas, like the screw. But, on the other hand, the screw can not, or *does* not, produce the same speed of vessel, as the side paddle-wheel; and the result is, a division of our steam marine into paddle-wheel and screw steamers, each possessing independent advantages—the paddle-steamer being preferréd for those uses where speed is of supreme importance, and the screw being preferred for those in which sails are largely depended on as aids in producing the motion. For naval purposes, also, the screw possesses great advantages in being entirely immersed, and thus out of the reach of shot. The proportion of vessels, in which the screw is preferred over the paddle-wheel, has been increasing for many years.

From the facts that the screw vessel is but partially impelled by steam, and that a lower rate of speed is considered satisfactory, the gross power of the engines is usually less than is applied by the engines of a paddle vessel of corresponding size. Aside from this, however, a most important distinction exists in the character of the mechanism. The engines of a paddle-wheel steamer are large and slow; those of a screw vessel of similar size, must be relatively small and quick-acting. The difficulty of working with sufficient speed to impel the screw, lies at the foundation of the differences between screw and paddle-engines. The English very frequently introduce gearing which allows the use of engines more like the ordinary style; but the friction and rattling sound, as well as many other disadvantages thus involved, has prevented its general introduction in America.

The Griffith Patent Propeller is a French invention, the form and construction of which is shown in Plate XXXIII., Figs. 44, 45, 46, and 47. The central boss is very large and spherical, and the arms, or blades, are capable of being feathered, or adjusted to different angles, and are, furthermore, broader at the junction with the boss than at their extremities. It has been quite extensively introduced in large war vessels, both in Great Britain and this country; and, as this form of propeller is, in several important points, opposed to all previous ideas of the true form and proportions of an efficient screw-propeller, the following elaborate analysis of its several points, and comparison with the commoner forms of screw, although rather strongly drawn, will be found of much practical value. It is contributed by Mr. Charles W. Copeland, of this city.

The peculiarities of this propeller, as compared with all those previously known, are:

1st. Instead of continuing the blades down to the shaft, and keeping the hub as small in diameter as the requisite strength will permit, about one-third of the entire diameter is occupied or filled by the spherical hub.

In the experiments made by the inventor, it was ascertained, that the central part of the blades of the ordinary screws *absorbed* 20 *per cent.* of the power, and had little or no propelling effect in consequence of that part of the blades (more especially in coarse pitched screws) being nearly in a line with the shaft, the effect when in operation being to throw the water off (by its flapping and centrifugal action) at right angles to the line of the vessel's motion, thus seriously disturbing and breaking up that portion of the water upon which the more effective part of the blades should act.

The great vibration of the stern of all screw-steamers is also to be attributed, in a greater or less degree, to the action of this central portion of the blades, which is, at the same time, so inefficient for the purposes of propulsion, and so wasteful of power.

It will be readily understood, that the power required to overcome the friction only of the central sphere must be but a mere fraction of the power wastefully expended in overcoming the resistance of the central portion of the ordinary propeller.

Another advantage resulting from the use of the spherical hub is, that the water not being violently agitated and broken up by the central portion of the blades, the more effective portions of the blades are permitted to work in water more quiescent and solid, producing a better result, and working with a decreased amount of "slip."

2nd. The second important feature of this propeller is, the *form of the blades*, which, instead of being largest at the extremity, is *directly the reverse.*

A twofold advantage results from this change of form: 1st, the resistance of the different portions of the blade is more nearly uniform, the width of the blade being decreased at the distance from the shaft, and consequently its velocity is increased: being narrowest at the periphery, where its rotative velocity is greatest.

3rd. By thus reducing the blade towards the periphery, the clearance between the blade and the stern and rudder posts is increased, giving the greatest clearance where the rotative velocity is greatest, and, by this increased clearance, the vibrating motion of the stern of the vessel, which is so severe with the ordinary screw, is entirely prevented.

4th. Another result of this form of blade is, that the water in which the blades work is so much less disturbed than with ordinary screws, that it is found by experience that less *surface* of blade is required to produce an equally efficient result, than with the ordinary screw.

When this propeller was first introduced by Mr. Griffiths, it was expected that much advantage would result from an arrangement in its construction, whereby the "pitch" or angle of the blades could be increased or diminished, or even "feathered" during the voyage, and whilst at sea, with facility, to suit the varying exigencies of sea-going steam-vessels, in regard to wind and weather.

Experience, however, obtained from many experiments, and careful observation, has shown conclusively, that the risk of derangement incident to the machinery requisite for this purpose, is too great to admit of practically useful results; also, that the advantages anticipated by such an arrangement are not practically realized.

It should be borne in mind, that this propeller is not the result of *mere* theorizing, but of extensive experiments of great number and variety; not upon models only, but with screws of practical size, and applied to vessels in actual service, and extending over a period of several years. Its advantages, as set forth, are *proven* and established by actual use on a large number of vessels in Europe. It has, as yet, been applied to but few vessels in this country, and its peculiarities and advantages are here but little known; but there is no doubt that ample trials will soon place it here in advance of all other propellers, as it has already done in Great Britain.

ADVANTAGES.

1st. Increased speed of vessel, relative to the power employed.

This has been proved in so many trials that have taken place with this screw against others, both in experiments made by the British Admiralty, and by private parties, that it is unnecessary to do more than allude to the results as fairly establishing its superiority for speed.

A great saving of power is effected by this propeller, because its central portion is filled up by a sphere, whereas other screws have their centres made as *small* as the necessary strength will permit; their blades, consequently, at the part near the hub, are nearly at right angles with the plane of revolution, and much power is, therefore, lost by driving the water *outwards* instead of *astern*, thus seriously disturbing the water upon which the more effective portion of the blades operates.

A screw-propeller is a rotating *fan*, which drives the water through its disc, and whether the vessel is moored or under weigh, that column passes through in proportion to the power exerted upon it; therefore, when any portion of the water is moved in any other direction than fairly astern, *power is absorbed* without giving out a corresponding useful effect in the propulsion of the vessel.

2nd. Less vibration of the vessel.

The tremulous motions, and disagreeable and injurious vibrations which are almost invariably observed in vessels fitted with the ordinary screw, and more particularly at the stern of such vessels, are so completely removed by the use of the Griffith propeller, as scarcely to excite attention.

This small vibration is the result of the combination of the central sphere—which dissipates the action of the blades at right angles with the line of propulsion—with the form of the propeller blades, which, by their diminished length and increased clearance as they approach the periphery, where the rotative velocity is greatest, aids in producing the desirable result.

That the vibration alluded to is not merely unpleasant to passengers and crew, but decidedly *injurious* and destructive to the whole framework of the vessel, is sufficiently established by the results produced upon iron-built steamers, where, as is well-known, it becomes frequently necessary to rivet and fasten anew all the after-end of the vessel near the propeller, in consequence of the whole having become loosened by the incessant vibrations produced by the ordinary screw.

The same destructive action is, without doubt, constantly going forward on board wood-built vessels; but, by reason of the greater elasticity of the wood over the iron, the evil results are not so soon developed or so readily discovered.

3rd. The facilities for ascertaining the "best pitch" to suit the vessels and engines to which it is fitted.

Up to the present time, no correct *a priori* rule has been laid down by which the "best pitch" (that is, that pitch which best fulfills *all the requirements* of the vessel and her machinery) can be ascertained, and it is very doubtful if this will ever be accomplished.

The present practice of engineers is, to make propellers of such a pitch as, from precedent and experience, they are led to *believe* will hold their engines at the required speed.

Should the result, on trial, prove satisfactory, the screw remains; but if this fortunate issue is *not* obtained, *other* screws are tried, of different pitches, or amount of surface, until the desired result *is* obtained.

This is a very *tedious and expensive* process, which, in many cases, cannot be resorted to, in consequence of the necessary delay to the vessel; and there is no doubt, that many screw steam-vessels have been running for years under the disadvantage of an improperly-constructed propeller, at a clear loss of speed to the vessel and of economy to the owners.

Experience of many years in adapting the screw propeller to vessels of varied models and dimensions, has shown conclusively, *that nothing short of actual trial* of different pitches can determine the "best pitch" for a propeller in each particular case.

The points to be considered in reference to this question are so numerous and complicated in their bearings upon each other, as to defy *reliable* previous calculations of their effects. For instance, the quantity of power employed, relative to the *size* of the *vessel* and of the *screw*, the *model*, or *form*, of the vessel below water, particularly at its after-end, all influence the results obtained by screw propulsion, and these results in practice often vary so much from what previous experience in the trials of other vessels would lead us to suppose, that actual trial of different pitches for each particular vessel is the *only* means of arriving at the best results.

As an instance of this difference: Two vessels of the British Navy were fitted with engines of the same power, and with screws exactly similar to each other; one vessel, however, had *full* after-body lines, whilst the other was finer in form, or, in other words, had a sharper and finer stern. On the trial, the engines of the former made 54 revolutions, and those of the latter 75 revolutions, per minute.

The speed of the former was, 6.3 knots per hour, and of the latter, 7.8 knots per hour.

In another case, a screw steamer was originally built with a fine stern—on trial, the engines made 32 revolutions per minute, and the vessel had a speed of 9.1 knots per hour.

As an experiment, the stern was then filled out, so as to give what would be called a *very full* stern; the result, upon trial, was, that the engines made 24 revolutions per minute, and the vessel had a speed of 3.25 knots per hour.

A *portion* of the filling at the stern was then taken off, and another trial made, when the engines made 26.5 revolutions per minute, and the vessel had a speed of 5.75 knots per hour.

The *whole* of the filling was then removed, and another trial made, when the vessel attained the original speed of 9.1 knots per hour.

An instance has also occurred in which an increased speed of *one knot per hour* was obtained by reducing the pitch of the screw, although the speed of the *engines* remained the same as before.

These anomalies may be attributed chiefly to the difference which exists in almost all vessels in the amount and velocity of the back water, or *eddy*, following the vessel—which must largely influence the efficiency of the propeller working in it; and hence the necessity of using a propeller which can have its pitch adjusted to suit these varying circumstances.

With the "Griffith" screw, the pitch can be altered, either when the vessel is in dry dock or beached, or even when afloat in many cases, by the use of a diving dress, or by weighing the vessel down forward and lightening aft; and where the propeller is arranged to lift in a well, as in many naval steamers, this change of pitch is readily effected, when afloat, and even when at sea, without in any manner disturbing the cargo or stowage of the vessel.

4th. No spare screw need be carried by a vessel fitted with "Griffith's" propeller, as the central sphere is never destroyed or injured, and spare blades only are necessary, which are comparatively readily handled and conveniently stowed.

When an ordinary screw is broken, the vessel must be docked, or beached, to replace it, and the whole instrument is useless, and of no value, except as old material; but with the "Griffith" screw, if one of the blades is broken, it is replaced in but a fraction of time required to fit a *new propeller*, and it is frequently done with the vessel afloat.

In this case, instead of the *whole* propeller being condemned and thrown aside as old material, only the broken blade is thus thrown away—the value of which is, probably, not more than one-fourth the total value of the propeller.

5th. The "Griffith" propeller possesses advantages over the common screw, which render it peculiarly applicable to auxiliary steam navigation.

The resistance which an ordinary screw occasions, when the vessel to which it is attached is under canvas only, and the serious difficulty which is experienced in steering and manœuvring the vessel, are well-known facts: the resistance to the vessel's progress is variously estimated at from one to three knots per hour.

There have been *three* remedies devised for these evils: 1st. To hoist the screw out of water. 2nd. To feather the screw; and 3rd. To adopt the "Griffith" propeller.

The first-named remedy has been largely adopted for war-steamers, because it is desired that they shall be essentially and primarily *sailing* vessels—the steam power being merely auxiliary, and to be used only when the sails cannot be depended upon to accomplish the purpose desired.

The great number of men attached to vessels of war, enables them to accomplish the raising and lowering of a propeller on board such vessels within a comparatively short period of time; and, therefore, this mode of overcoming the evils of a *common* fixed screw is not open to *one* of the objections to adopting it on merchant steamers.

The other objections to hoisting the screw, apply with equal force to both war and merchant steamers. They are: 1st. Great increase of weight upon overhanging and unsupported portions of the stern, combined with decreased strength of the stern frame, in consequence of the well, or hole, for hoisting the propeller, being cut vertically through the *whole* of that frame. 2nd. Very large additional cost for the frame, slides, cross-heads, pulleys, &c., necessary for a *hoisting* propeller. 3rd. The power of control of the rudder over the movements of the vessel, is very greatly reduced when the propeller is *hoisted*, in consequence of the large opening left in the dead wood, through which the currents from the two sides of the ship meet and counteract each other, instead of acting directly upon the rudder, thus increasing its efficiency.

The *feathering* screw was first brought into practical use about six or seven years ago, when the entire fleet of an English company—eight or ten vessels—were fitted with them.

As they were brass screws, and made by a most eminent engineering firm, we may conclude that no time or expense was spared to render them perfect feathering screws; but the results obtained from their use has not encouraged the owners of other vessels to adopt this plan.

It would appear, on the contrary, that the examples thus offered have deterred others from adopting it; and we believe that two of the smaller vessels referred to, have had a common *fixed* screw substituted for their former feathering propeller.

As stated above (see page 140), experience with screws having shifting or feathering blades, has shown that the risk of derangement incident to the machinery required for this purpose, is too great to admit of practical success; and the small number of steamers which have been fitted with the feathering screw, would seem to indicate that this is also the general opinion.

When the "Griffith" propeller (with two blades) is fitted to a steam-vessel, her engines can be stopped in such a position that the blades are perfectly vertical, and locked in that position, when it will be found that their resistance to the vessel's *sailing* is scarcely appreciable, and that the power, or *control, of the rudder* over the vessel's movements is *not at all* affected.

Similar results are obtained when the screw is disconnected from the engines, and permitted to revolve freely; and, no doubt, in *some* vessels, rather less resistance would be offered in this way, than by locking the blades in a vertical position; but *not in all* vessels, as cases have been reported where the experiment has been tried, to ascertain in which the least resistance was presented to the vessel's progress, and no advantage was found either in sailing or steering from disconnecting their propeller—which cannot be said of the ordinary screw.

Experiments have also lately been made with the "Griffith" propeller, fitted in the hoisting-frame of a wood-built steamer, and it was found, from repeated trials, that *less* resistance to her sailing was experienced when the screw was in its place in the water, and the blades locked vertically, than when it was *hoisted out of the water*.

The stern frame of the vessel being thick—as it was of wood—nearly covered the screw blades, and from the *aperture being filled* when the screw was in its proper position, and the blades fixed vertically, there was *less* resistance than when it was hoisted, and the aperture left open to the conflicting currents of water.

6th. Another advantage possessed by the "Griffith" screw is, that when the ship is opposed to strong head winds, with rough sea, and pitching heavily, she will make better progress, and the engines will *race* less, with the "Griffith" than with the common screw, *because* the chief propelling surface of blade of the "Griffith" screw is about midway of the radius of the screw, and is not, by pitching, thrown out of water; but, in the common propeller, the chief propelling surface of blade is near the periphery, which is the part first thrown out of water, and left by the sea, and when this occurs, the engines are relieved of a large proportion of their resistance, and *run off*, or *race*, very badly, in consequence of the relief, and not unfrequently to such an extent, as to cause serious derangement, and sometimes the *breaking down* of the engines.

Experience has shown, that, under such circumstances, with the "Griffith" propeller much less attention is required to guard against *racing*, than with the common screw.

THE OPERATION OF THE SCREW-PROPELLER, WITH REFERENCE TO ITS SCIENTIFIC PRINCIPLES.

RESISTANCE OF FLUID.

The scientific investigation of bodies in moving through a fluid, is still involved in much obscurity, from the want of independent research on the part of the various authors, who have undertaken the elucidation of the subject; and the mistakes incidental to the researches of Newton, and other eminent philosophers, have overrun various departments of physical science, and are now found most difficult of eradication. The circumstances in connection make it expedient to investigate the matter in a practical way, and to illustrate a few of the leading principles of Mechanics, from Bourne's Treatise on the Screw-Propeller, which relate to the question before us.

Mechanical power is pressure acting through space; and the amount of mechanical power developed by any combination is measurable by the amount of the pressure, multiplied by the amount of space through which the pressure acts. A pressure of 10 pounds acting through a space of 1 foot, represents the same amount of mechanical power as a pressure of 1 lb. acting through a space of 10 feet; and 10 lbs. gravitating through 1 foot, or 1 lb. gravitating through 10 feet, represents ten times the amount of mechanical power due to the gravitation of 1 lb. through 1 foot.

In the same way, 1000 lbs. gravitating through 1 foot, is equivalent to 1 lb. gravitating through 1000 feet; and in general terms, the weight or pressure, multiplied by the space through which it acts, represents the power universally. If, therefore, a body falls freely through space by the operation of gravity, since it parts with none of its power during its descent, the whole power must be accumulated in the falling body in the shape of momentum; and, at the instant of reaching the ground, the body must have such an amount of mechanical power stored up in it as would suffice to carry it up again to the position from which it fell, if the power were directed to the accomplishment of that object. The amount of mechanical power, therefore, in any moving body, is measurable by the weight of the body, multiplied by the space through which it must have fallen by gravity, to acquire the velocity it possesses; and this fundamental law, if distinctly apprehended, and kept constantly in recollection, will ensure exemption from the fallacies which prevail so generally among authors in reference to such subjects. In Newton's "Second Law of Motion," it is maintained that "the change or alteration of motion produced in a body by the action of any external force, is always proportional to that force," from whence it is inferred, that to produce twice the quantity of motion in a body, will require just twice the power; and this is the doctrine maintained by Robison, in his "Mechanical Philosophy," and by Hutton, Gregory, and most other English authors, who have undertaken to illustrate such questions. Nevertheless, there is no doubt whatever that this doctrine is altogether erroneous, as was shown by Leibnitz at the time of its promulgation; and subsequently by Smeaton, who, by a series of carefully executed experiments, proved very clearly that to double the velocity of a moving body, it required four times the amount of mechanical power that was necessary to put it into motion at first; and consequently, that the momentum of moving bodies of the same weight varies as the squares of their respective velocities. The soundness of this conclusion, is made manifest by a reference to the law of falling bodies, by which it will be found that it is necessary a body should fall through four times the height to double its ultimate speed; nine times the height, to treble its ultimate speed, and so on; showing that the height, and therefore the power exerted in creating the motion, must be as the square of the ultimate speed; and consequently, that the ultimate velocities of all falling bodies will be as the square roots of the heights from which they have respectively descended. In the case of two bodies of equal weight, therefore, moving in space, but of which one moves with twice the velocity of the other, the faster will have four times the amount of mechanical power stored up in it that is possessed by the slower; for it must have fallen from four times the height, to acquire its doubled velocity; and the relative quantities of power capable of being exerted by bodies of the same weight, are measurable in all cases by the spaces through which the weight or pressure acts. A cannon ball moving with a velocity of 2000 feet a second, has four times the momentum of a cannon ball of equal weight, moving with a velocity of 1000 feet a second; and every particle of a stream of water moving with a velocity of 10 miles an hour, has four times the momentum of every particle of a stream of water moving with a velocity of 5 miles an hour. Every particle of the faster stream, therefore, will exert four times the effect in impelling any body on which it impinges, that is exerted by every particle of the slower stream. But in the faster stream, not only will every particle impinge with four times the force, but there will be twice the number of particles impinging in a given time, and a quadrupled force for each particle; and twice the number of particles striking in a given time, gives an effect eight times greater, in a given time, with a doubled velocity of the stream. Accordingly, it is found that in water or wind mill, when the velocity of the current is doubled, the power exerted is about eight times greater than before; and it is also found that a steam-vessel, to realize a double velocity, requires about eight times the amount of power; but these results, it is obvious, have reference, not merely to the increased velocity of the particles of matter, but to the large number of them brought into operation; and any given quantity of water, if flowing with a doubled velocity, would only exert four times the power exerted before. In the same manner, a steam-vessel, to accomplish any given voyage in half the time, would require four times the quantity of coal previously consumed; for although eight times the quantity of coal would be consumed per hour, yet only half the number of hours would be occupied in accomplishing the distance. The number of particles of water to be displaced by a vessel in performing any given voyage, is the same, whatever the velocity of the vessel may be; but the number of particles displaced in the hour differs with every different velocity, and the power expended must consequently vary in a corresponding proportion. It may hence be asserted, generally, that the power or dimension of engine necessary to propel a vessel, increases nearly as the cube of the velocity required to be attained; but the consumption of fuel will only increase in about the ratio of the square of the velocity, looking to the number of miles of distance actually performed by a steamship.

It will be very important, to compare with these doctrines the following statements of the most eminent authors who have treated of Theoretical Mechanics. Robison, in his "Treatise on Mechanical Philosophy," Vol. II, page 269, gives the following, as the fundamental proposition of the doctrine of the resistance of fluids: "The resistances and (by the third law of motion) the impulsions of fluids on similar bodies, are proportional to the surfaces of solid bodies, to the densities of the fluids, and to the squares of the velocities jointly." In Tredgold's work on the Steam-engine, there is an Appendix on Paddle-wheels, by Mr. Mornay, where the same doctrines are propounded. At page 122, Mr. Mornay writes as follows: "In order to be able to calculate the absolute amount of power required to produce a given effect, it is necessary to be acquainted with the laws which govern the resistance of fluids to the motion of solid bodies in them, which are generally admitted to be based on the following theorem: "If a plain surface move at a given velocity through a fluid at rest, in a direction perpendicular to itself, the resistance is proportional to the density of the fluid, and to the square of the velocity of the plane." He adds, "It is assumed that the resistance to a plane moving in a fluid at rest, is equal to the pressure of the fluid on the plane at rest, the fluid moving at the same velocity and in the contrary direction to that of the plane in the former case; on which hypothesis, the ratio of the square of the velocity is explained in two very different ways. The first is, that 'the resistance must vary as the number of particles which strike the plane in a given time, multiplied into the force of each against the plane; but both the number and force are as the velocity, and consequently, the resistance is as the square of the velocity.' The second explanation is, that the force of the fluid in motion must be equal to the weight or pressure which generates that motion, which, it is known, is equal to the weight of a column of the fluid whose base is equal to the area of the surface, and altitude the height through which a body must fall to acquire the given velocity.'" These explanations, Mr. Mornay adds, are extracted from Dr. Gregory's "Treatise on Mechanics;" and in the works of Hutton, and most other English writers on Theoretical Mechanics, similar statements are to be found: yet it is quite certain that they are altogether erroneous, and their original promulgation is traceable to the accident of the mechanical force resident in moving bodies, having been set down by Newton as measurable by the velocity, instead of by the square of the velocity, as is now known to be the case. If, therefore, the assistance varies as the number of the particles multiplied by the force of each particle, it must vary as the cube of the velocity; for the number of particles varies as the velocity, and the force of each particle varies as the square of the velocity; and the velocity multiplied by the square of the velocity, is obviously the cube of the velocity. It consequently cannot be true that the impact of a fluid in motion will produce a pressure only equal to that due to the head of fluid that will produce the motion, as was long perceived by Daniel Bernouilli. For, as water issuing from a reservoir has the same velocity as any heavy body would acquire by falling freely from the level of the surface of water in the reservoir to the level of the issuing stream, and as, by the laws of falling bodies, the ultimate velocity of a falling body is just double its mean velocity, it is clear that a jet issuing horizontally, after having acquired the ultimate velocity due to the head, will pass through a distance equal to twice the distance that a body would pass through in descending from the level of the water-surface to the level of the orifice. Hence, Bernouilli inferred that the accumulated hydraulic pressure by which a vein of heavy fluid is forced out through an orifice in the side or bottom of a vessel, is equal to the weight of a column of the fluid, having for its base the section of the vein, and for its height twice the fall productive of the velocity of efflux.

Bernouilli's theory was adopted and still further developed by Euler, who gives a formula for ascertaining the percussive effect of a jet of water on a plate, which is as follows:

Let R = force of impact in permanent percussion.

a = area of vein.

H = height due to actual velocity of reflected water.

ϕ = angle of reflected water to axis.

Then,
$$R = 2aH\left(1 - \frac{\sqrt{h}}{\sqrt{H}} \cos \phi\right).$$

Morosi and Bidone's experiments lend material confirmation to the doctrines of Bernouilli and Euler on this subject. Euler says, that the theoretical value of the percussion of a fluid vein may increase until it is equal to the weight of a fluid column of the same base as the section of the vein, and of a height four times greater than that due to the velocity of the vein. Bidone found that the sudden shock of a jet upon a plate is to the force of the jet, when permanent, as 1.84 to 1; but this effect may, perhaps, be in some measure attributed to the momentum acquired by the parts of the instrument by which the percussive force was measured. Colonel Beaufay has made experiments for ascertaining the resistance of bodies moving through water, from the results of which it appears that at low speeds, such as two knots an hour, the weight necessary to draw the body varies as the square of the velocity; the weight necessary to overcome the friction of a body moving in water appears to vary as about the 1·7th power of the velocity; but the proportion appears to diminish slightly with an increase of speed. The friction upon a square foot of plank, moving through the water in the manner of the bottom of a ship, was found to be equal to a weight of 0.14 lbs., with a velocity of one nautical mile an hour; or, in other words, it would require a weight of ·014 lbs. acting on a string passing over a pulley, to overcome the friction of a square foot of plank, when passing through the water at a velocity of one nautical mile per hour. At a speed of two nautical miles per hour, the friction was found to be equal to a weight of ·0472 lbs.; three miles an hour, .0948 lbs.; four miles an hour, ·153 lbs.; five miles an hour, ·2264 lbs.; six miles an hour, ·3086 lbs.; seven miles an hour, ·4002 lbs.; eight miles an hour, ·5008 lbs.; and by carrying the law up to thirteen nautical miles per hour, the weight necessary to overcome the friction varies as the 1·823 power of the velocity. At eight nautical miles per hour, the weight necessary to overcome the friction varies as the 1·713 power of the velocity.

In general, it may be asserted that the weight necessary to draw any body through the water varies nearly as the square of the velocity; but the distance through which the weight descends varies also as the velocity, so that the power expended in any given time, varies nearly as the cube of the velocity. Whatever doctrines relative to fluid resistance may be eventually adopted, it is at least certain that in the case of ordinary vessels and ordinary velocities, the power necessary to accomplish any particular speed which may be prescribed, is *ascertainable by the equation*

$$\frac{S^3 A}{C} = \textit{horse-power},$$

where S is the speed in miles per hour, A the immerged sectional area of the vessel in square feet, and C a certain number, or coefficient, which varies with the form and also with the size of vessel employed.

This coefficient, as set down in the table, is obtained by multiplying the cube of the speed, in nautical miles per hour, by the immerged midship section of the vessel in square feet, and dividing by the indicated horse-power of the engine; and a number is thus obtained, by the aid of which the power necessary to accomplish other speeds, with a similar size and form of hull, may be approximately found.

$$\frac{(\text{speed})^3 \times \text{midship section}}{\text{nominal horse-power}} = \text{the coefficient};$$

or,

$$\frac{(\text{speed})^3 \times \text{midship section}}{\text{indicated horse-power}} = \text{the coefficient}.$$

If a steam-vessel be tied at the stern, and the engines be then set into revolution, their velocity will go on increasing until the resistance at the centre of pressure of the paddle wheels just balances the pressure on the piston—the centre of pressure being a point in the depth of every float at which the pressure above and below it is the same, or at which the aggregate pressure may be supposed to be collected. Now, as the resistance at the centre of pressure must just balance the pressure upon the piston, it follows that the pressure urging forward the vessel will be the same, whether the vessel is at rest or in motion, supposing always that the engines are adequately supplied with steam; and the resistance created at the centre of pressure will be the same, whether the paddle-floats are large or small—only, if they be small, a greater velocity of revolution will be necessary to create the resistance requisite to balance the pressure upon the pistons, and a large consumption of steam will be occasioned, without any countervailing advantage. If, however, the wheel be diminished in size, the pressure upon every paddle-float will be increased, for a larger resisting area will then be necessary to balance the pressure upon the pistons, in consequence of the diminished length of leverage acting against that area; and when a small diameter of wheel is employed, rather a large area of float is necessary, or else the centre of pressure will pass with a greater velocity through the water, which is tantamount to saying that the slip will be increased. In the case of the screw similar results will be found to ensue. Setting aside the loss of power occasioned by the friction of the screw when revolving in the water, and the resistance occasioned by its cutting edge, it will be obvious that, with any given pitch, the forward thrust of the screw-shaft will be the same, whatever the area of the screw-disc may be, for the velocity of rotation will go on increasing until the resistance which the screw encounters balances the pressure on the pistons; and if the pressure on the pistons be considerable, so will be the thrusting or pushing-force of the screw. If the screw, however, be of inadequate dimensions, then the velocity of its rotation will be much greater than what answers to the speed of the vessel, and there will be a larger consumption of steam by the engine than would be necessary, if the screw were of larger size. It is hence obvious that a very small diameter of screw, relatively with the midship section, or with the resistance to be overcome, is inadvisable, just as a small area of float-board is inadvisable in the case of a paddle-wheel. To diminish the pitch of a screw is tantamount to a diminution of the diameter of a paddle-wheel; but, unlike the paddle, the propelling area may be made too large; and this point will be attained when the friction consequent upon the increased diameter, the resistance arising from the extension of the cutting edge, and other analogous sources of loss, more than balance the loss arising from the slip. From some experiments which were made by Mr. Brunel, at Bristol, in 1840, with half a disc of metal, 5 feet 9 inches diameter, set on a shaft revolving in water, it appears that it took 6·4 horses power, by the indicator, to give the shaft a velocity of 101 revolutions per minute, when the semi-disc revolved in air without the contact of water; and that it took 9 horses power by the indicator to give the shaft a velocity of 100 revolutions per minute, when the semi-disc revolved in water. Hence it was inferred, that the resistance to the semi-disc which a screw with an equal amount of surface, and an equal length of cutting edge, suffers from the water at a speed of 100 revolutions per minute, will be overcome by about 3 horses power of the engine. This is equivalent to a weight of about 55 lbs. at the end of the arm, or at the circumference of the disc, hindering its revolution; for 5 feet 9 inches, or 5·75 feet.

$$5{\cdot}75 \text{ feet} \times 3{\cdot}146 \times 100 = 1806{\cdot}42 \text{ feet per minute, and 3 times 33000 lbs.; or,}$$

$$99000 \text{ lbs.} \div 1806{\cdot}42 = 55 \text{ lbs. very nearly.}$$

In steam-vessels of the usual form, and with the ordinary rates of speed, the resistance of the vessel, or what comes to the same thing, the amount of thrust, necessary to be imparted by the paddle or screw-shaft, increases very nearly as the square of the velocity; and as in order to communicate twice the velocity to the vessel, the engines must not merely be able to work against four times the load, but must also move with twice their previous speed. Contrarywise, if the engine-power of a vessel be increased, while her immersion and other elements remain without alteration, her speed will be increased in the proportion of the cube root of the increased power. If, therefore, the engine-power of a given vessel be doubled, her speed will be increased in the proportion of the cube root of 1 to the cube root of 2; or, in other words, in the proportion of 1 to 1·25. If the original speed of the vessel, therefore, were 10 knots an hour, the effect of doubling the power would be to raise the speed to 12½ knots an hour. While, however, this result may be confidently expected in the case of such speeds as 10 or 12 miles per hour, it does not follow that the law will apply in the case of such speeds as 18 or 20 miles an hour, supposing the same form of vessel to be retained. Indeed, it is well known that at high velocities, the resistance of any given vessel increases in a higher ratio than the square of the speed. The main cause of this accelerated increase in the resistance, in the case of high speeds, is traceable to the inability of the water to close in at the stern of the vessel with sufficient rapidity to impart its proper pressure thereto; and in addition, therefore, to the ordinary resistances, the vessel has under circumstances to encounter the hydrostatic pressure due to the deficient gravitation of the water against the stern. At high speeds it is consequently indispensable to make the stern very fine, else the vessel, in passing through the water, may leave a vacant space behind her, and the resistance will be enormously increased thereby. Each different speed, indeed, has a corresponding form of vessel, which will make the resistance a minimum. A vessel with any given amount of power, and with any given displacement, may be sharpened so much, that an additional sharpening would increase the resistance by increasing the friction of the bottom in a greater ratio than the bow and stern resistances were diminished. And when by adopting such an amount of sharpness as gives the best result, the total resistance is brought to a minimum for one particular amount of power, it will be found that a further sharpening is necessary to make the resistance a minimum for an increased amount of power. In practice, cases have occurred where a vessel has been made too sharp, since with the same engine-power, placed in a blunter vessel, a better speed was obtained; but with an increased power the sharper vessel would have afforded the best result.

In considering the amount of power necessary to be given to a screw-vessel, to propel her through the water with any given velocity, it is necessary first to settle the type of vessel, and next her size. When these points are determined, it will be easy, by selecting from any screw-vessel of the shape most nearly resembling that of the intended vessel, to tell the amount of power necessary to propel that vessel at any given speed, on the supposition that the two are of the same size. The coefficient of performance of the vessel which the new vessel resembles, will be found, as heretofore stated; and the number of horse-power necessary to accomplish any different speed, either of this vessel or of the new vessel, may be ascertained by multiplying the cube of the intended speed by the number of square feet of immersed section, and dividing by the coefficient proper for this particular case. The result thus obtained, however, supposes that the new vessel is of the same size as the similar vessel; but if she be smaller than the model vessel, the speed will be less, and if she be larger, the speed will be more in the proportion of the square roots of the length, or some other linear dimension of the two vessels.

If S be the speed of the vessel in knots, A the area of the immersed midship section in square feet, C a numerical coefficient, varying with the form of vessel, and P the indicated horse-power, then

$$P = \frac{S^3 A}{C}, \quad C = \frac{S^3 A}{P}, \quad \text{and} \quad S = \sqrt[3]{\frac{PC}{A}}.$$

By means of these equations, therefore, the power P, necessary for the accomplishment of any prescribed speed, and the speed, S which will be realized by the application of any given power, may be approximately determined.

If the new speed be higher than the old, however, then the actual speed will be somewhat less than the theoretical speed as thus ascertained, since this rule proceeds upon the assumption that the resistance varies as the square of the speeed; whereas in the case of vessels of a moderate sharpness, the resistance varies in a somewhat higher ratio than the square of the speed.

CONFIGURATION AND PROPORTIONS OF THE SCREW-PROPELLER.

FORM AND PITCH OF SCREW-PROPELLER.

The screw-propeller, as now commonly applied to the propulsion of vessels, consists of two or three helical or twisted blades set upon a shaft or axis, revolving beneath the water at the stern. The shaft where it protrudes through the stern of the vessel is surrounded by a stuffing box, containing hemp-packing, whereby the entrance of the water into the vessel is prevented, and the extremity of the shaft in the rear of the screw is supported in a socket or bearing attached to the rudder-post. This part rests upon the keel, and from it the rudder is suspended. The screw revolves in that thin part of the stern of the ship, which is called the dead wood, in which a hole of suitable dimensions is cut for its reception; and the thrust or forward pressure caused by the action of the screw upon the water is transmitted to some point within the vessel which can be amply lubricated. The most perfect lubrication of this point is indispensable, to counteract the friction caused by the combined thrust and rotation of the shaft, and cases have occurred in practice in which the end of the shaft became white hot, even with a stream of water playing upon it, and actually welded itself to the steel plate against which it pressed. It is the thrust of the shaft which is operative in propelling the vessel, and the amount of this thrust can be measured by means of a dynamometer, applied to the end of the shaft within the vessel. The diameter of the screw is the diameter of the circle described by the arms; and the length of the screw is the length which the arms occupy upon the revolving shaft. If a string be wound spirally upon a cylinder, it will form a screw of one thread. If two strings be wound upon a cylinder with equal spaces between them, they will form a screw of two threads. Three strings similarly wound will form a screw of three threads, and so of any other number. If instead of strings, flat blades be wound edgeways round the cylinder, and if each blade has one of its edges attached to the cylinder by welding, soldering, or otherwise, then if a slice be cut off the end of the cylinder, there will be only one piece of blade attached to that slice, if the screw be of one thread; two pieces of blade, if the screw be of two threads; three pieces of blade, if the screw be of three threads, and so of any number. The number of blades, therefore, of any screw determines the number of threads of which it is composed, and this indication equally holds, however thin the slice cut off the end of the screw may be.

The pitch of a screw is the distance measured in the direction of the axis between any one thread, and the same thread at the point where it completes its next convolution. Thus, a spiral staircase is a single-threaded screw, and the pitch of such a screw is the vertical distance from any one step to the step immediately overhead. Ordinary screw propellers are not made nearly so long as what answers to a whole convolution; and in speaking of their pitch, therefore, it is necessary to imagine the screw to be continued through a whole convolution at the same angle of inclination with which it was begun. Of this whole convolution any given proportion may be employed as a propeller, and the length of a screw, therefore, cannot be determined from the pitch, neither can the pitch be determined from the length.

The form of a screw most frequently employed in this country is a screw of two blades or threads, sometimes three and four blades are used. The pitch of the screw is not made less than its diameter, sometimes made near twice the diameter, and in some instances made over twice the diameter of the screw. The length of the screw is usually made equal to one sixth of the pitch. The thrusting of the screw is measured by the area of the circle described by the arms, which is termed the area of the screw-disc. The screw-disc has generally about 1 square foot of area for every $2\frac{1}{2}$ or 3 square feet in the immersed transverse-section of the vessel. Thus, a vessel with 226 square feet of immersed section will have a screw of such diameter, that the disc will have an area of about $75\frac{1}{2}$ square feet. This answers to a diameter of screw of 10 feet. The pitch of such a screw should be about 11 feet, and the length of the screw about 1 foot 10 inches. These proportions are those proper for screws with two plates fitted to vessels of 800 or 1000 tons; but they will also apply to screws with three blades; and with small vessels a larger screw in proportion should be employed.

NEGATIVE SLIP OF THE SCREW.

By "slip" it will be recollected, is meant the difference between the actual advance of the propeller through the water, and the advance which would be accomplished, if there were no recession of the water produced by the pressure of the propelling surface.

A screw of 10 feet, if working in a stationary nut, would advance 10 feet for every revolution it performed; but, when such a screw acts in the water, it may only advance 9 feet or less for every revolution—the water being, during the same time, pressed back 1 foot, from its inertia being inadequate to resist the moving force. In such a case, the slip is said to be 1 foot in 10, or 10 per cent. With every kind of propeller which acts upon water, there must be a certain amount of slip; for any force, however small, will overcome the inertia of the water to a certain extent; but, by so proportioning the propelling apparatus that it will lay hold of a large quantity of water, the backward motion of the water will be small relatively with the forward motion of the vessel—or, in other words, the slip will be reduced to an inconsiderable amount.

One of the most remarkable phenomena connected with the action of the screw is, that under some circumstances its apparent progress through the water is not only as great as that of the ship, but greater. In some of the early voyages of the "Archimedes," when the vessel was proceeding under the joint action of steam and sails, it was found that the progress made by the vessel through the water was greater than if the screw worked in a solid nut. It was from hence inferred that the ship must be overrunning the screw; yet that, it was also plain, could not be the case, as the engine was all the while well supplied with steam, and had the usual load upon it. The engine was, therefore, evidently driving something, and it was certain that the mere friction of the machinery and of the screw in the water could not consume all the power. There was also the usual thrust upon the screw-shaft; so that the screw, although moving slower than a patent log would do, if put over the stern, was nevertheless propelling the vessel. Shortly afterwards, the vessel was fitted with a number of different screws, by Mr. Brunel, and it was ascertained that with some of these screws the vessel went faster without the aid of sails, than if the screw had been working in a solid nut. In various other vessels the same action has since been observed; and if the pitch of the screw be made much less than the diameter of the screw, this action is very likely to follow. At first the phenomenon appeared so paradoxical as to be pronounced incredible; but it is now known to be a fact; and the action has been ascribed to the circumstance of the screw working in a column of water which follows the ship, instead of in the stationary water of the sea.

When a strong current of water runs through the arches of a bridge, the water may be observed to curl around those ends of the piers which stand lowest in the stream; and if a chip of wood be thrown into that spot, it will not be carried of by the stream, but will remain at rest, showing that the water is not in motion in that place. Now, if we suppose a screw to be placed in this stationary water, it will be obvious that any movement of rotation given to it will produce some thrust upon the screw-shaft; whereas if the screw were placed in the stream, it would require to revolve faster than the stream runs, before any thrust upon the screw-shaft could be produced. If, now, we suppose the pier to be a ship, the other circumstances we have specified will not be altered thereby; and it is conceiveable that a screw acting in this dead water, might aid the vessel to stem the current, even though the screw moved with a less velocity than that of the current itself. That the screw will exert some reacting force upon this dead water, even with any speed of rotation, is obvious enough; but whether with a speed inferior to that of the stream, it will produce a sufficient thrust to enable the vessel to stem the current, will depend very much upon the shape of the vessel and the dimensions of the screw employed. If the pitch be fine, and the number of revolutions answering to a given speed of vessel be great, there will be a tendency to pile up the water at the stern, owing to the adhesion of the water to the rapidly revolving blades, and the consequent acquisition of a considerable centrifugal force by the water. When this action occurs, the vessel will be forced forward, to some extent, by the hydrostatic pressure produced by the elevation of the water at the stern, and this pressure will aid the thrust of the screw. If, then, by such an arrangement, a vessel could be made to stem a current, she could obviously, under like conditions, be made to move through still water. All vessels carry a current in their wake, which answers to the dead water in the case of the bridge; and if the screw acts in this current, then the apparent slip will be positive or negative, just as the real slip, or the velocity of the current, may preponderate. In every case, the screw must have some slip relatively with the water in which it acts, but if that water has itself a forward motion, the result cannot be the same as if the water were stationary, and it will be necessary to reckon the forward motion of the current as well as the forward motion of the ship. Thus, if the real slip of the screw be three miles an hour, and the following current runs at the rate of three miles an hour after the ship, then there will appear to be no slip, if the comparison be made with the open ocean on each side of the vessel; or there will appear to be a negative slip, as it is termed, of one mile an hour, if the following current runs at the rate of four miles an hour. The whole perplexity vanishes, if we consider that a current follows the ship at a rate which may be either greater or less than the slip of the screw. This current is confined to the water very close to the ship, so that a log, whether of the ordinary or the patent kind, will not take cognizance of it if thrown over the stern. But if a patent log were to be set in the spot where the screw revolves, it would show the velocity with which the vessel leaves the current, and the real slip of the screw would then be ascertained. It appears not improbable, moreover, that the centrifugal action of the screw, besides piling up the water at the stern, and thus forcing the vessel on with a velocity which may be greater than that of the screw, also causes a current of water to flow radially from the centre of the screw to its circumference; and this stream of water, by intervening between the surface of the screw and the nut of water in which it works, may assist in making the vessel travel faster than the screw itself. In all screw vessels, I believe the slip to be greater than it is commonly reckoned, for in all of them there is a following current in which the screw works; and as, in some cases, the current conspires to make the apparent slip to disappear altogether, so it will, I believe, in every case, reduce the visible slip to a less amount than the real slip, and it is the real slip which it concerns us to determine. There is no benefit derived from the existence of a following current in screw vessels; for to produce the current requires a large expenditure of power; and in screws so proportioned as to produce a negative slip, a worse performance has been obtained, than in cases in which screws producing an apparent slip of 10 to 20 per cent. have been employed.

EXPERIMENTS ON SCREW-PROPELLERS, AND THE PRACTICAL RESULTS.

A number of experiments have been made with the French steam vessel *Pelican*, and their main object was to determine the relations which it is expedient to establish between the diameter, the pitch, the number of blades, and the length of a screw propeller. A subsequent series of experiments upon screws of larger diameter was made on board the same vessel in 1849, and the results then obtained corroborated those which had been arrived at in the previous trials. These several trials were conducted by M. Bourgois, a lieutenant in the French navy, and M. Moll, naval engineer; and in 1850, a committee of the French Institute, composed of MM. Arago, Dupin, Poncelet, Duperrey, and Morin, examined and reported favorably upon the results which had been thus obtained. The *Pelican* is a vessel of 120 nominal horse power, 131 feet long, 22 feet 4 inches wide, and the mean immerged midship section is 109¾ square feet. The displacement of the vessel during the experiments was 258 tons. The engines consist of two vertical oscillating cylinders, with the necessary supplementary apparatus. The diameter of each cylinder is 44 inches, and the length of the stroke is 37.8 inches. The expansion valves were so adjusted as to be capable of cutting off the steam at 0.08, 0.15, 0.30, 0.50, 0.70, and 0.80 of the stroke. The maximum pressure of the steam during the experiments did not exceed 15 lbs. on the square inch above the pressure of the atmosphere. The total weight of the machinery of the *Pelican*, including water in the boilers, was 80 tons. The diamter of the screws tried were 8 feet 2½ inches, 6 feet 10¾ inches, and 5 feet 6 inches; forming thus a geometrical progression, of which the ratio is 1.22.

By comparing the speed of the vessel with the number of revolutions of the screw, the advance of the vessel for each revolution of the screw becomes readily ascertainable, and the ratio of the excess of the pitch over this advance, constitutes a quantity, which MM. Bourgois and

Moll have termed the coefficient of slip. If the screw, instead of acting upon a fluid, worked in a stationary nut, the coefficient of slip woud become nothing; or if the screw were set into revolution when the vessel was at anchor, the coefficient of slip would become equal to 1. If, in like manner, a screw vessel in trying to stem a very strong head wind was driven bodily backward, the advance would become negative, and the coefficient of slip would become greater than unity. It is easy to conceive that the coefficient of slip is a quantity which must exercise a marked influence upon the efficiency of the vessel; and MM. Bourgois and Moll, by taking the relative velocities for the abscissæ of a curve, and the coefficients of slip for the ordinates, have shown that the coefficient of slip increases with the relative velocity, and that consequently the advance diminishes in like manner with this velocity.

In the experiments made with the screw of 66 inches diameter, pitches were tried of 76.18 inches, 92.95 inches, 113.38 inches, 128.3 inches, and 159.7 inches, and with each of these pitches screws were tried of a length answering to the following fragments of the pitch, 0.300, 0.375, 0.450, 0.600, and 0.750, making in all 30 series of experiments for this one diameter. To make the results accruing from these experiments more readily intelligible, they were laid down in a curve, in which the pitches were taken for abscissæ, and the coefficients of slip for ordinates. This comparison was made both in the case of the experiments performed when the resistance of the hull was aggravated by the plane immerged in the water at the bow, and in the case of the hull when unobstructed by this addition; and two series of curves were thus obtained, which showed very clearly that the coefficient of slip diminished with the pitch, and diminished also as the fraction of the pitch increased. In other words, it was thus made plain, what, indeed, had been known before, that there was less slip with a fine pitch, than with a coarse one, and less also when the screw was tolerably long, than when it was very short. The amount of slip, however, is not the only question to be considered in such a case, for the increased friction produced by the expedients which diminish slip may more than compensate for the advantage gained; and in another series of curves, in which the areas of the indicator diagrams were taken as the ordinates, the pitches being still the abscissæ, it was found that the performance increased, both as the pitch diminished, and as the fraction of the pitch diminished.

It appears from these, in common with other experiments upon the screw, that the slip increases the more the immerged area of the midship section exceeds that of the screw's disc. With a given midship section, it is, therefore, advisable to make the screw as large in diameter as possible, consistently with the observance of other conditions. Screws with two blades have a larger slip than screws of four blades constructed of the same length in the direction of the axis, and of the same pitch, and screws of six blades have about the same amount of slip as screws of four blades. *As regards the mere question of slip, therefore, the screws with four blades appeared to be preferable to those with two blades; but as regards efficiency, the screws with two blades appear to be equally effective. The increased slip with a screw of two blades appears to be compensated by its diminished friction.* In employing screws of a larger and larger diameter relatively with the immerged area of midship section, the following consequences are found to ensue: 1st. The efficiency of the engine power in propelling the ship increases. 2d. The ratio of the pitch to the diameter, which produces a maximum effect, goes on increasing. 3d. It becomes proper to employ smaller and smaller fractions of the screw or of the total pitch.

Thus, in the case of the *Pelican*, when fitted with screws of four blades, and of the diameters of 98.42 inches and 66 inches, the results have been found to be in the ratio of 1 to .823. The most advantageous ratio of the pitch to the diameter was found to be 2.2 in the case of the large screw, and 1.384 in the case of the small. Finally, the fraction of the pitch found to be most advantageous was .281 in the case of the large screw, and .450 in the case of the small screw. These results show that there are no absolute proportions of screws which are properly applicable to all vessels alike, but that the proportions and configuration must vary with the form of the vessel, with the draught of water, and with the amount of the engine power to be employed. Screws of two blades, in order that they may realize the same results as screws of four blades, should have a finer pitch, and screws of six blades appear to act very efficiently in the case of large diameters. According to these conditions, a vessel being given, and the area of the midship section being known, so that the limit of the screw's diameter—when taken as large as possible—is determined, the ratio of the square of the screw's diameter to the area of midship section becomes at once ascertainable. Multiplying this ratio by the coefficient K of the resistance of hull, and which MM. Bourgois and Moll take at the mean value of 13.23 lbs. avoirdupois per square métre, or 10.75 square feet English of immerged section, at the speed of one métre, or 3.28 feet per second, they obtain a product which they term the *Relative Resistance*, because it expresses the relative resistances of the vessel and the screw; and putting for abscissæ the values of this quantity in the case of the *Pelican*, and, successively, for ordinates the fractions of the pitch and the values of the ratio of the pitch to the diameter corresponding to the maximum performance, they have constructed curves from which they have deduced, for values equi-distant from the quantity which they have termed the relative resistance, the fraction of the pitch which should be employed, and also the proper proportion of the pitch to the diameter, in the case of screws with two, with four, and with six blades. They have next proceeded to classify the different vessels of the French navy according to the ratio of area of the immerged midship section to the square of the diameter of the screw, and they have thereon deduced the value of the relative resistance for those vessels, and have shown how to determine the pitch, and the fraction of the pitch proper to be employed in each particular case. For war vessels, which require to be capable of proceeding either under sails or under steam, they recommend the use of screws with two blades.

SIZE OF SCREW.

We have to learn, so far as experiment is concerned, what increase in the immersion of a screw will be equivalent to a given enlargement of its diameter, and what alteration in the thrust is produced by a given alteration of the pitch. In the case of superficial screws, however—and these comprise most of the examples which occur in practice—the experiments of MM. Bourgois and Moll enable us, with any given diameter, to specify the best pitch and the best length of screw that can be employed, whether the screw is formed with two, four, or six blades. For, taking K as the resistance per square foot of immersed section, B^2 the area of immersed midship section in square feet, d the diameter of the screw in feet, and p the pitch of the screw in feet, then $\sqrt{\frac{B^2}{K}} = d$, and d multiplied by the ratio of pitch to diameter given in the second column of the following table, will be equal to p.

TABLE OF THE PROPER PROPORTIONS OF SCREW-PROPELLERS.

Class of Screw-Vessel.	Usual relative resistances.	Screws of Two Blades.		Screws of Four Blades.		Screws of Six Blades.	
		Ratio of pitch to diameter.	Fraction of pitch.	Ratio of pitch to diameter.	Fraction of pitch.	Ratio of pitch to diameter.	Fraction of pitch.
———	$K \times 5.5$	1.006	0.454	1.342	0.454	1.677	0.794
———	$K \times 5.0$	1.069	0.428	1.425	0.428	1.771	0.749
———	$K \times 4.5$	1.135	0.402	1.513	0.402	1.891	0.703
Auxiliary line-of-battle ship.	$K \times 4.0$	1.205	0.378	1.607	0.378	2.009	0.661
Auxiliary frigate.	$K \times 3.5$	1.279	0.355	1.705	0.355	2.131	0.621
High-speed line-of-battle ship.	$K \times 3.0$	1.357	0.334	1.810	0.334	2.262	0.585
High-speed frigate. . . .	$K \times 2.5$	1.450	0.313	1.933	0.313	2.416	0.548
High-speed corvette. . . .	$K \times 2.0$	1.560	0.294	2.080	0.294	2.600	0.515
Dispatch-boat.	$K \times 1.5$	1.682	0.275	2.243	0.275	2.804	0.481

Finally, $\frac{p \times \text{fraction of pitch}}{\text{number of blades}}$ = length of screw.

It will be useful to illustrate this question by an example, and I shall take the case of a high-speed corvette, or screw steam packet, of full power. The same mode of procedure is applicable with any other class of vessels, and I shall take the dimensions in English measure. Referring to the table of the proper proportions of screw propellers, it will be seen that the relative resistance of a high-speed corvette is $K \times 2$; or, taking K as unity, the relative resistance may be put as equal to 2. Let the immerged midship section, or B^2, be 416.79 square feet; then $\sqrt{\frac{416.79}{2}}$ = 14.43, which is the diameter of the screw in feet. Supposing a screw with four blades to be employed, then the ratio of pitch to diameter proper for a screw of that kind, when applied to a high-speed corvette, being by the foregoing table 2.080, we have for the proper pitch of the screw, 14.43 × 2.080 = 30.01 feet. The proper length of the screw being the pitch multiplied by the fraction of the pitch, and divided by the number of blades, and the proper fraction of pitch for a screw of this kind being by the table .294, we have 30.01 multiplied by .294, and divided by 4, which gives a length of 2.2047 feet. By a similar procedure there will be no difficulty in fixing, for any given class or size of screw vessel, the proportions of screw which will give a maximum performance, whether the screw selected be one of two, of four or six blades. The main difference which will be produced by increasing the number of blades, will be to diminish the speed of the engine; for a coarse pitch is proper for a screw of many blades, and a coarse pitch involves a slow engine, in order that there may not be an injurious amount of slip. In cases, therefore, in which it is apprehended that the engine, if coupled immediately to the screw shaft, would move with an inconvenient velocity, there are two alternatives which present themselves whereby the velocity may be diminished. The one is the use of gearing; the other is the employment of a screw of several blades and of a coarse pitch. The latter expedient appears to be the preferable one, except in cases in which the screw has to be drawn up through a trunk, and under those conditions screws of two blades must of course be employed.

MM. Bourgois and Moll, describing the manner in which the experiments were made, and discussing fully the nature of the results which had been obtained, gave one of the most elaborate investigations of the action of the screw which has yet appeared, and upon the whole is considered a safe and useful guide in practice, and it will be useful, therefore, to recapitulate here such of the leading topics of its information as may be available for the more complete elucidation of the subject.

The object, in experimenting upon the *Pelican*, was to determine the specific efficiency of all kinds of screw propellers in vessels of every size, proceeding at every speed, and under all circumstances of wind and sea, to the end that the particular species of propeller most proper for a given vessel might be readily specified. It was also a leading object to determine the value of the revolving force that it was necessary to bring to act upon the screw-shaft to make the screw perform a determinate number of revolutions in a given time, supposing, of course, that the form of the vessel was known as well as the dimensions and form of the propeller; or rather, having once determined the law of the revolving force in functions of the number of revolutions, to assign the value of the power consumed in a single revolution per unit of time, and the solution of this double problem evidently involves the elucidation of the question in all its generality.

By the term utilization, or efficiency, is meant the ratio of useful effect to the power transmitted by the engine to the screw-shaft, or, in other words, it is the ratio of the engine power to the aggregate resistance, multiplied by the distance through which the vessel passes. This, therefore, is the same as the ratio of the indicator power to the dynamometer power. The value of this ratio, as has been seen by the experiments made with vessels to which a dynamometer has been applied, depends not merely on the proportions of the screw, but also on the size and form of the vessel, and upon the action of the winds and sea.

By elementary resistances, are meant the resistances of two hulls per unit of speed, both brought into relation to the respective speeds of the two hulls; and in order to make the laws ascertained to exist in the case of the *Pelican* applicable to any vessel of a different form, it is only necessary that in both vessels the ratio of the elementary resistance of the hull to the area of the screw's disc, or the square of the screw's diameter, should be the same. It is obvious that if a vessel be propelled with twice the difficulty, she should have twice the propelling area in the screw's disc, or in the square of the screw's diameter, if the same velocity has to be maintained; and the elementary resistance of the hull is merely an expression of the ease or difficulty with which the vessel is propelled.

The ratio of the square of the diameter of the screw to the elementary resistance of the hull, gives the relative resistance of the screw and hull, and this is a quantity which enters largely into the subsequent investigations, and the precise nature of which it is necessary to apprehend. If a vessel of a given form, and with a given number of square feet of immerged section, be propelled through the water at a given speed, there will be a certain resistance per square foot of immersed section, which is ascertainable. A blunter vessel, if driven by the same power, at the same speed, must have a less immersed section; she will, therefore, present more resistance per square foot of immersed section, while a sharper vessel will present less. If, therefore, these several vessels be driven at the same speed, with the same power, and with the same screw, then as the screw exerts the same pressure or thrust in each case, and remains unchanged in diameter, the ratio of the square of the screw's diameter to the total area of immersed section must be different in each of the vessels, and this ratio is indicative of the relative resistance of the screw and ship. The thrust of the screw will always be just balanced by the resistance of the ship.

Having the immerged midship section of the *Pelican* taken at 109¾ square feet—which immersion appears to have been maintained throughout the experiments with little variation—

and the diameters of the screws tried, at 66.14 inches, 80.7 inches, and 98.4 inches, or 5.51 feet, 6.72 feet, and 8.2 feet, with the common geometrical ratio of 1.22, then the different relative resistances corresponding to the series of diameters will be expressed by

$$K \times \frac{109.75}{(5.51)^2}, \qquad K \times \frac{109.75}{(5.51)^2(1.22)^2}, \qquad K \times \frac{109.75}{(5.51)^2(1.22)^4},$$

K being the value of the resistance of a square foot of immersed section of the vessel at a speed of 3.2808 feet per second. It is obvious from these figures that the relative resistances are as the numbers $(1.22)^4$, $(1.22)^2$, and 1; or, as 2.21533456, 1.4884, and 1.000; or, say, as 2.215, 1.488 and 1.000.

The efficiency being the ratio of the useful effect to the amount of engine power transmitted to the screw, and the useful effect being nothing more than the resistance of the vessel multiplied by the space through which she passes, or, in other words, being just the dynamometer power, it is easy, when the engine and dynamometer powers are known, to tell what the efficiency is, whatever may be the speed of the vessel. If we knew with precision the law of the increase in the resistance which a vessel experiences when her speed is increased, we should be able, by knowing the resistance at any one speed, to tell what it would be at any other, and also what would be the useful effect at that increased speed. Thus, if it were the fact that the resistance increased as the square of the velocity, then a constant elementary resistance multiplied by the cube of the speed would give the useful effect at the increased velocity; and for speeds differing but little from one another, this mode of computation may be adopted. In the case of dissimilar speeds, however, such a method of estimation will give fallacious results, since it is known that the resistance of vessels increases more rapidly than the square of the velocity, in the case of considerable speeds. Thus, when the speed of a screw propeller was increased from 6½ to 9½ knots per hour, or about one-third, the resistance rose, not as the square, but as the 2.28th power; so that, calling B^2 the immerged section of the vessel, V the velocity of headway, and K the resistance of a square foot of immerged section at the speed V, it will only be permissible to take the expression KB^2V^3, as representative of the resistance, on the understanding that K varies in functions of the speed according to the approximate law, $K = K' \times V^{0.28}$, where K' is a constant coefficient. In the following investigations, it will not be necessary to adopt this notation, the formula KB^2V^3 having the advantage of superior simplicity, and being, moreover, sufficiently exact in the case of similar speeds. But in the case of speeds of a different order, it is necessary to understand that the ordinary formula does not give exact results; and when employed in connection with such speeds, therefore, a correction of the nature here indicated must be applied. If we put h to denote the pitch of the screw, γ the coefficient of recoil or of slip, and n the number of revolutions of the screw in a unit of time, then the expression KB^2V^3 may, it is clear, be put under the form $KB^2(1-\gamma)^3h^3n^3$, or under the form $KB^2(a^3n^3)$, supposing a to be understood to represent the distance through which the vessel is advanced through the water by each revolution of the propeller. The area of the immersed midship section B^2, is readily ascertainable when the draught of water is known; and the advance a, the coefficient of slip γ, the speed of the vessel through the water V, and the number of revolutions n, are all to be determined by experiment. As regards the coefficient K, its value for different speeds was fixed approximately by the aid of experiments directed to that object;

but its value in functions of the speed is also deducible from the slip, and the efficiency of ratio of the dynamometer to the indicator power.

By the absolute speed of the vessel, is meant the speed over the ground, determined by dividing the length of each run by the time of its duration. From the absolute speed the relative speed, or speed through the water, may be derived, by adding or subtracting the velocity of the current; and as this speed raised to the third power enters into the formula which expresses the efficiency of a propeller, it is highly important that it should be accurately determined, since any error which exists in the determination of this element will be multiplied by the subsequent operations.

The precaution taken to insure accuracy in determining the speed of the *Pelican*, is to make four runs with the whole power of the engines acting, so as to give the full speed; four runs with only one boiler in operation, if two are applied for the engines, so as to produce a reduced or medium speed; four runs with only one boiler, and with the steam throttled, or rather worked with much expansion, so as to produce a low speed; and four runs with one boiler, and with the engines using all the steam which that boiler produced, but with a board or plane of 13.988 square feet lowered into the water at the bow, and fixed across the stem, so as purposely to increase the resistance of the vessel. The mean values of the speed thus obtained were, with the simple hull, 9.5, 7.7, and 6.5 knots; and if N be the number of strokes made by the engine during the run, t the duration of the run in seconds, and r the ratio of the gearing wheels, then the number of revolutions, n, made by the screw in a second will be represented by the formula $n = \frac{Nr}{t}$.

To determine the speed of the vessel through the water, from the speed over the ground, the vessel running with and against the current, and if a be the advance of the screw through the water made by each revolution, then it is clear that the value of this quantity will remain unchanged, or nearly so, whether the vessel proceeds with the current or against it, so long as there is an absence of wind. Putting, then, U, $U + a$, and $U + 2a$, to represent the mean speeds of the current during each succeeding run, and n, n', and n'', the corresponding number of revolutions of the screw per second, then na, $n'a$, and $n''a$ will represent the speed of the vessel through the water during each of the runs in question; and if we designate the absolute speed, or the speed over the ground in each of the runs, by v, v', and v'', we shall have, $v = na \pm U$, $v' = n'a \pm U \pm a$, and $v'' = n''a \pm U \pm 2a$. We hence find the value of a to be expressed by the equation, $a = \frac{v + 2v' + v''}{n + 2n' + n''}$; and when the distance that the screw advances through the water by each revolution is known, it is easy to tell the speed of the vessel through the water per second, or per hour, by multiplying the advance by the number of revolutions per second or per hour made by the screw.

These conclusions are only quite accurate under the supposition that the vessel is sailing in a perfect calm; but vessels are generally more or less affected in their speed by the influence of the wind, and this influence will not be properly eliminated in an experimental trial in running the vessel first before the wind, and then against it, nor will it often happen that the current and the wind proceed in the same direction. MM. Bourgois and Moll have made experiments to determine the effect of winds of different strengths upon the progress of vessels, and also upon the coefficients of slip; and by the aid of these experiments, they have constructed a table which gives the corrections, in functions of the speed, proper to be applied to the coefficient of slip in the case of vessels proceeding at full speed, and aided or opposed by winds of different strength. These corrections are as follows: With the winds on end, or at two points of deviation from that direction, the correction proper for a strong breeze is, 0.024; for a pretty strong breeze, 0.021; for a light breeze, 0.018; for a gentle or feeble breeze, 0.012; and for light airs, 0.006. With the wind at six points of deviation from the course of the vessel, the corrections are just half the foregoing; and for a deviation of four points the corrections are intermediate between the other two. Besides the corrections which have reference to the influence of the wind, there are two others to be taken into account, and they relate to the immersion and speed of the vessel. The mean draught of water of 8.2 feet, which corresponds with an immerged section of 109.75 square feet, was adopted as the normal immersion of the *Pelican* during the experiments.

From the general tenor of the experiments upon the *Pelican*, it appears that with every kind of screw, the difference in calm weather between the coefficients of slip at speeds of 9.5 knots and 6.5 knots does not exceed 0.03. A correction, therefore, of 0.01 per knot, in the case of such speeds as 9.5 knots per hour—which is adopted as the normal full speed in these experiments—will be a near approximation to the truth; and taking the coefficient of slip as indicated in the following tables, in the case of the experiments made at full speed, and in the case also of those made with the resisting plane, the several corrections proper for the force of the wind, the immersion, and the speed have been applied to them. The slip is the same in the case of all vessels of similar forms, and which have similar screws and a resistance of hull proportional to the squares of the diameters of the several screws. By similar screws, is meant screws which have the same number of blades, the same fraction of pitch, and the same ratio of the pitch to the diameter; and similar forms of hull and similar screws are merely hulls and screws constructed on a different scale from the original, but in all other respects the same. Taking B^2 as the immersed midship section of the vessel, and D as the diameter of the screw, then $\frac{B^2}{D^2}$ will express the ratio of the immersed section to the square of the screw's diameter. Substituting for this expression, the letter b, then Kb will represent the ratio between the resistance of the hull per unit of speed and the square of the screw's diameter; or, in other words, it will represent the relative resistance already referred to. Putting R for the resistance of the hull at the speed under consideration, then K being the coefficient of the resistance of the hull, we have, $K = \frac{R}{B^2 V^2}$. But K, according to the experiments upon the resistance already recited, varies slightly with the speed, and therefore, putting $Kb = \frac{KB^2}{D^2}$ for the same screw, it is clear that the quantity Kb, and the slip which answers thereto, will vary also, and will in fact increase with the speed. The amount of increase, however, is not considerable, and is scarcely appreciable in speeds under 7 knots; and it appears probable that the increased proportion of slip at high speeds is due altogether to the increase in the resistance beyond that which the common law supposes, or in other words, to the increase in the exponent of the speed which answers to the resistance in the case of high speeds. Thus, while with the simple hull the difference in the slip at a speed of 6.5 knots, and at a speed of 9.5 knots, is 0.03, as has already been explained, the difference in the slip, if the comparison be made between the simple hull and the hull with the resisting plane, is, with equal speeds, from 0.10 to 0.11, and this quantity varies but little with the pitch of the screw. Bearing then in mind that the resisting plane occasions an increase of resistance of 0.5, while the vessel passes from the speed of 6.5 knots to 9.5 knots, we may discover by interpolation the ratio of the increase of the coefficient of resistance which shall produce an increased slip of 0.03, and that ratio we shall find to be 1.11. If, then, we call V' the high speed of the vessel, and V^2 the low speed, and R' and R the respective resistances of the hull, we have,

$$R = KB^2V^2, \qquad \text{and} \qquad R' = 1.11KB^2V'^2.$$

But $1.11 = \left(\frac{V'}{V}\right)^{0.28}$ from whence it would appear, that while the vessel passes from the low speed to the high speed, the resistance varies as the 2.28 power of the speed. This result, deduced from the experiments upon the *Pelican*, presents also a satisfactory agreement with the results of experiments made upon a smaller boat, when brought into relation to the speeds corresponding to the difference of dimensions.

POWER OF THE ENGINE, SCREW, AND RESISTANCE OF THE SHIP.

The power of the engines of the *Pelican* was ascertained by means of an indicator which was applied both to the top and bottom of the cylinders. The mode of reckoning the power exerted by the engines, was not the same as is usually adopted in this country, but the measures employed are nevertheless convertible. In the United States and England the practice is to take the mean pressure in pounds per square inch exhibited by the Indicator, less a pound and a half, which is deducted as an allowance for the power consumed in overcoming the friction of the engine itself; then multiplying the residual pressure by the number of square inches in the area of both pistons, by the number of feet travelled by the piston in the minute, and dividing by 33,000, and we have the measure of the dynamical effort of the engine in actual horse power.

In the French experiments the pressure upon the piston is taken in centimètres of mercury; and a centimètre is .39371, or somewhat more than one third of an English inch, so that a square centimètre is .155 of an English square inch. Now a column of mercury, an inch square and an inch high, weighs, at 60°, about .491 lbs.; hence, a column of mercury, a centimètre square and an inch high, will weigh .155 of this amount, or .0761 lbs.; and a column of mercury, a centimètre square and a centimètre high, will weigh .39371 of this last amount, or .02996 lbs. A kilogramme is 2.2055 lbs. *avoirdupois*, so that a cubic centimètre of mercury will weigh .01358 kilogrammes. A linear mètre contains 100 centimètres, therefore, a square mètre contains 10,000 square centimètres; and 10,000 times .01358 kilogrammes, or 135.8 kilogrammes, will be the pressure exerted on a mètre of surface by a column of mercury a centimètre high.

If now we call D the diameter of the cylinder in mètres, C the stroke of the piston in mètres, N the number of strokes made per minute, and p the mean pressure exerted upon the piston in centimètres of mercury; then, bearing in mind that there are two cylinders, and that the stroke of the piston both ways has to be reckoned, the power exerted, measured in kilogrammes, raised one mètre high in the minute, will be represented by

$$135.8 \times 2D^2 \times .7854 \times 2C \times N \times p; \quad \text{or, in other words,} \quad 426.6D^2CNp.$$

The power exerted per second will be $\frac{1}{60}$th of this, or $7.11D^2CNp$; and MM. Bourgois and Moll have adopted the expression, $7.117D^2CNp$, to represent the gross power exerted by the engine per second in kilogrammètres, or in kilogrammes raised one mètre high. From this, however, some deduction has to be made for friction, before it can express the power transmitted to the screw; and it is clear that the friction consists of two parts, of which one is nearly constant, whatever be the pressure put upon the piston, while the other varies with the amount of strain transmitted through the working parts. For, overcoming the constant friction of the engine, MM. Bourgois and Moll, from some experiments they made, consider that 5 centimètres of mercury, or .9666 lbs. per square inch, must be accepted as a sufficient allowance; while the friction, which varies with the strain, is designated by the coefficient A. The power actually operative in turning round the screw shaft will, therefore, be expressed by the formula, $A \times 7.117D^2CN(p-5)$, and this quantity will always be proportional to the quantity $7.117D^2CN(p-5)$, so that the latter expression will serve as a measure of the power transmitted to the screw. This, then, gives an expression of the power exerted by the engine. The power utilized by the ship, or, in other words, the dynamometer power, is approximately represented by the expression,

$$KB^2V^3 = KB^2a^3n^3 = KB^2(1-\gamma)^3h^3n^3;$$

K, being the resistance in kilogrammes per square mètre of the immerged section of the vessel, at a speed of one mètre per second; B^2, the immerged midship section in the square mètres; V, the speed of the vessel through the water in mètres per second; a, the advance of the screw through the water per revolution; n, the number of revolutions of the screw per second; h, the pitch; and γ, the slip of the screw. The utilization, therefore, u, or in other words, the ratio of the dynamometer to the indicator power, will be represented by the expression,

$$u = \frac{KB^2V^3}{7.117CN(p-5)} = \frac{KB^2a^3n^3}{7.117CN(p-5)} = \frac{KB^2(1-\gamma)^3h^3n^3}{7.117D^2CN(p-5)};$$

and calling r the ratio of the gearing wheels interposed between the screw shaft and the engine, we may put the expression of the utilization, or ratio of dynamometer to indicator power, under the following forms:

$$u = \frac{KB^2r^3}{7.117D^2C \times 60^3} \times \frac{N^2}{(p-5)} \times a^3 \quad \ldots\ldots\ldots \quad (1.)$$

$$u = \frac{KB^2}{7.117D^2C \times 60} \times \frac{n^2}{\left(\frac{p-5}{r}\right)} \times a^3 \quad \ldots\ldots\ldots \quad (2.)$$

$$u = \frac{KB^2}{7.117D^2C \times 60} \times \frac{1}{\frac{\left(\frac{p-5}{r}\right)}{n^2}} \times a^3 \quad \ldots\ldots\ldots \quad (3.)$$

The efficiency of all the screws tried in the *Pelican*, has been calculated by MM. Bourgois and Moll by the aid of these formulæ, and especially by the aid of Formula (1). When the engines are connected immediately to the shaft, the effect is obviously the same as if the gearing wheels were of equal diameter. The ratio of the gearing would, in such a case, be expressed by the fraction $\frac{1}{1}$, which is the same as 1, so that the ratio of the gearing would spontaneously disappear in the formulæ, since the multiplication or division of the other quantities by 1 would not in any way alter their value. The whole of the quantities entering into the composition of these formulæ have been derived from direct experiment, except the quantity K; and as that quantity is involved in some uncertainty, so far as the experiments in the *Pelican* are concerned, it will be useful to investigate the values of the other quantities comprehended in the formulæ, independently of K, so that they may continue to be applicable even if a new value for K be finally adopted.

Now if, as appears by Formula (1),

$$u = \frac{KB^2 r^3}{7.117 D^2 C \times 60^3} \times \frac{N^2}{(p-5)} \times a^3;$$

then,

$$\frac{u}{K} = \frac{B^2 r^3}{7.117 D^2 C \times 60^3} \times \frac{N^2}{(p-5)} \times a^3;$$

and putting u' for $\frac{u}{K}$, we have,

$$u' = \frac{B^2 r^3}{7.117 D^2 C \times 60^3} \times \frac{N^2}{(p-5)} \times a^3;$$

or, by Formula (8),

$$u' = \frac{B^2}{7.117 D^2 C \times 60} \times \frac{1}{\left(\frac{\left(\frac{p-5}{r}\right)}{n^2}\right)} \times a^3.$$

Now, it has been already explained that $p-5$, with the addition of a constant factor, expresses the mean effective pressure, in centimètres of mercury, of the steam urging the piston, and that it is permissible, therefore, to take $p-5$ as representative of the mean effective pressure exerted by the engine. If, then, the screw travels twice, three times, or any other number of times faster than the engine, it is plain that, by connecting the engine immediately to the screw shaft, a half, a third, or other fraction of the pressure corresponding to the new speed of piston, would impart the same power as before to the screw. If r be the ratio of the gearing, it is clear that, neglecting the constant factor already mentioned, the fraction $\frac{p-5}{r}$ will represent either the reduced pressure, which, when applied directly to the screw shaft, will suffice to give the same power, or it will represent the diminished power expended in producing a revolution of the screw compared with that necessary to produce a revolution of the engine. The fraction, therefore, $\frac{\left(\frac{p-5}{r}\right)}{n^2}$, will, still neglecting the constant factor, represent the effective pressure acting on the screw shaft, divided by the square of the number of revolutions of the screw. Now in paddle vessels it has been found that the pressure in the cylinders increases as the square of the number of revolutions, supposing of course that the immersion remains invariable, and that the vessel is not affected by the wind. The experiments in the *Pelican* show that this law also obtains in the case of screw vessels, and it hence follows that the fraction $\frac{\left(\frac{p-5}{r}\right)}{n^2}$ will express the effective pressure necessary to be employed in cylinders coupled immediately to the screw shaft, in order to cause the shaft to make one revolution per second; but the result has to be corrected by a constant factor, the value of which depends on the size of the screw, and the size and shape of the vessel. It is time, however, to give some of the promised citations from the Memoir of MM. Bourgois and Moll, to which the foregoing remarks have reference; but as this production stands greatly in need of compression, I cannot undertake to give in all cases a literal translation, but shall abridge the verbiage as much as seems practicable without obscuring the sense:

"ELEMENTARY STROKE OF ROTATION.—The abstract value of the abstract ratio $\frac{\left(\frac{p-5}{r}\right)}{n^2}$, which we shall henceforth, for more simplicity, call P_e, expresses, with the exception of a constant factor—which depends solely on the proportions of the screw, and the size and shape of the vessel—the effective pressure requisite to be exerted in the cylinders, when coupled immediately to the screw shaft, to produce one revolution of the screw in the second. As regards the elementary stroke of rotation capable of producing this speed of revolution of one turn of the screw per second, and which we shall denote by C_e, it would be determined by the relation,

$$C_e = A \times 1.0330 \times \frac{D^2C}{2 \times 76} \times P_e;$$

or putting, $A_1 = A \times \left(\frac{1.0330}{2 \times 76}\right)$; then, $C_e = A_1D^2CP_e$,

the kilogramme and the mètre being taken as the measures of the forces, and A being the general coefficient of the reduction of the power of the engine, as before explained. It is easy now to see that the complete resolution of the problem rests substantially on the determination of the values of the second members of the equations,

$$u = Ku', \qquad \text{and} \qquad C_e = A_1D^2CP_e;$$

and principally on the determination of the values of

$$u' = \frac{B^2}{7.117D^2C \times 60} \times \frac{1}{P_e} \times a^3, \qquad \text{and} \qquad P_e = \frac{\left(\frac{p-5}{r}\right)}{n^2}.$$

The precise values of K and A, though interesting to ascertain, are of inferior importance to the determination of the values of u and C_e, and in the *Pelican*, the elements from which the values of u and C_e were computed remained without alteration in the several experiments, so that the relative results must be correct, even if we suppose the absolute values to be in some degree uncertain.

"UNKNOWN QUANTITIES OF THE PROBLEM.—Besides the two principal unknown quantities u' and P_s, we have called in the aid of an auxiliary unknown quantity, the intervention of which, if not indispensable, is at least of much utility in facilitating some of the calculations. The coefficient of slip has already been expressed by γ, and the combination of γ and P_s leads immediately to the value of u'; so that if we discover the practical and rational formulæ, or in default of this, if we draw a series of curves, to express γ and P_s in functions of the defined variables on which those quantities depend, we shall have discovered the law of the efficiency, and consequently will be able to specify the circumstances which make the efficiency a maximum.

"GENERAL ESTIMATE OF THE RESULTS OBTAINED.—It will be apparent that the value of the expression $u = Ku'$, or of the coefficient of efficiency u', diminishes as the speed increases. This is inevitable from the increase, in a higher ratio than the square of the speed, of the coefficient of resistance K; for, with an aggravated resistance, there is an increased slip, and consequently an inferior efficiency. This increase in the resistance, in a more rapid ratio than as the square of the speed, is imputable to the difference in the level of the water at the bow and stern, whereby the vessel is forced back or resisted by a hydrostatic pressure. If our attention be turned to the values of P_s with the three different speeds at which each screw has been tried, we shall find that those values remain constant in each case in which the experiment has been made during a calm. The wind will of course either diminish or increase the resistance of the hull, as its direction may be favorable or adverse; and this effect will be the more conspicuous at the low speeds.

"CURVES OF THE COEFFICIENT OF SLIP AND EFFICIENCY.—In investigating the values of the coefficients of slip and of efficiency, we have represented the results obtained with the screws of the diameter of 5.51 feet English, or 1.68 mètres, by two clusters of curves, the one relating to the experiments with the simple hull, and the other to the experiments with the resisting plane. They have all the pitches for abscissæ; and in the case of the screws with four blades, three series of curves have been traced for each of the fractions of pitch, 0.30, 0.375, 0.45, 0.60, and 0.75. The ordinates of the first series represent the coefficients of slip cleared from all disturbing influences; the ordinates of the second series represent the values of $\frac{\left(\frac{p-5}{r}\right)}{n^2}$; and the ordinates of the third series represent the values of the efficiency already designated by u'. This last curve is naturally the product of the other two. The same mode of procedure has been adopted in the case of the screws with two blades, but only with the fractions of the pitch 0.30 and 0.45. The elements of the curves, which refer to the simple hull, have been brought to the speed of 9.5 knots, in preference to the medium or the low speed, so as to reduce as much as possible the disturbing influence of the wind, and of the irregular variations of the current. The maximum efficiencies, in the case of the simple hull, which answer to different fractions of the pitch, with screws of four and of two blades, are as follows:

"SCREWS WITH FOUR BLADES, DIAMETER 5.51 ENGLISH FEET.

	PITCH IN FEET.	FRACTION OF PITCH.	RATIO OF PITCH TO DIAMETER.	EFFICIENCY, OR UTILIZATION.
Simple hull	9.012	0.300	1.035	$K \times 0.07495$
	9.606	0.375	1.743	$K \times 0.07485$
	10.154	0.450	1.842	$K \times 0.07465$
	10.942	0.600	1.985	$K \times 0.07405$
	11.394	0.750	2.067	$K \times 0.07350$

"SCREWS WITH TWO BLADES, DIAMETER 5.51 ENGLISH FEET.

	PITCH IN FEET.	FRACTION OF PITCH.	RATIO OF PITCH TO DIAMETER.	EFFICIENCY, OR UTILIZATION.
Simple hull	8.376	0.300	1.520	$K \times 0.07500$
	9.311	0.450	1.689	$K \times 0.07690$

"It would appear, from these figures, that screws of two blades are a little more efficient than screws of four blades; but at the time the experiments upon the two-bladed screws were made, the engine was working with somewhat less friction than when the experiments upon the four-bladed screws were made. The two kinds of screws, therefore, must be accepted as about equal in efficiency, but the screws with two blades should be made with a somewhat shorter pitch.

"If we bring a straight line parallel to the axis of the abscissæ or pitches, but distant therefrom 0.97 of the ordinate, which represents the maximum efficiency in the case of screws with four and with two blades, and whether combined with the simple hull or with the addition of the resisting plane, it is clear that the intersection of this straight line with the curves of efficiency for each fraction of the pitch will answer to an equal number of accurate solutions of the problem. Each of the curves of efficiency which rises high enough to be met by a straight line, will furnish two intersections, and, consequently, two values for every fraction of pitch interpolated between those of the experiments. We are thus led to two sets of solutions that are equally satisfactory—the one with longer, and the other with shorter pitches. The constituents of each series are as follows:

"SCREWS WITH FOUR BLADES, DIAMETER 5.51 ENGLISH FEET.

SIMPLE HULL.	PITCH IN FEET.	RATIO OF PITCH TO DIAMETER.	FRACTION OF PITCH.	SLIP.
1st series, shortened pitches	6.045	1.097	0.300	0.2351
	6.773	1.229	0.375	0.2485
	7.865	1.427	0.450	0.2710
	8.970	1.628	0.600	0.2865
	9.800	1.779	0.750	0.2960
2d series, elongated pitches	11.152	2.024	0.300	0.3885
	11.555	2.097	0.375	0.3810
	11.890	2.158	0.450	0.3735
	12.283	2.229	0.600	0.3630
	12.447	2.259	0.750	0.3550

"SCREWS WITH TWO BLADES, DIAMETER 5.51 ENGLISH FEET.

SIMPLE HULL.	PITCH IN FEET.	RATIO OF PITCH TO DIAMETER.	FRACTION OF PITCH.	SLIP.
1st series, shortened pitches	7.068	1.283	0.300	0.2905
	6.461	1.173	0.450	0.2445
2d series, elongated pitches	9.167	1.664	0.300	0.3575
	11.266	2.045	0.450	0.3825

"These figures show, that with the same fraction of pitch, and the same relative resistance, the ratio of the pitch to the diameter ought to increase when we pass from a screw of two blades to a screw of four blades; and this rule also holds in passing from a screw of four blades to a screw of six."

RIVER-BOAT BEAM ENGINE OF THE STEAMBOAT FRANCIS SKIDDY.

DESCRIPTION AND DRAUGHTING.

The engine of the *Francis Skiddy* was designed and constructed by Mr. Joseph Belknap, now of the Neptune Iron Works, of New York, and the hull built by Mr. George Colyer. The following are the dimensions of her engine and boat:

Four iron boilers, extreme length of each . .	. 28 feet 1 inch.
Diameter of shell	. 8 "
Whole amount of fire surface for the 4 boilers .	. 5132 square feet.
Whole amount of grate surface " " .	. 208 "
Area of lower flues for one boiler . .	. 7.48 "
Area of upper flues " " . .	. 6.97 "
Area of chimney " " . .	. 9.61 "

She is provided with a single-beam engine.

Diameter of cylinder	5 feet 10 inches.
Length of stroke of piston	14 "
Diameter of air-pump	3 " 10 "
Length of stroke of air-pump	4 " 4 "

With old hull, running as passenger boat.

Length on deck	322 "
Breadth of beam	38 "
Depth of hold	10 " 4 "
Tonnage	1235 tons.
Average draught of water without load	7 feet.
Average draught of water when loaded	7 feet 6 inches.
Pressure of steam in boiler	45 pounds.
Cutting off steam from the commencement of the stroke of piston	7 feet 6 inches.
Mean pressure of steam in cylinder	39.1 pounds.
Average number of revolutions at the highest speed	21½ per minute.
Greatest speed of the *Francis Skiddy* through the water	22$\frac{8}{10}$ miles per hour
Consumption of anthracite coal per hour, upon a square foot of grate surface, at the highest speed	26 pounds.
Consumption of anthracite coal at an average speed	23 "
Consumption of anthracite coal for the whole trip, from New York to Albany and back, at the highest speed, running the blower to supply the wind for the furnaces	32 tons.
Consumption of anthracite coal at an average speed for the whole trip	25 tons.
Diameter of paddle-wheel	40 feet.
Length of paddles	11 "
Depth of paddles	3 "
Number of paddles	28

With enlarged hull, 14 ft., as a freight boat.

Average draught of water without load	5 " 4 inches.
Average draught of water loaded	6 " 4 "
Pressure of steam in boiler	33 pounds.
Mean pressure of steam in cylinder	30.92 "
Average number of revolutions	16 per minute.
Cutting off steam from the commencement of stroke of piston	9 feet.
Consumption of anthracite coal per hour, upon a square foot of grate surface	11.1 pounds.
Consumption of anthracite coal for the whole trip, from New York to Albany and back	23 tons.
Diameter of paddle-wheel	41 feet 4 inches.
Average speed of the *Francis Skiddy* per hour, with enlarged hull	13$\frac{3}{10}$ miles.
Tonnage	1500 tons.

ENGINE OF THE FRANCIS SKIDDY.

Engines of the character of that of the *Francis Skiddy* are no novelty to engineers in the Eastern States and in the vicinity of the Great Lakes. The mechanical construction of the engine is not intricate, and we refer to it as a practical illustration of the proper proportions in the several parts, developing the greatest amount of propelling power with the use of the least weight of metal. These considerations involve, first, the proper application of the urging force upon the piston at one end of the working beam, and the transfer of this force or power at the other end of the beam to the crank and the paddle-wheel. Referring our readers to pages 136 and 169, for the speed of the *Francis Skiddy* as a freight and-passenger boat respectively, we shall now describe generally the engine with reference to the operation of the valves.

The engine is of that description called a condensing engine. The steam having entered the cylinder, A, and forced the piston from the commencement to the end of the stroke, escapes at once through the lifting of the exhaust valves, and rushes with great velocity through the steam chests, A^6 and A^7, and escape pipe, A^8, on the left, into the condenser, A^1, where the steam, by the injection of water passing through the injection pipe, A^{10}, is condensed into water at a temperature of about 100° or 120° Fahr. (see page 76, explaining condensation of steam). The condensed steam and injected water now collects in the hollow bed-plate, A^4, and is drawn off by the air-pump, B^0, to be raised into the hotwell case, B, whence it is forced to run into the ocean or river, as the floating top placed upon the air-pump prevents it running back into the air-pump. The valves which regulate the admission and liberation of steam are placed within the steam chests, A^6 and A^7, and are termed steam and exhaust valves. Sectional drawings of them are shown in Plate XVIII., Figs. 4, 5, 6, and 7. In explaining below the mechanism for the lifting of the exhaust valves, the same mode is applied to the steam valves, and will therefore be appropriate to the movement of the latter as well. The exhaust valves are raised from their seat by the movement of the eccentric, F, in connection with the eccentric rod, F^0F^0, and F^3; the latter having on its end the eccentric hook, F^8, to cause, through the revolving motion of the eccentric and through the medium of the rocking lever, the rocking motion of the rock shaft, F^B, on which is secured by a key the exhaust wiper, F^6, raising and lowering the toe, F^7, and the lifting rod, I. These two are secured fast to each other by two bolts. The lifter, F^9, secured by two bolts near the upper end of the lifting rod, I, has an arm which extends to the valve stems, F^{12}. The valve stems being made fast to the valves, raise and lower the valves by the lifting rods as required, at the proper moment.

The cut-off arrangement applied to this engine, and which we find in general use in river and lake-boat beam engines, is the invention of Messrs. R. L. and F. B. Stevens, of Hoboken. Its principal feature is the use of long arms secured to the rock shaft, so arranged as to use only a portion of the rocking motion of these arms in lifting and lowering the toes attached to the lifting rods; giving, at the extremities of the arms, in consequence of their length, a very quick action to the lifting rods, permitting the steam to be cut off at any desired point in the stroke of the piston.

The lead of the steam valves on this engine is $\frac{3}{16}$ of an inch, or in other words the steam valves are raised $\frac{3}{16}$ of an inch from their seats when the piston is on the upper or lower termination of the stroke. The whole lift of the steam valves from their seats is five inches. The lead of the exhaust valves is 1½ inch when the piston is on the upper or lower termination of the stroke. The whole lift of the exhaust valves from their seats is six inches.

It is common, in slide-valve engines, to close the passages by what is termed the "lap" of the valve, before the termination of the stroke of the piston. The term "lap" is familiar to all steam engineers, as denoting those portions or edges of the working faces of the valves which extend past or beyond the ports. There is not, and can not be any "lap" on a poppet valve, but the same effect is obtained in engines of this character by so adjusting the eccentrics and the lifting toes, that the valves will be lowered into their seats at the right period, or at the time when the passages require to be closed, to produce the effect. The early closing of the steam valves allows the steam to expand itself, and thereby to work with economy in the cylinder. The early closing of the exhaust valves prevents the escape of the weak steam which is before the piston, and compresses it. By closing the exhaust valves very early, this compression may be carried to such an extent as to induce a pressure equal, or even superior, to that in the boiler. It is never desirable, in engines of this character, to "compress" to this extent, but by judiciously closing the exhaust valves, and producing a gentle compression, an effect somewhat equivalent to "lead" is produced. In some engines, this compression is substituted for lead altogether, and the piston, at the moment of commencing its stroke, is only actuated by the steam which has been thus imprisoned. This mode of operating, works very economically, because it tends to save the steam which would otherwise be required to fill the valve passages, and the little space at the end of the cylinder, generally known as the "clearance." In other words, the steam, on flowing in at the commencement of the stroke, usually has to fill up a certain amount of empty space at the end of the cylinder, before it begins to act with its proper effect upon the piston; but when this space has been previously filled, to some extent, by the compression above described, a smaller quantity of steam is required to be drawn from the boiler for this purpose. In the Corliss engine, and some others in which the valves are opened and closed very rapidly, this mode of working is almost invariably practiced.

In the *Francis Skiddy*, the exhaust valves are just closed when the piston is about ten inches from the termination of its stroke, and the steam valves are commencing to raise from their seats when the piston is about four inches from the end of its stroke.

DRAUGHTING OF THE ENGINE.

To design and draw an engine of this character, it is common to determine: 1. The general dimensions of the hull; 2. The diameter of the cylinder; 3. The stroke of the piston; and, 4. The size and dip of the wheel. The draught of water, or, in other words, the position of the surface of the water in relation to the hull, can usually be nearly ascertained, also, by calculations. It is usual to receive these points from the owners or builders of the vessel. Having been determined, and accurately noted, the designer will stretch a sheet of stout drawing-paper, sufficiently large to admit of a scale of about three-fourths of an inch to the foot. The general working drawing of the engine of the *Francis Skiddy*, on a scale of half an inch to the foot, is on a sheet forty-two inches long and twenty-nine inches wide, and presents, besides the side elevation, the front and back views. Commencing with the side elevation, a few inches above the lower edge of the sheet, he will draw a horizontal line, to indicate the upper surface of the engine keelsons, *N*, being the level at which the engine work proper is commenced. A red line will next be drawn perpendicularly in the centre of the sheet, to indicate the centre of the frame, *G*, usually termed the gallows frame. It rests upon, and is bolted to the keelsons, and supports all the principal portions of the engine. The designer will next fix upon any convenient position on the perpendicular line, as a temporary centre for the working beam, and determine, by calculation, the length of this beam. In American river-boats, the length of the working beam is usually somewhat less than double the length of the stroke of the piston. In the *Francis Skiddy*, the stroke of the piston is fourteen feet, and the length of the beam, between the end centres, is twenty-five feet.

A horizontal cross line will then be drawn temporarily in pencil, to indicate the position of the beam when at half-centre, or when the piston is at the middle of the cylinder. Then, towards the side on which the cylinder is to stand, two similar horizontal lines will be drawn, one at half the length of the stroke *above* the temporary beam line, and another an equal distance *below* it; these lines indicating the position of the ends of the beam when at extreme points.

The designer will now take, in compasses, one-half the length of the beam, and, planting one leg upon the temporary centre, will describe a curve on each side of it, which will indicate, temporarily, the path of the end centres of the beam. The ends of each of these curves are, of course, nearer than the middle of the same to the perpendicular central line of the engine; and the distance of these several points will be very carefully laid off on the horizontal line. The horizontal difference between these several positions of the end of the beam, will now be very accurately divided off, in order to arrive at the mean position of the end of the beam. Permanent red lines will now be drawn through these points, vertically across the entire sheet, one of which is the central line for the cylinder; the other will aid in locating the shaft. The principal object of the temporary drawing of the beam is to fix these important centre-lines. The actual height of the beam centre above the keelsons not having been determined, the temporary pencil lines will not set forth that particular with reference to the position of the beam.

Having determined the centre-lines for the cylinder and the shaft, the designer will next ascertain the thickness of the bed-plate, A^4, Pl. XVII. ($1\frac{7}{8}$ inch), with compasses, and make a line-mark of the same. Also, the height of the condenser, A^1 (6 ft. 6 in.), making a line-mark of that. It has been ascertained, by experience, that the thickness of the bed-plate, and the height of the condenser, above set forth, are in proper proportion to a cylinder of fourteen feet stroke. The height of the condenser is indicated by the decimal 0.4714 of the length of the stroke; or, in other words, multiply that decimal by the length of stroke in feet or inches, and the product will be the height of the condenser in feet or inches, for an engine of this character. Of course the height of the condenser is often increased or diminished from this rule; but this has been found to be the true proportion.

Having drawn the condenser, the designer will next proceed to the cylinder bottom, A^8, ($11\frac{3}{4}$ inches high), which rests upon the condenser. The height of the cylinder bottom in proportion to the stroke of the piston, is indicated by the decimal 0.07; multiplying the length of stroke in feet or inches by this decimal, the product is the height of the cylinder bottom in feet or inches. The position of the cylinder bottom having been drawn by the designer, he will next proceed to ascertain the height of the cylinder. This is done by adding to the stroke of the piston, the thickness of the piston at the circumference, the clearance above and below the piston at each end of the cylinder, and the depth of the cylinder cover. Thus, in the case of the *Francis Skiddy:*

	Feet.	Inches.
Stroke of the piston . . .	14	0
Thickness of piston at circumference	0	7
Clearance above piston . . .	0	1
Clearance below piston . . .	0	1
Depth of cylinder cover . . .	0	9½
Total height of cylinder	15	6½

The next objects for consideration are the proportions of thickness of piston, Figs. 8 and 9, Pl. XVIII., at the centre and at the circumference, with reference to the bore of the cylinder. The thickness of the piston at the circumference is indicated by the decimal 0.1, and the thickness of the piston at the centre is indicated by the decimal 0.2285; multiplying these decimals by the inside diameter of the cylinder, which is seventy inches, will give seven inches as the thickness of the piston at the circumference, and 15.8 inches, nearly, as the thickness of the piston at the centre. For the depth of the cover and stuffing-box, no established rule can be stated, as it depends upon the height of the steam-port.

Having found the thickness of the bed-plate, the height of the condenser, the height of the cylinder bottom, the height of the cylinder, the thickness of the piston at centre and circumference, the designer will next proceed to ascertain the dimensions of the socket of the piston rod. The length from the centre of the cross-head to the lower extremity of the socket is indicated by the decimal 0.31, multiplied by the diameter of the cylinder (70 inches), which will give nearly 21¾ inches. The next thing to be done is, to draw the lower line of the cylinder cover; next, the outlines of the stuffing box and gland—the latter at its greatest height, that is, when the stuffing box is filled with packing to its greatest capacity.

Now, sufficient clearance must be given between the upper extremity of the gland and the lower part of the socket of the piston rod, in order that the screws holding the gland in its place may not come in contact with the socket of the piston rod; say 4¾ inches. The height from the top of the cylinder to the cross-head may also be found by multiplying the decimal 0.5 by the diameter of the cylinder (70 inches), giving thirty-five inches, which, under ordinary circumstances, is sufficient for the thickness of the cylinder cover, the height of the stuffing box and gland, the necessary clearance, and the depth of the socket of the piston rod, reaching to the centre of the cross-head.

Having now drawn in a cross line the height of the centre of the cross-head, when the piston is at the lowest extremity of its stroke, the designer has now to measure off from the centre of the cross-head in its lowest position, the stroke of the piston (14 feet), which will indicate the centre of the cross-head in its highest position. The length of the front links is next to be determined. No definite rule has been established as a guide on that point. However, a decimal may be approximated from practical results, by adopting the proportion of 10 ft. 1½ in. to 14 ft. (the stroke of the piston), that is, in decimals, 0.723, which, multiplied by 14 ft., gives 10 ft. 1½ inch, nearly.

In the early stage of the drawing, lines were made indicating the horizontal positions of the centre of the beam, and of the end centres, at their highest, middle, and lowest points. Draw a line parallel to the centre line of the cylinder, running through the end centre of the working beam in its lowest position. Now, measure off 10 ft. 1½ in. (the length of the front links), from the centre of the cross-head in its lowest position, to a point on the line which passes through the end centre at its lowest position. This point will indicate the *actual position* of the end centre of the working beam at its lowest position; and, of course, seven feet (half the stroke of the piston) in a right angle above the end centre in its lowest position, will be the horizontal line of the centre of the working beam.

The height of the centre of the main shaft is next to be ascertained. The draught of water of the boat, and the diameter of the wheels, are given by the builders, or can be calculated by rules that will be made the subject of future illustration. The dip of the paddles of the paddle-wheel must also be considered, and is subject to various theories. The prevailing rule is, in river-boats, to bring the interior line of the paddles, when the vessel is light, at or a little below the surface of the water.

The draught of the boat being given by the carpenter or builder (5 ft. 4 in. light or without load), the thickness of the planking, engine-keelsons, &c., under the bed-plate, must be drawn, before the water line, in its relation to the engine, can be shown. The position of the water line to the engine should be indicated by a horizontal line. The dip of the paddles (3 ft.) and the diameter of the paddle-wheels (40 ft.) being now given, determine at once the location of the centre of the main shaft.

The gallows frame next requires the attention of the designer. The size of the main pillow block should be arrived at, giving on the bottom and sides iron of sufficient strength to endure all the heavy strains that naturally fall upon it, or may follow an accident to the engine. A full drawing of a main pillow block is presented in Plate XX., Figs. 117, 118, which give the dimensions of that of the *Francis Skiddy*. The proportions here adopted may be applied to a larger or smaller engine of the same general character. The beam pillow block must next be drawn, the general dimensions of which are given in Plate XX., Figs. 115, 116. The beam pillow block must also be of great strength, but little inferior to that of the main pillow block. These points having been arrived at, the designer will draw such lines of the main and beam pillow blocks as are necessary as a basis for designing and drawing the gallows frame. The position of the gallows frame must be so arranged as to allow the side flanges of the main pillow block to fit exactly to the side of the gallows frame (G and G^1, Plate XVII).

A circle must now be drawn, taking as a radius half the stroke of the piston (7 ft.), from the centre of the main shaft (E, Plate XVII.), showing the path of the centre of the crank pin. The points occupied by the crank pin in its highest and lowest positions—that is, at the extremes of the stroke of the piston—are where the lines of the path of the crank pin cross the perpendicular line passing through the centre of the main shaft. The position of the crank, E^1, when the beam, D, is at its level, must also be represented; and that can be arrived at by drawing a circle from the end centre of the beam, passing through the centre of the main shaft, and crossing the path line of the crank pin. Now draw a line from the centre of the crank pin to the end centre of the beam, and the length of this line will be the length of the connecting rod, D^{12}.

The air-pump, B^6, must now receive the attention of the designer. Its position must be determined with a view to leave sufficient space between the hotwell case, B, and the condenser,

A^1, to afford room for the valves connecting with the condenser, A^4, and for the cover (A^5) of those valves. The bore of the air-pump for the *Francis Skiddy*, being forty-six inches, has been demonstrated to be in the proper proportion to a cylinder of seventy inches, and is represented by the 0.657, which, multiplied by seventy, will give forty-six, nearly. The stroke of the air-pump bucket has been fixed at 4 ft. 4 in., as the proper proportion for a cylinder of 14 ft. stroke, represented by the decimal 0.321, which, multiplied by 168 inches (the stroke of the piston), will give fifty-two inches, nearly (the stroke of the air-pump bucket). The actual position of the centre of the air-pump, between the main pillow block and the cylinder, depending upon the two other considerations, variable in themselves (*i. e.*, the space for the valves in the bed-plates and their cover, and the stroke of the air-pump bucket), will, of course, be subject to them, and cannot be accurately determined until those two points are settled. In the *Francis Skiddy*, the centre of the air-pump is 7 ft. 8 in. from the centre of the cylinder.

The designer will next draw the centre of the air-pump cross-head, B^D, Fig. 3, in its highest and lowest positions. At its lowest extremity, it is placed in the *Francis Skiddy* 2 ft. 2 in. from the top of the hotwell case. A circular line must next be drawn from the centre of the working beam towards the end connecting with the cylinder, which will show the path of the air-pump centre in the working beam. The distance of this point from the centre of the beam depends, of course, upon the stroke of the air-pump bucket (4 ft. 4 in.). Divide the stroke of the air-pump bucket equally above and below the horizontal line of the working beam. Draw horizontal lines at those distances, and the points at which these lines cross the centre-line of the working beam in its extreme inclined positions, should be traversed by the circular line above mentioned.

We are now enabled to find the length of the air-pump rods, B^2, measuring from the air-pump cross-head in its lowest position, to the air-pump centre in the working beam in its lowest position.

On a working drawing like that under consideration, centre-lines of the various parts of the engine will be sufficient for practical purposes. It is necessary, however, to fully draw the gallows frame, in order that no part of the engine shall come in collision with it. Such a gallows frame is fully represented in Pl. XVII., Figs. 1, 2, 3—giving side, front, and back elevations.

In speaking of the main and shaft-pillow blocks, we briefly set forth some of the principal lines of the gallows frame; and we now give a more complete description, as follows: The gallows frame work consists of four main frames marked *GG*, two on each side, resting, at the lower end, on the engine-keelsons, *P*, *N*, and G^1, and secured by knees, G^2, which are bolted to the frames and keelsons. On the top of the main frames, *G*, rests the main pillow block D^8, to support the beam *D*. Two strong braces, G^7, strengthen the gallows frames on the top, extending down to the engine-keelson *N*, and are connected with the framework of the main pillow block, K^2, to which they are secured by two iron bolts, reaching from the outside of the brace, G^7, through the main pillow block framing, and main frames of the gallows frame, thus uniting, strongly, these three points—the top of gallows frame, the framing of main pillow block, and the engine-keelson. The four stays, G^6 G^6, are bracing the main frames, *G*, and pillow block framing from the top of engine-keelsons. The upper four horizontal braces, two on each side, connect the main frames, *G*, with the gallows frame brace, G^7, which are held together by iron bolts along the horizontal braces, reaching from one end to the other. The diagonal braces, G^3 G^3, fit exactly into the corners of the horizontal braces, G^4 G^4, will strengthen the gallows frame to some extent. Figs. 2 and 3, representing front and back views of the gallows frame, will show the cross braces, G^3 G^3, and the diagonal braces, G^6 G^6, strengthening the gallows frame sidewise.

Having obtained the exact sizes of all the frames, braces, and stays of the gallows frame of the *Francis Skiddy*, we can recommend them as true proportions for an engine of this character. For any other size of engine, the designer can be guided by these proportions as to strength of the framing, by applying similar proportions with reference to the power required for the propelling force of the boat. The designer, however, must exercise good judgment in determining the proportions of the various parts of the gallows frame, to secure sufficient strength to withstand all the strain which naturally attaches to the framework.

In regard to the draughting of this gallows frame, it may be said, that the designer or student has but to follow closely the description we have given, referring to the full drawing embraced in Plate XVII., illustrating completely the position and relative dimensions of the several parts, and they should be drawn in the order in which they are named in the above description.

More detailed directions to the student, in draughting the gallows frame, are believed to be unnecessary to the purpose of this article. Much valuable information can, however, be derived from a careful examination of the drawings in Plate XVII.

DETAILED DESCRIPTION OF THE ENGINE OF THE FRANCIS SKIDDY.

At the commencement of this article, and in the directions which we have given for draughting an engine of this character, we have presented the principal points in a general description, to which we now refer, and subjoin the following detailed description of such parts of the engine as appear to us to present valuable and interesting features.

Plate XVIII.—The steam and exhaust chests, of which Fig. 6 presents longitudinal sections, and Fig. 4 a transverse section of the exhaust chest, are here shown. In Fig. 6, *G G* represent the steam valves, the lower valve of which is bolted with four bolts to the upper valve, and G^1 is the valve-stem, having a collar on the top, and a nut on the lower end of the lower valve, to keep the valves and valve-stems in close proximity. In fitting the steam valves to their places, care should be taken when grinding them to provide for the expansion caused by the high temperature of the steam. Place a sheet of the thinnest letter paper between the joints of the two valves; fit them to their place with this sheet of paper in that position, and when ground, remove the paper; the usual expansion of the valves, when in operation, will cause them to fit closely to their seats; otherwise the upper valve will leak. G^2 is a lower cover, and serves as a guide to the valve-stem. G^3 is the cover, and G^4 the gland placed on the top of the steam chest. They are also applicable to the exhaust chest. The exhaust and steam chests are separated by the partition *L L*. The exhaust valves, *E E*, are cast solid to each other. The appendages of the exhaust valves are the same as those of the steam valves; thus, E^1, valve-stems; E^2, lower cover and guide; E^3, upper cover; E^4, gland. The exhaust valves are placed in their seats through the upper opening in the steam chest; the upper steam valve, *G*, is placed in its seat through the upper opening in the steam chest; the lower steam valve, *G*, is put through the steam port and bolted to the upper, while both are in their seats. The steam and exhaust valves are balance puppet valves; the pressure upon the lower area of the lower steam valves, *G*, is upon a surface of seventeen inches diameter, and upon the upper area of the upper steam valve, upon a surface of the same diameter. The pressure of steam upon the exhaust valves is from the steam port, and would raise the upper valve from its seat if not counterbalanced by the pressure of steam upon an equal area of the top of the lower valve. In Fig. 5, to the right, is a top view of the lower steam chest, A^7, showing, I^0 and I, the bearings for the steam lifting rods, and, I^1 and I^2, bearings for the exhaust lifting rods. Fig. 7 is a side view of the lower steam chest, A^7, looking from the side where the steam chest is bolted to the cylinder bottom, A^8. The open chamber, A^9, is cast to the steam chest, A^7, and bolted by its flanges to the condenser, A^1; being the passage for the exhaust steam to enter the condenser.

Fig. 16 is a vertical longitudinal section of the condenser, showing the injection pipe and nozzle, A^{10}, to conduct the water near to the highest part of the condenser, which is then equally distributed by the strainer, *B B*, over the whole surface of the condenser, running down through the numerous holes, and condenses the exhaust steam which rushes from the nozzle, A^9, into the condenser. To the condenser are cast three flanges, 2, 2, 2, supporting the strainer, *B B*, in its proper height, which is secured by three bolts, 1, 1, 1, to these flanges. *C* is a manhole plate. Fig. 18 is a horizontal section view of the condenser, of which 3, 3, are heavy flanges, keeping the condenser to the front of the gallows frame, and *D D* brackets for bolts, securing the condenser to the bed-plate and engine-keelson. The top flanges, 4, 4, hold the ends of bolts which extend to the flanges of the beam pillow blocks, with a nut on each end, to keep the condenser and beam pillow block firmly connected. The lower flanges, 4, are intended for placing holding-down bolts, reaching to the lower end of the engine-keelson, fastening the condenser strongly to it. A^9 is the exhaust passage, the flanges of which are bolted to the lower steam chest. Fig. 17 is a front view of the condenser; the several parts having been described in Figures 16 and 18. The bed-plate, Figs. 14 and 15, is drawn in a top view and longitudinal section, *C* being the place for the cylinder, and *D* for the air-pump; E^2 is the foot valve made of brass (in some engines India-rubber is used for the foot-valve); B^1 is the foot valve cover. *F* is a projection cast to the bed-plate, allowing room for a wedge to hold the joints of the foot valve seat and bed-plate tight; the same means is used for the same purpose in the middle of the foot valve seat.

Plate XIX.—Fig. 33 is the longitudinal section view, and Fig. 34 the top view of the feed pump, to supply the boilers with water. *C* is the feed pump barrel, made of composition. C^1 and C^3 are the suction and valve chambers. The part C^8 is for the suction pipe, and C^1 for the feed pipe. C^5 is the valve for the feed pipe, and the suction chamber has a valve of the same character, the seat of which only is shown in the drawing. C^4, C^4, C^4, C^4, are flanges resting on the air-pump brackets. C^6, C^6, are the valve covers. *E* is the plunger. E^1 the cover to the same. *D* is the rod for plunger. D^1 is the upper end of the plunger rod, fitted to the extreme ends of the cross-head. D^2 a nut to hold D^1 to the cross-head. D^3 is that part of the plunger rod which is fitted to the plunger, and is held by two nuts, D^4 on the top, and D^5 on the bottom. Fig. 35 is a transverse section view of the suction and feed chambers.

Plate XX.—The air-pump illustrated in Figs. 92 and 93 is drawn in a longitudinal section and a top view, of which B^0 is the casing of air-pump, made of cast iron, and filled on the inside with brass segments one quarter of an inch in thickness; the segments are carefully fitted on the edges towards its outside circumference, but allowing a little clearance in their joints towards the inside for hammering, which operation will fill up the clearance in the joints, and the segments holding each to the other, as if made solid. The floating top, *D D*, has a groove where it rests on the top of the air-pump, for an India-rubber gasket, or wooden segments may be fitted into the grooves, to make a tight joint when the floating top is on its seat, the former being preferable. D^1, D^1, are handles, D^2 stuffing box, and D^3 gland of the same. The hotwell case, B^5, is a separate piece of casting from the air-pump, and is kept tight by rust-cement, and a few bolts. B^6 is the cover, with a stuffing box for the air-pump rod. C^4, C^4, are brackets supporting the feed pumps which supply the steam boilers. C^3, C^3, are flanges, to make the barrel of pump tight with a rust joint. A is the air-pump bucket, A^4 the valves, and A^5 the follower to hold the packing of the air-pump to its place. A^3 is the guide for the air-pump pocket valves. A^1 is the air-pump rod, inclosed in a copper casing, about one eighth of an inch thick. A^2 is the place for the air-pump socket. B^E is the air-pump cross-head.

The details presented in Plates XVIII, XIX, and XX, to which especial reference is not made in the above detailed description, are drawn with such clearness and accuracy, giving the names of the principal parts, that further description is not required.

PLACING THE ENGINE IN THE BOAT.

The manner of proceeding to place the engine in the boat is a subject of interest and importance, and we have now to offer some general observations upon it, together with specific directions in such particulars as seem to require them.

The carpenter, in erecting the gallows frame, fixes some of the temporary lines, such as the horizontal centre of the foot of the gallows frame, and a perpendicular line rising therefrom at right angles with the level of the boat in all directions. As we stated in the general directions respecting the drawing of the engine, the gallows frame must be so arranged as to connect with the flanges of the steam cylinder, and of the main shaft pillow block, as well as to clear all the moving part of the engine; to which observations we now refer the reader. The carpenter will bolt the gallows frame to a temporary proximate position.

The first business of the engineer will be to ascertain the precise beam centre of the boat. He will erect perpendicular straight edges towards each end of the boat, sufficiently far apart to clear the cylinder at one end, and the shaft and crank at the other. These straight edges must rise strictly perpendicular to the side level of the boat, because they are to serve as a guide in establishing the sidewise centre lines of the cylinder, gallows frame, walking-beam, and main connecting rod; and, indeed, must be kept in view in the whole operation of placing the engine in the boat. The manner of placing a straight edge in its true position is, to rest the lower end upon the centre keelson, in such position that one side of the piece forming the straight edge shall be exactly in the beam centre of the boat. To carry up the straight edge in strict perpendicular from this centre, a straight edge must be laid also across the boat, at the level of the deck, resting in an exact horizontal position by means of blocks placed under each end, at the sides of the boat. The exact position of the perpendicular straight edge may now be ascertained, either by means of a T square, or by measuring from the outside of the hull towards the centre. Having found the position of one perpendicular straight edge by the means we have described, the second will of course be fixed in an exact relative position to the first. The proper height of the gallows frame, where the pillow blocks rest upon it, must now be measured from the flooring of the boat, upon which the engine-keelsons rest. This height must always be specified in the working drawings of the engine, to which reference must be had;

and having been measured on either side, that side of the gallows frame must be cut off at the proper point to receive the beam pillow block. A T square applied to the side that has thus been cut off, will indicate the point of cutting off the other side of the gallows frame, bringing both sides to exactly the same height. In applying the T square for this purpose, care should be taken to keep the long or perpendicular arm of the square in exact line with the perpendicular straight edges above mentioned. The beam pillow blocks can now be placed in their positions, precaution being taken to see that these blocks are of equal dimensions from the point resting upon the gallows frame to the centre of journal, any difference to be obviated by the variation of the height of either side of the gallows frame, from the exact point heretofore reached. A beam main centre piece of wood must next be made, of the same dimensions as that of the beam centre itself. This wooden beam main centre piece must be placed in the journals of the beam pillow blocks. The middle of this centre piece, measured from each journal, and indicating the beam centre of the working beam, must be marked upon it, either by the person turning it, or by the engineer himself; usually by the former. A piece of small cord, of very perfect manufacture, and very strong (cat-gut is generally used), must now be employed, stretched from one straight edge to the other. A short straight edge may also be fastened, with screws, to the wooden centre piece, exactly at the middle thereof, indicating the beam centre of the working beam. This centre-line must correspond with the cat-gut line drawn from the two straight edges. The wooden centre piece should be brought to rest in exact right angle to the several centre-lines of the working beam; that is, in right angle to the horizontal, perpendicular, and beam centres of the working beam, as heretofore ascertained and described. Having described the manner of ascertaining the various points to be considered in fixing the beam pillow blocks in their places, the work becomes merely mechanical, and will be accomplished by such means as suggest themselves to the engineer; which done, the beam pillow blocks may be permanently bolted to the top of the gallows frame.

The laying of the bed-plate, A^4 A^4, Pl. XVII., Fig. 1, is the next object to receive the attention of the engineer. The bed-plate is laid on two oak planks, which may easily be adjusted to the necessary variations in the bed-plate. The planking which lies on the engine-keelson, and comes in immediate contact with the bed-plate, should be adjusted as nearly as possible to its proper position. This may be done with the use of a T square. The exact centre of the steam cylinder must now be taken into consideration, and keeping that point in view, the bed-plate can now be placed upon the planking prepared, as above referred to, for its reception. The centre points where the condenser and air-pump rest upon the bed-plate, must now be accurately ascertained, and again the T square must be employed to lay the bed-plate true in every direction. This being accomplished, a line must be stretched from the two upright straight edges before described, and the air-pump and condenser centres must be in accordance with this line. Now fix the true perpendicular centre-line of the gallows frame, and measure from that line to the centre-line of the cylinder, as established in the working drawing—(11 ft. 5½ in.) This will settle the exact position of the bed-plate. The relative positions of the beam centre—as indicated by the cord drawn from the two upright straight edges—and of the transverse centres of the cylinder, condenser, and air-pump, must be indicated by marks with a chisel upon the ends and sides of the bed-plate flanges, preparatory to the removal of the cord when the condenser is placed in position. Before doing this, however, four marks on the upper and lower flanges of the condenser must be made with a chisel, at right angles to each other, corresponding with the centre of the condenser. This centre is ascertained by means of a wooden cross placed in the condenser, the arms of the cross fitting closely to its inside diameter. The same marks, by the same process, must also be made on the upper and lower flanges of the steam cylinder. The condenser may now be fitted to its place, care being taken to bring the centre at the top in exact position, measuring from the centre-line of the gallows frame as before (11 ft. 5½ in.), and in accordance with the line drawn from the two straight edges; in other words, that the centre-line of the condenser is in exact perpendicular. The work of fitting the condenser to the bed-plate must, of course, be performed upon the "chipping-strips," in the lower flange of the condenser; and when perfected, the condenser may be permanently bolted to its place. The cylinder bottom is always bolted to the cylinder, and when thus joined, the two are placed together upon the condenser. The marks upon the outside of the flanges will assist in bringing the lower part of the cylinder to its exact centre point, The cylinder must now be fitted to its place, care being taken, as in the case of the condenser to maintain the perpendicular of its centre—the same rules governing both cases.

The slides for the cross-head must next be fitted to their places. To this end, a wooden cross must be placed in the lower extremity of the cylinder, the arms fitting closely to its inside diameter. A temporary platform, near the top of the gallows frame, must now be employed, to which a cord should be stretched from the centre point of the cylinder marked on the wooden cross in the cylinder. This cord should indicate a continuation of the true perpendicular centre-line of the cylinder, for the purpose of fixing the true position of the slides. In fixing the position of the slides, a wooden piece may be employed to represent the cross-head, with a point in the centre to show where the continuation of the centre-line of the cylinder should pass, and when the slides are accurately set they should be bolted to the flanges of the cylinder. A brace, A^{16} (made of cast iron), must now be fitted to the upper flanges of the slides, retaining them in their proper position; and wrought iron braces, A^{15}, extend from the slides to the gallows frame. The second brace from the top, A^{15}, is a diagonal cross brace, connecting the flanges of the slides with the other sides of the gallows frame. The four lower braces, two on each side, are of wrought iron, and extend directly from the slides to the gallows frame. All these braces are bolted to the flanges of the slides, and to the gallows frame. The piston, with piston rod, and the cylinder cover, may now be put in their places, and then the cross-head put in its place, and fastened to the piston rod. The working beam, also, may be laid in the beam pillow block journals.

The setting of the main or shaft and the out-port pillow blocks is the next work to which we direct attention. A cord must be extended, indicating the centres of these pillow blocks. These centres are to be ascertained by measuring the height on the working drawing, from the top of the flooring of the hull to the centre of the shaft, then drawing a horizontal line with a cord from the straight edge to the perpendicular centre of the gallows frame, at the height of the main shaft; a line must now be drawn perpendicularly through the centre of the shaft, parallel, in every direction, with the centre-line of the cylinder, when a T square can be employed in determining the centre-line of the main shaft, to be indicated by a cord drawn from one out-port pillow block to the other; the actual height, as well as the distance from the centre of the gallows frame, having been, as previously stated, ascertained by measuring the height in the working drawing, from the flooring of the boat, and in using the T square, to see that this line is in strict right angle to the cord drawn from the straight edge to the centre of the gallows frame, and also in right angle to the perpendicular line parallel to the centre of the cylinder. The height of the centre of the main shaft of the *Francis Skiddy*, above the flooring, is 22 feet 4½ inches, and its distance from the centre of the gallows frame is 11 feet 5½ inches.

The placing of the main and out-port pillow blocks must now be proceeded with. The engineer must, of course, measure the distance from the centre of the journals to the lower edges of the pillow blocks, to ascertain the exact height of the resting places of them above the flooring of the boat, when he will cut the timbers accordingly, with a view to the exact height of the centre of the shaft, fitting these timbers to the lower sides of the pillow blocks. The centres of the journals of the out-port pillow blocks must always be slightly higher than the centre of the journal of the main pillow blocks, on account of the great weight of the paddle-wheels, and of the fact that the sides of the boat will yield more than the centre to the weight of the engine. If the two parts of the shaft as usually employed in river boats, lie perfectly true, the cranks will show no variation in their distance from each other at any point in their revolution.

In placing the air-pump in its seat, reference must be had to the working drawing, following the centre-points, &c., as there laid down, subject, of course, to the formation of the bed-plate. Mechanically, the operation is precisely the same as in placing the condenser and cylinder.

We have to remark, before leaving this subject, that it is well to fasten a piece forming a straight edge along the engine-keelson, the upper or straight edge of this piece to be in strict right angle to the perpendicular centre-line of the cylinder. With the aid of a T square, this straight edge will supply the place of the perpendicular straight edges before described, in case their removal should become necessary from any cause.

It may also be observed, that the bed-plate is laid upon a mixture of white and red lead, spread carefully over the oak planks which come in immediate connection with the bed-plate. The object of this mixture of paint is to fill up all crevices or imperfections of any character which may exist, either in the surface of the bed-plate itself, or in the planking, causing the bed-plate to receive equal and substantial support in all its parts. The same process must be followed in laying the main and beam pillow blocks, these important parts of the engine having been broken in consequence of not being set firmly and accurately in their places. A gasket of red lead is also employed to secure a perfectly tight joint between the condenser and the bed-plate. Lay rolls of red lead, well hammered, upon the chipping-strip of the bed-plate, then place the condenser on it, and bolt it to the bed-plate; after which a rust cement, composed of cast-iron filings or turnings (dust) ten lbs., pulverized sal ammoniac one lb., pulverized sulphur half lb., as prescribed in page 26 of this work, may be employed in securing the joint perfectly air-tight. A rust cement, formed by the use of sea-water and cast-iron filings or turnings, is often employed, but it is inferior to that above described. Be careful that joints to come in connection with this cement are not greasy when placed upon it. The joints of the cylinder bottom and cylinder, and of the air-pump and condenser, are made air-tight in the same manner. The joint between the cylinder cover and cylinder is made steam-tight by the use of an India-rubber gasket. Some use hemp instead of India-rubber, as cheaper. The joints of the steam chest and the steam port flanges, are made tight with India-rubber; also those of the exhaust passages. A sheet of very thin copper, with red lead, is used for the joints between the steam and exhaust pipes, and the upper and lower flanges of the steam chest. All the other joints may be made with India-rubber.

STEAM FIRE ENGINE MISSOURI.

Plate XXI.—Fig. 1 is a side elevation of the new steam fire-engine "Missouri," built by Abel Sharok, of Cincinnati.

This engine possesses, in an eminent degree, all the qualities necessary in an efficient fire-engine, viz.: extreme lightness, simplicity, beauty, and power, and an enormous steam generative capability. One noticeable point is its quickness of action, occupying but four minutes in generating steam and throwing water—time taken from the moment smoke appears at the top of the chimney, until water is thrown from the nozzle, at the end of one hundred feet of hose. The steam engine is of the simple direct-action, reciprocating description, as seen from the engraving. Its most important feature is the motion of the valves. Both the steam engine and pump are fitted with balance piston valves, which receive an instantaneous motion directly before the commencement of each stroke, by the action of steam directly from the steam chamber, giving the whole great freedom and efficiency of action. The construction and mode of operation is described in reference to the sectional details. It is only necessary here to describe its general arrangement.

Fig. 2 is a front elevation. The same letters refer to all the figures. A is the steam cylinder. A^4 is the valve chamber. A^7 is the exhaust or eduction pipe leading to chimney. A^8 is the induction pipe. A^9 is the throttle (see details). B^3 is the outside casing of pump cylinder. B^6 valve chambers. B^9 water chamber. B^{10} air vessel. h, h, are "goose neck" discharge pipes, to which the hose is attached. These pipes are fitted on the inside face with sliding valves, worked from the outside by the handles B^{11}, B^{11}. B^{12} (front elevation), is the hose. B^{13} is the suction or supply pipe.

C is a small steam cylinder, fitted with piston valves to be moved by the tappet arm, C^1, in connection with the tappets, C^2, C^2, at the termination of each stroke, to admit and exhaust steam alternately at the ends of the cylinder, D. C^3 is the exhaust pipe from cylinder, C. A pipe, fitted with throttle, leading from the steam chamber, N, is admitted between the piston valves in the cylinder, C. The exhaust is effected on the outside (as shown), through the pipe, C^3. The supply steam pipe to this cylinder is not seen in the engraving, being on the opposite side to that from which it was taken. The steam cylinder, D, is fitted with a piston used for the purpose of shooting over the steam engine valves and pump valves, directly before the commencement of each stroke. E is the steam cushion (described hereafter in the reference to the sectional details). F is a rod connecting steam engine with pump valves, to which is also attached the small piston of cylinder, D, and the cushion piston.

G is a small cylinder, fitted in the ends with gum cushions, against which a dash, or plate, fitted on the rod, H, comes in contact, after receiving motion from the tappet arm. The stuffing boxes, t, t, t, t, contain metallic packing, as shown in section, and u, u, u, u, u, are packed with hemp or cotton wick. H^1 is a handle, by which the valve rod may be moved at pleasure. H^2, H^2, are oil cups for lubricating the valve chambers, A^4 and C.

H^3 is a small steam pipe for supplying cushion. H^4 is a pipe connecting the water chamber of pump with the water tank, to keep up a supply of water for the generator, and to communicate the pressure of the water in the pump to the water tank, which, being superior to the pressure in the steam generator, a continual supply of water to the generator may be insured, without the aid of the pump attached to the engine. H^5 is an air vessel immediately connected with the suction, for the purpose of decreasing the effect of the ram of the suction water, at the moment the valves pass over and close the ports.

H^6. H^6, are lugs on the steam cylinder and pump, to which the braces, H^7, H^7, are firmly connected. H^8 is the lamp. I is the fire-box, containing the steam generator. I^1 the fire-door, I^2 the fire-door frame. The pipe I^6 conducts the steam from the generator to the lubricator. K is the outside casing of the generator. L the chimney. M is the lubricator, resting upon the patform, M^1.

A portion of space in the lubricator contains water, through which the superheated steam is passed, and becomes charged with its due complement of water, before being conducted to the steam chamber. The usual working height of the water in the lubricator is eight inches. The remaining space is occupied by steam. The pipe, I^6, from generator, is conducted to within 1½ inch of the bottom of the lubricator. The steam therefore passes through 6½ inches of water, and is conveyed through the pipe, M^2, to the steam chamber, N. The steam in passing through the generator rapidly conveys jets of water into the lubricator, which is proportional to the supply the generator receives from the force pump. By this operation the lubricator receives a supply of water which can be easily regulated.

A steam pipe, M^3, and water pipe (not shown), from the top and bottom of the lubricator, respectively connect to the stand pipe, M^4. This stand pipe is fitted with a water gauge and gauge cocks, for the purpose of indicating the height of the water in the lubricator. M^5 is the safety valve, fitted to the lubricator. M^6 waste pipe of safety valve. To the lever of the safety valve is attached the spring balance, M^7, fastened at its lower end to the steam receiver, as shown. M^8 is a steam whistle. It is generally used for the purpose of whistling when a fire has been extinguished, as an indication that the presence of other engines is unnecessary. M^9, M^9, are steam gauges, having pipes connecting with the steam receiver. M^{10} is a speaking tube, through which the engineer gives orders to the fireman.

N^2 is the water-tank for supplying steam generator. O is the force pump, bolted to the steam cylinder, and worked by the tappet arm, C^1. O^1 is the stand pipe, or valve chamber of force pump, O. O^8 is the discharge pipe. P is a double-acting hand force pump, used before the engine is in motion, for supplying the generator with water. P^0 barrel of pump. P^1, P^1, valve chambers. P^2 pump piston rod. P^3 links or side rods. P^4 stand. P^5 hand lever. A long movable socket lever (see details) is used to fit on the lever, P^5, so that it may be easily worked by hand. P^6 is the air vessel. Both the hand and steam force pumps are connected at P^7 with the same supply pipe, P^8, and at P^9 with the same discharge pipe, P^{10}. O^8, front elevation, is a cock in the supply of steam force pump, to which is attached a finger and dial plate, to regulate the supply of water. Q, Q, Q, are brackets for supporting the suction, B^{13}. B^{14} is the strainer of suction. It is a cylinder of copper, having, at regular distances over its entire surface, a number of half inch holes. This cylinder is covered with copper net work, forming a fine strainer. The suction is made of gum, or vulcanized India-rubber, supported from the inside by thin flat iron rings, nearly close together in its entire length. It is covered with canvas. The suction always remains attached, and when not in action, is brought round and supported on the brackets, Q, Q, Q.

R is the hind axle of the wheel R^1, extending across and below the fire-box. The springs, R^2, are connected to the axle, as shown by the dotted lines. On the ends of the springs, the links of straps, R^3, connect them with the straps, R^4, which turn at right angles, and pass across the fire-box to support the weight. R^5 is an axle brace between the axle and end of fire-box. R^6 is a hog or brace connecting the fire-box with frame at R^7. R^8 is a pipe from pump, to which are attached three pipes, R^9, passing round three sides of the fire-box. In these pipes, facing the inside of the fire-box, a number of small holes are drilled, to admit of jets of water being thrown upon the ashes of the fire, to prevent them causing the wheels to take fire. A stop valve, R^{10}, is fixed in the pipe, R^8, to be opened only when necessary by the firemen, with the aid of the rod, R^{11}, and wheel, R^{12}.

At the centre, *S*, of the straps, S^1, S^1, rods or shafts pass across the fire-box, carrying on their extremities brake arms, S^2, S^2, and brakes, S^3, S^3. On the opposite side to that from which the engraving was taken, a lever, S^4, is attached to the brake arm. At the end of the lever is fitted a small chain pulley, S^5. A chain, from the axle of the hand wheel (worked by the brakesman), passes over this pulley, and connects to a similar lever on the brake-arm of the opposite side. By this arrangement, the brakesman can apply or release the brakes at one and the same moment.

The bolts, *v*, *v*, connect to the holding-down bolts, I^6, I^6, I^6, I^6 (see details). The outside casing of the generator, *K*, is secured to the fire-box by four bolts, *w*, *w*, *w*, *w*, on each side of the casing. *x*, *x*, *x*, *x*, are the chimney tie rods. *T* is a platform, or engineer's stand, with railing, T^1. T^2, T^2, is a circular platform, encircled by railing, T^3, T^3. The top of this railing, and that of the engineer's stand, is a tube, into which the vertical stands (also tubes) are screwed. The platforms, T^1, T^2, rest upon the brackets, T^5, T^5, bolted to the frame of water tank. From the ends of the brackets, stays are bolted, which connect with the circular disk, *U*, to which the frame of engine is bolted. A pin projects from the disk, *U*, on which the disk, *V*, and forward running gear turns. This arrangement allows of the axle of the forward running gear being turned to an angle of 45° with the central line of frame; thus the machine may be turned in a circle, of which its forward length is the radius, and the fire-box the centre of motion. *W*, *W*, are the forward wheels, constructed of wood, with wrought iron tires. W^1 is the axle. W^2, W^2, the springs connected to the frame by the links W^3, W^3, W^3, W^3.

Due attention has been paid to the strength and lightness of the frame, by which the weight of the forward part of the machine is communicated to the springs. Longitudinal bars, W^4, W^4, are connected to the disk, *V*. To the bars, W^4, W^4, the transverse bars W^5, W^5, W^5, W^5, are bolted. On the extremities of the two outside bars, the links connecting them with the springs are attached. To the transverse bars and disk, *V*, is bolted the reception frame, W^6, of the tongue, W^7. The tongue and reception frame are constructed of wood. To the inside transverse bars, a platform, *X*, is bolted.

DESCRIPTION OF DETAILS.

Fig. 3 is a longitudinal section of steam pump. *A* is the steam cylinder, 9 inches in diameter, and admitting of a 26-inch stroke. A^1 is the piston. A short description of this piston will suffice to show its simplicity of construction, and efficiency of action. *a* is the piston head, *b* the rod. The rod is turned down at the end from its full size, 2 inches diameter, to 1⅞ inch, to form a collar on the rod for piston head, and is also tapered down to 1¾ inch, a length of 2⅝ (the central thickness of piston). The piston head, *a*, is bored conically to fit the rod. *c* is the follower. The whole is secured by the nuts on the end of the rod. *d*, *d*, are the outside rings, $\frac{1}{4}$th of an inch in thickness in the parts between the head and follower, and $\frac{3}{16}$ths in the parts covering them. These rings are turned to exactly the same diameter as the cylinder. *e* is the inner ring, $\frac{3}{8}$th of an inch in thickness, turned rather larger in its outside diameter than the inside diameter of the outer rings. When the rings are cut and placed in their proper positions in the piston, the outside ring will be pressed tightly to the cylinder by the elastic force of the inner one, in its tendency to resume its original diameter. The rings are retained in the piston, and make a perfectly steam-tight joint, in the manner shown in the engraving. The outside rings cover the entire thickness of the piston, a thin portion of each covering the head and follower, not making a close joint, but leaving space for the admission of a thin annular body of steam between the head and follower and outside rings, which serve as an additional packing force. A^2, A^2, are alternately steam and exhaust ports. A^3 is the steam port connecting with the induction pipe, leading from steam chamber. A^5, A^5, are balanced piston valves. These pistons are constructed in the same manner as the one described. Annular recesses and bridges are formed in the valve chamber, in connection with the steam ports, to balance the pressure of steam upon the piston. A^6, A^6, are exhaust ports. A^7 is the exhaust pipe leading to chimney. B is the pump cylinder; B^2, the outside casing; B^3, the piston; B^4, B^4, ports for the admission and discharge of water; B^5, port directly connected with the suction, for the admission of water; B^6, valve chamber; B^7, B^7, piston valves, which are moved instantaneously, directly before the commencement of each stroke, for the admission and discharge of water. *f*, *f*, *g*, *g*, are bridges and recesses, as in steam valve chamber, to equalize the pressure upon the piston in passing over the ports; B^8, B^8, discharge ports; B^9, water chamber; B^{10}, air vessel. *h*, *h* (see front elevation), are "goose neck" discharge pipes, to which the hose is connected. These pipes are fitted on the inside faces with sliding valves, worked from the outside by the handles, B^{11}, B^{11}. C is a small steam cylinder, fitted with piston valves, *i*, *i*, to be moved by the tappet arm, C^1, when in contact with the tappets, C^2, C^2, at the termination of each stroke, to admit and exhaust steam alternately, on the ends of the cylinder, D. Steam is admitted between the piston valves, *i*, *i*; the exhaust is effected from outside the valves. C^3 is the exhaust pipe leading to chimney. The steam cylinder, D, is fitted with a piston, D^1, used for the purpose of shooting over the steam engine and pump valves, directly before the commencement of each stroke. E is the steam cushion cylinder, fitted with piston, E^1. *j*, *j*, are the steam ports, to the centre of which a steam pipe, leading from the steam chamber, is connected. The cushioning is effected by the piston passing over the ports, and compressing the steam in the ends of the cylinder. When the compressed steam has attained a sufficient elastic force to overcome the momentum of the valve rod, F (to which the steam valves, pump valves, steam piston, D^1, and cushion piston, E^1, are all connected), and is brought to rest, the piston valves will have arrived at the proper position in the valve chamber, for the admission and discharge of water. By the admission of steam alternately at each end of the cylinder, D, the piston valves are shut over almost instantaneously, and brought to rest by the cushion, directly on the commencement of each stroke of the main piston. By this arrangement, steam is admitted full port instantly, and continued throughout the stroke, and has an unusually free exhaust. The pump valves being changed by the same action, possess the same advantages. G is a cylinder, fitted with gum at the ends, *k*, *k*, to which a dash, *l*, fitted on the rod, H, comes in contact, after receiving motion from the tappet arm. Steam under full pressure, as before stated, is continued throughout the entire stroke, no attention being paid to the economy of steam by expansion, as it is necessary in a steam fire engine to have great power and simplicity, with the least possible amount of weight. The arrows illustrate the direction of water and steam in the position of the engine in the engraving. It is necessary, however, to add that in the engraved position, the tappet arm, C^1, has moved the piston valves, *i*, *i*; steam has been admitted through the port, *m*; the valves have shot over, and the main piston commenced its stroke in the direction of the arrow. The stuffing boxes, *t*, *t*, *t*, *t*, contain metallic packing, on the same principle as the pistons, the rings merely being reversed, in order to pack inward. The stuff boxes, *u*, *u*, *u*, *u*, contain hemp packing.

Fig. 4 is a longitudinal section of boiler, water chamber, and steam chamber. Figs. 5 and 6 are horizontal sections of boilers. *I* is the fire-box, formed of wrought-iron plates, firmly riveted together; I^1 is the fire door; I^2 the fire-door frame. In the fire-box, at the distance of an inch from the sides, is formed a continuous coil of 1¼ pipe, a section of which is shown on Fig. 6. I^8, I^8, are vertical malleable cast-iron pipes or boxes, to which the horizontal pipes are connected, leaving a space for firing (see Fig. 6). These vertical pipes or boxes have partitions in the inside, forming return for the horizontal pipes, so as to insure a complete circulation of water through every pipe.

Commencing with the pipe I^4, are five flat coils of 1 inch pipes put together, with return bends (see Fig. 5). By turning the last bend over, so as to rest on the bars, *n*, *n*, *n*, *n*, &c., shown in section, each layer of pipes intercepts the opening of the one immediately above and below (see Fig. 5). Above the 1 inch pipes, are four layers of 1¼ pipes, three of 1½, and three of 2 inches, the whole forming a continuous line of pipe through which water is forced and steam generated. The usual working height of the water is at the second coil of 1 inch pipes, I^5. Above this point is steam space, where the steam becomes highly superheated, and passes from the generator through the pipe, I^6, into the lubricator, *M* (see Fig. 1). I^7, are supports for the pipes, bolted to the angle iron of fire-box. The whole coils of pipes are bolted down by the four bolts, I^8, I^8, I^8, I^8, passing over the top, and joining the ring, I^9. A^7 is the exhaust pipe from steam valve chamber. M^6 is the waste steam pipe from safety valve. *K* is the outside casing of generator, resting upon the angle iron of fire-box. The casing is double, to the height of 28 inches, and admits of jets of air through the holes, *o*, *o*, *o* (see Fig. 1). *L* is the chimney. The fire-bars, *p*, *p*, rest upon the brackets and bars, *q*, *q*, *r*, *r*, the brackets being bolted to the angle iron of fire-box, and to the bars, *s*, *s*, which support the entire weight of the generator, upon the springs of the hind running gear. *N* is the steam chamber, separated from the fire-box by the plate, N^1. N^2 is the water tank. The plates, N^3, N^3, separate the water and steam chambers. The water tank is supplied with water from the main pump chamber, as previously described. The pipes forming the steam generator are screwed in reverse direction on the ends, or have what is termed a right-hand screw on one end, and left on the other. With this arrangement, the pipes are easily put together, and in case of any of the pipes giving way under the pressure, they can be replaced with little loss of time.

TENDER FOR THE LOCOMOTIVE "TALISMAN."

The "Talisman" is a locomotive built by the New Jersey Locomotive and Machine Company, and the following description of its Tender includes all the details that can be regarded as necessary, to which are added a few general observations:

Plate XXIII.—Side elevation, Fig. 1. Half of the front elevation, Fig. 2. Half of the back elevation, Fig. 3. Plan or top view, Fig. 4. View of a transverse section through the truck, Fig. 5. Wheels and axle, Fig. 6. Brake arrangements, Figs. 7 and 8. Water tank, occupying both sides and the rear part of the Tender, A, Fig. 1. Framing, A^1, Fig. 1. Buffing timbers, A^2 A^2, Figs. 1 and 4. Cast iron back coupling, A^3, Figs. 1 and 4. Frame beams, A^4 and A^5, Figs. 1 and 4, and A^{10}, Fig. 4. Centre bearing beams, A^8 and A^9 A^9, Fig. 4. Bumper beam, A^6, Fig. 4. Centre casting, A^{12}, Figs. 1, 4, and 5. Side bearing, B^2, Fig. 1. Equalizing levers, B^4. Spring hangers, B^5. Spring straps, B^6. Brake, C^{12}. Brake cross bar, C^{11}. Brake rod, C^1. Bracket wheel, C. Foot pin, C^3. Spring, C^5. Brake springs, I^1. Brake lever, C^8. Brake lever holders, C^9. Brake rod, C^{18}. Brake rod and chain, C^6 and C^7. Hanger for suspending the brake, C^{17}. Bracket for hangers, D. Truck pedestal, B^9. Truck frame, B^8. A wedge, A^7, to take up lost motion between Tender and locomotive. Manhole, F, to get into tank. Wheel, G, for tank valves. Tank valve, E. Four pedestal braces, B^{14}.

The tank is made of iron plates, No. 9, wire gauge, one-eighth inch full, riveted together with angle iron frames, top and bottom, and braced inside with angle iron also. The frame is made of oak timber, tenoned together, fastened with cross bolts, and bound all around with iron plates, three-sixteenths of an inch thick. There is a casting secured on each end of the Tender, to receive the draw pins. The space between the two wings of the tank is to contain the fuel. The frame and tank are carried on two four-wheeled centre bearing trucks. The rear truck has two side bearings, besides the centre bearing, fastened upon the frame, to steady the Tender, and also to take part of the weight off the centre pin, the rear part of the tank being considerably heavier than the forward part. The brake appliance is very simple, and is controlled with one hand wheel, C; at the same time it is very efficacious, acting on the eight wheels of both trucks at the same moment.

GEARED SCREW PROPELLER ENGINES, FOR THE STEAMSHIP "CAROLINE," OF HAVANA.

The geared screw propeller engines for the steamship "Caroline," of Havana, were designed and built by Reaney, Neafie & Co., at the Penn Iron Works, Philadelphia. They include two cylinders, 40 inches in diameter, 36 inches stroke, and are completely illustrated in Plates XXIV., XXV., XXVI., XXVII., and XXIX., the various parts being presented in minute detail. The steamship "Caroline" is a merchant vessel, and is designed to run with great speed, as a passenger boat, on the coast of Cuba. The cylinders are placed in the boat transversely, side by side, as seen in Plate XXIV. Below the cylinders are the condensers, upon which are placed the brackets, A^1 A^1, and B^1 B^1, Pl. XXIV., which support the cylinders. The brackets, A^1 A^1, are cast to the bottom of the cylinders, and B^1 B^1, are cast to the top of the condensers. The condensers rest upon *D D D D*, the engine keelsons. E^0 is a pedestal, which operates to ease the forward action, or "thrust," of the shaft upon the journals, by which, otherwise, the shaft may be displaced through the wearing of the journals. E^1 is the "thrust collar," bolted to the top of the pedestal, to prevent the propeller-shaft working to one side, as is always the tendency of screws. *C C C C* are the bed plates for the condensers and air-pumps. C^1 C^1 are the air-pumps. C^2 C^2 are the waste water cases, cast to the air-pumps; on the top of these cases are the hot wells, C^3 C^3, with the air-chambers, C^4 C^4. The waste water is conducted from the hot wells through the pipes, C^5 C^5, and the outboard delivery valves, C^6 C^6, into the ocean. The steam is carried from the boilers to the steam screw stop valve, A^{10}, and enters the steam chests, A^{12} A^{12}, through the intermediate pipe, A^{11}. A^{14} A^{14} are the steam chests, for the cut-off valves, of which H^6 H^6 are valve stems. I^7 I^7 are valve stems for the steam and exhaust valves. A^{13} A^{13} are the escape pipes for the exhaust steam, conducting the exhaust steam to the condensers. A^7 is a cylinder brace, bolted to the flanges of the cylinder, A^9 A^9. A^8 A^8 and A^8 A^8 are the cross-head guides, placed upon the cylinder flanges, two upon each. A^5 A^5 A^5 A^5 are diagonal braces, for the protection of those guides. Two shafts, F^5 F^5, have their bearings on the condensers, and have, at their extreme ends, spur wheels, F^4 F^4, to operate the pinions, E^5 E^5, the latter being keyed to the propeller shaft, *E*. (See Fig. 1, Pl. XXV.)

At the extreme ends of the main cross-heads, F^1 F^1, (Pl. XXIV.) are placed the connecting rods, F^3 F^3, and F^3 F^3, one on each side of each cylinder, and connected with the cranks, F^6 F^6 F^6 F^6, of the spur wheels, F^4 F^4 F^4 F^4. The cranks are cast to the spur wheels. The upward or downward movements of the pistons, connected with the piston rods and cross-heads, and in the extreme ends of the cross-heads, the connecting rods, reaching to the spur wheels, cause the revolving motion of the wheels, and, through them, that of the shaft, by means of the pinion fastened to it. The movements of the air-pumps are brought about through the main cross-heads, operating by means of small rods upon the air-pump levers, F^{10} F^{10}, which turn on the centres, F^{11} F^{11}. At the other end of the levers, F^{12} F^{12}, small rods, F^{13} F^{13}, connect with the air-pump cross-heads, and give them the desired motion. *G G G G* are the feed-pumps. G^1 G^1 are the flanges to which are bolted the feed-pipes connecting with the boilers. G^2 G^2 are the suction pipes, connecting with the hot well cases, C^2 C^2. G^3 G^3 are the flanges, bolted to the hot well cases. G^4 G^4 are the glands for the stuffing boxes. G^5 G^5 are the ends of the air-pump cross-heads, which hold the force-pump plungers, G^6 G^6. G^7 G^7 G^7 G^7 are the air-pump guides, two for each pump.

A very ingenious arrangement is shown in the manner of applying the eccentric *I*, in connection with two eccentric hooks, I^3 I^3, and the eccentric rods, I^1 I^1, the latter being fastened to the eccentric. Another eccentric, which is placed on the opposite side of the engine, is not shown in this Plate (XXIV.). The main valve rock shafts, I^5 I^5, have double arms, I^4 I^4, made solid to them. The double arms are for the purpose of reversing the engine. The eccentric hooks, I^3 I^3, are placed at either end of the arms, and admit the steam to the cylinders, in such a manner that the motion of the engine will be reversed. Two levers, I^6 I^6, are fastened, one to each main rock shaft, for the purpose of operating the steam valves, which are connected with these levers by means of links with the steam valve stems, I^7 I^7. The motion of the expansion valves is produced by the eccentrics, *H H*, in connection with the rods, H^1 H^1, the levers, H^2 H^2, and the expansion valve rock shafts, H^3 H^3. The arms, H^5 H^5, are fastened to the rock shafts last named, and connected, by means of two links, to the expansion valve stems, H^6 H^6, the latter being fastened to the expansion valves.

Plate XXV, Fig. 1, is a view of the transverse elevation of the starboard engine, showing the propeller shaft, E; the pedestal, E^0; the collar, E^1, and friction rings, E^8. Also, the shaft pillow blocks, E^3 E^3; the pinions, E^5 E^5; the condenser, B; the nozzle for injection pipe, B^3, &c. The letters used in the view of the side elevation are used in the same connection in this view.

The boilers are illustrated by Plate XXIX, Figs. 98, 99, and 100. They are of such size and construction as admit of an abundance of steam when driving the engines up to thirty-five revolutions. Their form and design are of such nature as to entirely preclude any danger from explosion, with ordinary care; safety being the first and principal feature consulted by the well-known builders of these engines.

The details of the engines and boilers are illustrated in Plates XXV, XXVI, XXVII, and XXIX.

PATENT ROLLING MILL OF THE TRENTON IRON COMPANY.

Plate XXVIII.—This plate represents an upright rolling mill, for making heavy beams for building purposes, to the manufacture of which the Trenton Iron Company gives special attention. This mill was built at the works in which it is in operation. The dimensions of the principal parts are as follows: Distance between centres of rolls, 24 inches; length of outside rolls between journals, 27½ inches; diameter of journals, 11 inches; height of framing, 10 feet 8 inches from the point of support at bottom; length of framing from outside to outside, 8 feet 5 inches; width of framing, 4 feet 2 inches; dimensions of uprights, 8 inches by 9 inches; diameter of propelling shaft, 11 inches; diameter of cog wheel, 4 feet 2 inches, measured on pitch line; diameter of driving wheel, 3 feet 3 inches, measured on pitch line; diameter of upright shafts of spur pinions, 10½ inches; diameters of spur pinions, 24 inches.

The ground line of the mill coincides with the bottom of that part of the frame marked D^3, and is shown by the extended line in Fig. 1. The frame-work rests upon the timbers C^1 C^1 C^1 C^1, which are supported by substantial stone walls. The plate represents only the finishing rolls, with the parts connected therewith, which is ample for the purpose of fully illustrating the mill. It is proper to understand, however, that the mill has another frame and set of rolls, corresponding in all respects, excepting in the shape of the grooves, with the one represented by the plate. In each set of rolls are four grooves. Of these all are in use in the roughing rolls, and three in the finishers. Their shapes may be seen in Fig. 19, and the form of the pile in Fig. 18.

This mill is the invention of William Borrow, who in its construction had chiefly in view the attainment of means by which rolls of large diameter and corresponding strength might be employed, with the advantage of rolling in both directions, and with less lift from the lower to the upper grooves than if the rolls were placed in a horizontal position. In connection with this he made another important innovation upon the common system of rolling, by introducing the friction rolls E E, placed upon the horizontal stationary axes, E^1 E^1, and set in motion by contact of the bar in process of manufacture. This arrangement is most valuable in the manufacture of deep T bars, as through its agency direct rolling action takes place at the same time upon the web and the stem of the bar. The necessity of cutting deep grooves to receive the stem of the bar is also avoided, and the strength of the rolls thereby remains unimpaired. A T bar, with web six inches wide and stem nine inches deep, has been successfully manufactured by this mill in large quantities, and it is believed that no other mill in existence is capable of producing a T bar of equal dimensions.

Figs. 14 and 15 represent side and end elevations of a convenient device for combining the effect of two hooks in one. The plate B is duplicated, as seen in the end view, and the hook plays between the two plates. When the iron is on the level of the lower grooves, it is sustained and moved on the point of the hook A, which is then permitted by the workman to drop below the position shown by the figures, the pin 4, which is firmly attached to the plate B, acting then as the fulcrum, and the pin 5, which is attached to the lever, rising from its seat and playing freely in the slot B^2. When the iron is upon the level of the upper grooves, it is supported and moved upon the part A^1, which is then elevated sufficiently to throw the pin 5 upon its seat at the end of the slot B^2, thus changing the fulcrum from 4 to 5, and securing for the workman nearly the same leverage in both cases.

The cast-iron ring, Fig. 12, is marked K in Figs. 1 and 2, and represents the form of the bearings under the ends of the vertical rolls, the projections represented in the detailed view serving to prevent the rings from turning when the mill is in motion. These rings are chilled on their upper surface, and are supplied with oil from the upper end and through the body of the vertical rolls, which are cast hollow and made to act as oil reservoirs for the rings and lower journals.

The cross-head is shown in detail by the side and end elevations, Figs. 10 and 11. A^3, of Fig. 10, represents the bearing for the upper journal of the middle vertical roll, E^1 E^1 the axes for the friction rolls, and E^2 E^2 the parts by which the cross-head is secured to its place in the framing. To these the bolts D^9 D^9 are attached. By their agency in raising, and the reverse action of the keys D^{11} D^{11}, the position of the cross-head and friction rolls may be accurately adjusted at will. Referring to Fig. 1, the key D^{11} will be seen to bear upon the casting D^{10}, and through it to act upon the cross-head at E^2. The casting, D^{10}, is shown in section, and is weakened at the back, for the purpose of breaking under unusual strain, and thus relieving the pressure upon and preventing the fracture of more important parts of the mill. This block is of little value, is never fitted with care, and if broken can be easily replaced.

The present superintendent of the Trenton Iron Works, by whom the hook we have already described was devised, has also applied springs to the guides of this mill, by which they are kept constantly in contact with the rolls, and the danger of breakage or delay from the iron running under them and around the roll, which was at one time a serious difficulty, is now no greater than in a common horizontal mill, where the contact of the guide is maintained by its own weight.

With the mill we have described were produced the first solid rolled beams made in this country. Its products have been extensively applied by private parties, and by the United States government in public buildings, in nearly every State of the Union.

The mill makes thirty-five revolutions per minute. With regard to the necessary power to drive it, we will suppose that the engine be connected directly upon the principal shaft, and that the fly-wheel be upon the same shaft, when the dimensions and other points of the engine will be as follows:

Diameter of steam cylinder, ...	30 inches.
Length of stroke (cutting off at five-eighths of stroke)	42 "
Pressure of steam to square inch, ...	80 lbs.
Revolutions of engine fly-wheel, per minute, . .	50
Weight of fly-wheel, ...	26 tons.
Diameter of fly-wheel, ...	20 feet.
Number of plain cylinder boilers, ...	6
Diameter of boilers, ...	36 inches.
Length of boilers, ...	40 feet.

DIRECT-ACTING SCREW-PROPELLER ENGINES FOR THE UNITED STATES FRIGATE "NIAGARA."

The engines of the United States frigate *Niagara* are illustrated in Plates XXX., XXXI, and XXXIII, representing a plan, side and front elevations, and detailed drawings of her engines. She is provided with three cylinders and condensers. The cylinders are 72 inches internal diameter, and allow 36 inches stroke for the piston. The diameter of the screw is 18 feet 3 inches, and the blades of the screw are set at 30 feet pitch. The total heating surface of the boilers is 17,960 square feet, and 448 square feet of grate surface. The boilers have twenty-four furnaces: each is 7 feet 6 inches long and 35 inches wide.

The engine shaft makes on an average 46 revolutions per minute, cutting off steam at 17 inches from the commencement of the stroke, with a pressure of steam in the boilers of 15 pounds upon a square inch above the pressure of the atmosphere. The speed of the frigate *Niagara* is on an average 10 knots per hour. One knot is equal to 6,080 English feet. The slip of the screw is 36 per cent. The immersed midship section of the ship is 910 square feet.

Figs. 1, 2, and 3, Plate XXX., represent respectively a plan and side and front elevations, to which the following is a literal reference:

The bed-plates, B, B^1, and B^2, 3 inches thick, are secured to the keelsons, B^3, by 2⅞ holding-down bolts, and nuts, B^4, as shown in section, Fig. 3. At one end of the bed-plates are the cylinders, A A A, and at the other or opposite extreme, in the cases D D D, are arranged by a series of diaphragm, the condenser, the reservoirs or hot wells, the air-pumps, E^9, and the feed pumps, E^{11}, with their appendages, as shown in dotted lines.

The cylinders, the front ends of which are cast solid, are 72 inches internal diameter, and 36 inches stroke, and rest upon the upper part of the bed-plate, B^1, and are secured to it by bolts and nuts, through the flanges, A^5.

The condensers, reservoirs or hot wells, air and feed pumps, contained in the castings, D D D, are secured to the bed-plate, keelsons, and flooring by the flanges, D^1.

The guides, in which the ends of the cross-heads, E^3, furnished with boxes, work or slide, are formed by the upper part of the bed-plate frame, B^1, and a piece parallel to it bolted to a corresponding projection cast on the solid cylinder head to receive it.

The piston-rods, E^4, of which there are two to each cylinder, placed relatively as indicated by the conical chamber, A^4 (Fig. 2), in the cylinder cover, A, made to receive the nuts which secure the follower and fasten the cones to the pistons, are fitted to and secured by means of cones and nuts to the upper and lower arms, E^3, and connected by couplings to the rods E^5, and with similar couplings at the opposite end to the air-pump piston-rods, E^8. The connecting rods, E^6, coupled to the lower steam piston-rod, attach by similar means to the feed-pump plunger, E^7.

The screw shaft, C, 19½, 17½, and 16 inches diameter, with bearings 17⅛, 15⅝, and 14⅝ inches diameter, and 27, 25½, and 24½ inches long, is of wrought-iron, and revolves in its bracketed pillow blocks, furnished with gun metal boxes and caps, B^3, secured by powerful bolts, 3, and nuts, 4.

The main connecting rods, $E\ E\ E$, also furnished with composition boxes, are strapped, keyed, and bolted to the cross-head and crank journals, C^2, respectively. The diameter of cross-head journal is 10½ inches, and of the crank, 15½ and 13½ inches.

The valve geer consists of a rock shaft and its levers, connected, by means of rods, to the eccentrics and valves.

The rock shafts, F^{21}, are supported by brackets, F^{27}, bolted to the solid cylinder heads, while they rock in the capped bearings, F^{22}. The rock shafts are armed with two levers, F^{23}, studded to receive and carry the links, F^{24}, which attach to the wrought-iron valve cross-head, F^{25}, and are fastened to the coned end of the valve rod by nuts. F^{20} are levers at the ends of the rock shafts, to which the blocks in the curved link, F^{11}, are fastened. F^{12} and F^{17} are straps on the driving and backing motion eccentrics, bolted at F^{14} to their rods, F^{13} and F^{18}, and further connected by eyes and bolts to the curved link, at F^{15} and F^{19} respectively.

The reversing geer consists of a hand-wheel, F, fastened to the rod, F^1, which, according to the motion imparted by exigency or necessity, also imparts a corresponding motion to the curved link, by means of beveled geer, F^2 and F^3 (best seen in Fig. 3), and shaft, F^4. On the other end of the same shaft is a screw, represented in dotted lines, which works a corresponding wheel, F^7, keyed to the reversing shaft, F^8. This shaft is also armed with a series of levers, F^9, connecting them by the rods, F^{10}, to the lower extremities of the curved link, which is kept in its radial position, while moving up or down, by the rod F^{28}, fastened to the curved link at one end, and to the frame, at F^{29}, on the other.

The "slotted," or curved link motion, as applied to valves, is generally known; there may be instances, however, when a few remarks would render its comprehension more facile.

The curved link, as has been already described, is capable of moving up or down by the appliances of wheels and levers. It will be clear, therefore, that the valve rod will take its motion from the nearest eccentric. Thus, when the link is lowered, the top eccentric, when adjusted as is the case here, becomes the one for a forward motion; and since the top is for that purpose, when lowered to the block which is attached to the valve rod, the lower one must when raised be the reverse. It is conclusive, therefore, that between these two points, the valve can have little or no motion.

Bolted to the steam chests are brackets, G^2, the lower sides of which admit of the heads, fastened to the steam cut-off valve rods, G^1, sliding in it as a guide.

The escape valves, A^3, for priming or condensed steam, are fitted to both ends of the cylinders and worked by levers and rods, I^1, I^2, I^3.

The steam chests, A^6 and A^9, which are supplied through the apertures, A^{18}, have covers, A^7 and A^{10}, strengthened by ribs.

The cylinder covers also are strengthened by ribs, A^9, and an auxiliary cover, A^1, bolted to a corresponding flange.

The exhaust pipes, A^{12}, are bolted to the nozzle, A^{11}, Fig. 2, and extending over all, finally bolted to the condenser by a flange at D^8.

The bilge pumps, L, L, with their appendages, L^1, L^3, L^4, and L^7, are fastened to a projection cast on the bed plate to receive them, and worked by the rods L^2, attached to the plunger and to the stem, L^5, screwed into the main shaft at C^6.

The main injection valves, H^4, H^4, H^4, are worked by a combination of hand wheels, H, H, H; small beveled gear, H^8, H^8, H^8; rods, H^2, H^2, H^2; upright spindles, H^6, H^6, H^6, as seen in Figs. 3 and 4, Plate XXX., representing a front elevation and plan of the engine. The water for the main injection valves, H^4, H^4, H^4, is obtained from the ocean and conducted through a pipe connecting with a hole made in the hull of the ship. This pipe is carried from the hull of the vessel underneath the condenser to the lower flange of the main injection valves, H^4, H^4, H^4. The water, inflowing through the injection valves, is regulated by means of a screw valve, and enters the strainer placed horizontally in the condenser. The strainers have numerous holes on their top sides, to let the water enter the condenser and come in contact with the steam which is to be condensed into water. The condensed steam and injected water, now on the bottom of the condenser, is carried off by the air-pumps, which are placed within the condenser, and is forced into the ocean or river through the waste water pipes, D^8, D^8, D^8, connecting with the hull.

The bilge water injection valves, H^7, H^7, H^7, are used in case of leakage of the vessel, and then the condensers receive their supply of water partly or entirely through the bilge injection valves instead of the main injection valves. To the bilge injection valves are attached pipes placed within the condenser, with an upright elbow, widened on the upper extremity, to allow the water to be distributed in the condenser. The bilge injection valves are raised or lowered to their seats by screws and hand wheels. The outside bilge pipe is bolted to the lower flange of the bilge injection valve, and carried to the lowest point of the ship within the hull.

The force or feed pumps, E^{11}, E^{11}, E^{11}, are cast solid to the condenser, and extend through the whole length of it. The suction valve chambers, E^{12}, and stop wheels, E^{13}; the feed valve chambers, with air vessels, E^{14}, for the pump, are bolted on the back of the condenser, as shown in Figs. 1 and 3 of Plate XXX.

The three steam cylinders, A, A, A, receive their supply of steam for the action of the piston through the connection of pipes from the boilers to the steam chest branches, A^{18}, A^{18}, A^{18}, and the steam exhausts or escapes from the cylinders into the condensers, D, D, D, through the exhaust pipes, marked near the cylinder, A^{12}, A^{12}, A^{12}, and D^9, D^9, D^9, nearest to the condensers.

The Griffith Screw Propeller is represented in Plate XXXIII., Figs. 44, 45, 46, and 47. The advantages of this propeller have been fully explained on pages 139, 144. But we have to make a few additional remarks concerning the setting and securing of the plates for the purpose of altering the pitch of the blades. As may be seen in Figs. 44 and 45, showing a longitudinal section of the hub and blade, A^4, representing a boss cast on either side of the blade (which is fitted into the hub), against which wedge blocks are fitted for the purpose of changing the angle or pitch; they are held in place firmly by the keys, A^8, and the angle is made greater or less as the blocks are longer or shorter. A^6 is a key driven at right angles to A^8, and by bearing on the boss, A^4, secures the blade to the hub.

The Kingston valve, as shown in Figs. 48, 49, 50, and 51, is applied to supply water for various purposes.

The bilge delivery valve, as represented in Figs. 52, 53, and 54, is bolted by the side flange to the hull of the ship, and the lower flange connected with the bilge pump. At each supply from the bilge pump the valve is raised and the water forced into the ocean. Without such supply the valve remains on its seat.

The main, or steam valves (Fig. 56) are represented in three views, one view being a longitudinal section through the valve, and the others a top and side view of the same.

The cut-off valves are shown in Figs. 59, 60, and 61. A longitudinal section through the cylinder is shown in Fig. 55, and of the steam whistle placed on the boiler, in Fig. 62.

The boxes or brasses for the main shaft are fully represented in Figs. 63, 64, 65, 66, 67, and 68.

SUGAR MILL,

AS BUILT AT THE NILES IRON WORKS, CINCINNATI, OHIO.

Plate XXXII.—Fig. 1 represents a plan or top view of the sugar mill and gearing. Fig. 2 a front elevation. Fig. 3 a side view of the gearing. Fig. 4 a side view of the sugar mill and cane carrier. Fig. 5 is axle and wheels for the cane carrier.

Figs. 1 *and* 2.—A, A, are the side rollers, and A^0 the top roller of the mill. A^1, A^1, are the caps for the framing of the mill. A^2, A^2, are the mill frames. A^3, A^3, the bed plates. A^4, A^4, are the cane carrier frames. A^5, A^5, A^5, are the spur wheels connecting the three rollers, and giving them their rotary movement. A^6 and A^7 are chain wheels, attached to the ends of the journals of the side rollers. A^8 is another chain wheel, receiving its rotary movement through a chain connecting with the chain wheel A^7. A^9 is a rope pulley. A^{10}, A^{10}, are two chain pulleys, to operate one side of the cane carriage. A^{11} is the axle to the chain wheel A^8, the rope pulley A^9, and the chain pulleys A^{10}. A^{12} is a chain wheel, receiving its motion through a chain from the chain wheel A^6. A^{13} is the axle of the same. A^{14}, A^{14}, are two chain pulleys, fastened to the axle A^{13}. A^{15} is a spindle passing through the hollow shaft A^{13}, to operate the coupling clutch in connection with the chain wheel A^{12}, for the purpose of loosening the chain wheel A^{12}, which when loosened will turn on its own axle, and in consequence the chain wheel A^{14}, the axle A^{13}, and the cane carrier connected therewith, will stop. To put these parts in motion again, move the coupling clutch, by means of the lever, A^{18}, and intermediate spindle, A^{15}, to the chain wheel A^{12}, from which they will receive motion. A^{16}, A^{16}, are guards or guides for the cane carrier, supported by the brackets, A^{17}, A^{17}, and bolted through these brackets to the sides of the mill framing. A^{19} is a bracket holding a cast-iron piece, A^{20}; to the latter is fastened a fork, A^{21}, which serves as a support for the cane carrier beam, A^{23}. A^{24} is a heavy cast-iron scraper, fitting close to the roller which brings the cane to the mill. B (Fig. 2), is the steam cylinder, and C is the crank, the latter receiving its motion through the action of the piston and connecting rod. The crank, C, and pinion, B^1, are fastened on the engine shaft, $B\,B$, by means of keys; the pinion, B^1, receiving its motion from the shaft, $B\,B$, and the crank, C, operates the spur wheel B^2, the latter being fastened to the shaft, B^5. The pinion, B^3, is also fastened to the same shaft, by keys, and rotates in the same direction as the spur wheel B^2, operating the spur wheel B^4. The latter is fastened by keys to the shaft, B^6, on one end of which, towards the mill, is a coupling box, B^7, fitted to the shaft, B^6, for the purpose of receiving the breaking spindle, B^8, and connecting with another coupling box, B^7, which is fastened to one end of the upper roller, A^0. The shafts, B^8 and B^6, are supported in their places by the pillow blocks, B^9, B^9, and B^{10}, B^{10}, which rest upon the solid bed plate, B^0. One end of the engine shaft, $B\ B$, rests with its journal and pillow block upon the bed plate, B^0, the pillow blocks being bolted to the bed plate; the other end of the shaft resting upon the engine bed plate. The engine and sugar mill are by these devices fully connected together and made ready for operation.

SIZE AND SPEED OF THE MILL.

The size of the engine for the sugar mill is 12 inches bore of cylinder, and 3 feet 6 inches stroke of piston, with two steam boilers, each being 26 feet long and 48 inches diameter, with two flues for each boiler. The size of the mill is 24 inches diameter of the rollers, and they are 4 feet long. This sugar mill is geared 1 to 8. The speed of the rollers of the sugar mill is from 18 to 20 feet per minute on the circumference, or from three to four revolutions per minute by the rollers.

Manufacturers of sugar mills in the Northern states gear their mills of this size: Not less than 1 to 12, up to 1 to 15—sometimes 1 to 16—with a pressure in the steam boilers of 60 pounds upon a square inch. The speed of the rollers on the circumference is from 18 to 20 feet per minute.

A. C. POWELL'S PATENT SCREW CUTTING MACHINE.

Plate XXXIV.—The Patent Screw Cutting Machine, the production of Mr. Archibald C. Powell, which is the subject of the present illustration, is in use at the New York Central Railroad machine shop, at Syracuse, and presents many novel and valuable features, and similar machines are in use in several other important shops.

The Plate illustrates the machine in front and side elevations, and a plan or top view. Fig. 1, front elevation. Fig. 2, plan or top view. Fig. 3, front elevation. The main framing of the machine, A, A, is a solid cast-iron piece, connected by cast-iron cross braces, A^1, A^1, and A^5, A^5, with the auxiliary portion of the framing, also marked A, A, which is cast-iron likewise. The two cast-iron cross braces, A^1, A^1, both have journals, A^4, A^4, for the main spindle, B^{11}. On one end of this main spindle, is tightly fitted the die-box, marked B, secured by a bolt. On the main hollow spindle, B^{11}, B^{11}, is cut a thread for the friction roller, B^8, which is fastened to the brass nut, B^9, B^9, to work on its screw backward or forward. Two wheels, B^{15} and B^{16}, are running loose on this spindle, the latter wheel having a long hub, B^{18}. The coupling, C^6, slides only sidewise, on the main spindle, and when moved to the wheel B^{15}, and coupled with it, will make the return motion. If the coupling, C^6, is moved to the wheel B^{16}, it is coupled with it for cutting bolts, nuts, or bars. The sliding motion for the coupling in the main spindle is performed by the lever, C, moving on its fixed centre, C^0, and connected by a wrought-iron lever, C^1, with the cast-iron lever, C^2 (see Fig. 2). The cast-iron lever, C^2, has a stationary centre, C^3, and a pin, C^4, fitting into the groove of the coupling, C^6. The motion for the main spindle is obtained through a belt connecting the main shafting to the pulley, B^{12}. The pulleys are fastened by keys to the shaft, B^{13}. The shaft is supported by the journals, A^{9} and A^{3}, and has two pinions, fully shown in Figs. 19 and 20, marked B^{14} and B^{17}—B^{14} running into the wheel B^{15}, and B^{17} into the wheel B^{16}. If the coupling, C^{5}, is in the centre of its sliding motion, as shown in Fig. 2, then both wheels, B^{15} and B^{16}, will run loose on the main spindle; but if moved into one of the wheels, one wheel will run loose, and the other drive the spindle in the intended direction. A horizontal section through the main spindle, B^{11}; spur wheels, B^{15} and B^{16}; coupling, C^{6}; friction wheel, B^{5}; brass nut, B^{9}; and die-box, B, is shown in Fig. 4. The die-holders, B^{3}, B^{3}, slide in or outward in the die-box, B, receiving their motion and position from the angular levers, B^{8}, B^{8}, the latter receiving their action by the forward or backward motion of the brass nut, B^{9}, which is firmly secured to the friction wheel, B^{5}. A thin iron band, fastened on one end to the side of one of the main cast-iron frames, is laid over the friction roller, B^{5}, extending on the other end of the friction roller to the foot lever, B^{6}, held up by a spring marked 9, where, by the operation of the machine, a slight downward pressure is given to the lever connected with the iron band, which is forced to the surface of the friction roller, B^{5}, to prevent its rotary movement with the main spindle, B^{11}. This will produce the effect of setting the angular levers, B^{8}, B^{8}, in motion, one end of which being placed in the groove of the nut, B^{9}, and the other fitted on the ends of the die-holders, B^{3}, B^{3}. The forward or backward motion of the brass nut, B^{9}, in connection with the friction roller, B^{5}, if kept stationary, is regulated to move backward, until the attachment to the friction roller, B^{6}, secured to its place by a bolt, will come in contact with the pin, B^{7}, placed on the side of the die-box; by this operation, the dies are now in their extreme inward positions, if the thread can be cut in one cut, and every bolt will therefore be cut to the same size in diameter, until the position of the attachment, B^{6}, for another size bolts is changed. The machine must be reversed by the lever, C; clutches, C^{5}, C^{5}; and gear, B^{14} and B^{15}, whenever the bolt is cut, for the purpose of causing the outward movement of the dies, which is done, as heretofore stated, to facilitate the immediate removal of the bolt from the dies. If the dies have sufficiently moved apart from each other, and the operator ceases with the pressure with his foot upon the lever, B^{6}, a friction ring, B^{10}, illustrated fully in Fig. 23, reaching the angular lever, B^{8}, is applied, tightened by a bolt on its groove, sufficient to move the friction roller, B^{5}, back on its screw, B^{11}. The bolts, bars of iron, or nuts, are held between the holders, D^{4}, D^{4}, which are screwed together or apart by the hand wheel, D, fastened to the spindle, D^{3}, the latter having two threads working in the nuts, D^{2}, D^{2}, as shown in Fig. 3. D^{1}, D^{1}, is a wrought-iron guide, on which slide the holders, D^{4}, D^{4}; these bars are connected by the round slides, D^{5}, D^{5}, secured by the bolts, D^{6}, to the bar, D^{1}. The slides, D^{5}, D^{5}, with their appendages, move in or outward, according to the thread of the screw to be cut, and are supported near to the die-box, B, by two spring bearings, as illustrated in Figs. 16, and by the back cast-iron brace, A^{1}, having two holes through it. E is a cast-iron table, to collect the oil to run into the cast-iron sliding oil-box, E^{1}. This oil was used for cutting bolts, and would have been wasted, if no provision had been made of this kind. The details not yet referred to are represented in Figs. 5, 6, and 7, showing the various views for the bolt or bar holders; Figs. 8, 9, and 10, side, front, and top views of the die-box; Figs. 11, 12, and 13, views of the die-holder; Fig. 14, view of the dies; Fig. 15, a section through the pulley; Fig. 17, a spindle for the bolt holders; Fig. 18, the spindle for the bolt holder, with a right and left thread; Figs. 19, 20, pinions; and 21, driving shafts for the pulleys and the two pinions, B^{14} and B^{17}; and Fig. 22, view of the wrought-iron guide, D^{1}, for the bolt holder.

BOLT MACHINE,

INVENTED BY WILLIAM SELLERS, OF PHILADELPHIA.

This machine is intended for use in cutting and threading bolts, and appliances for this purpose enter so generally into the outfit of shops, that the improvements presented in this machine are worthy of especial attention.

In machines employed for cutting bolts, previous to the invention of Mr. Sellers, some caused the bolt to revolve, others the dies; but all required either the reversal of the machine to cut the bolt, the reversal to get the dies off when the bolt was cut, or the stopping of the machine to put in a fresh bolt. In cutting long bolts, it is very inconvenient to revolve the bolt, and for this reason the revolving die is preferred; but when the motion of the die is reversed, either to get the die off after doing its work, or to take a fresh cut in the unfinished bolt, an amount of injury is done to the edge of the die greater than that caused by the cutting, as by this operation it is very sure to break off the fine edge, so as in a short time to make a defective thread upon the bolt. In many revolving die machines, the dies are made in a solid block, so that in case of wear there is no means of adjustment. The tool must then be thrown aside before it is worn out, making it impossible by such a process to maintain a uniform standard of size. When the dies are stationary and the bolt is revolved, it is necessary to stop the machine for every fresh bolt. On small work the time lost by this arrangement will be almost as much as that required for cutting the bolt, so that on all kinds of work it is desirable to revolve the dies, providing they can be as easily adjusted and operated as when they are stationary.

The object of this invention is to avoid the necessity of reversing the motion of the cutting dies, or of stopping the machine to change the bolts, and at the same time to so arrange the dies that they can be readily adjusted to the size of the required bolt; to be adjusted so as to compensate for wear, and to admit of the greatest facility in changing from one thread to another, or to tap nuts. To this end the dies, B^{14}, B^{14}, B^{14}, Fig. 7, Plate XXXVIII., are placed in a cylindrical piece of metal, B^{9}, which is called the die-box, and which is provided with radial grooves of such size as to receive the dies and to allow them to slide freely in a radial direction. The die-box, B^{9}, is securely attached to the face plate, B^{4} B^{4} (shown in Fig. 1), of the hollow spindle, which carries the spur wheel, B^{11}, and its hub, B^{10}, that receive motion from the pulley shaft, C, through the medium of a pinion, B^{8}. Surrounding and sustaining the hollow spindle, B^{4}, B^{4}, B^{4}, is placed another hollow spindle, B, having a face plate or flange, B B, at one end. This spindle, B, is bolted to and supported by the tubular piece, A^{8} A^{8}, of the frame. On the face plate, B B, there are three eccentric or scroll cams, B^{1}, B^{1}, B^{1}. of equal eccentricity, securely bolted to the face plate, leaving spaces on the edges between them of sufficient width to admit the dies, B^{14}, B^{14}, B^{14}, to pass freely. On the outer surface of the cams, B^{1}, B^{1}, B^{1}, a covering plate is bolted, having on its inner surface three cams, 6, 6, 6, parallel with the cams, B^{1}, B^{1}, B^{1}, and three springs which are so placed as to form one side of the openings between the scroll cams, B^{1}, B^{1}, B^{1}, through which the dies are inserted, and having also projections which form a continuation of the cams, 6, 6, 6. These springs are for the purpose of keeping the die from sliding out of the openings between the scroll cams, whenever the dies are opened for the purpose of removing the bolt, and also to govern the position of the dies when first entered, so as to guide them upon the cams, 6, 6, 6. The face plate, B; the cams, B^{1}, B^{1}, B^{1}; and the cover, B^8, with its cams, 6, 6, 6, are called the cam box. On the wheel, B^{11}, and moving freely, a suitable distance around the hub thereof, is provided an adjustable stop, B^{12}, which may be securely attached to the wheel, B^{11}, by means of the bolt, B^{18}, Fig. 1, and having near its centre a projection, which is of such length as to come in contact with a similar projection on the wheel, B^6, thereby conveying motion to the wheel, B^6, and spindle, B, from the pulley shaft, C. To put the machine in operation, it is necessary—first, to move the die-box, B^8, into such a position that the grooves, to receive the dies, shall correspond with the openings between the cams, B^1, B^1, B^1. Into these openings the dies are then inserted, and pushed forward until the projection on the spring falls into an opening in the edge of the die, which is made to receive one of the cams, 6, which cams are for the purpose of withdrawing the die from the centre. The dies now being in the die-holder, the next operation will be to adjust them to the size of the bolt to be cut. This is accomplished by slacking the set bolt, B^{18}, Fig. 1, on the adjustable stop, B^{12}, Fig. 3, and then moving the die-box, B^8, by means of the pulley shaft, C, pinion, B^8, and wheel, B^{11}, the wheel B^6, with the cams, B^1, B^1, B^1, being held stationary; as this is done, the dies, B^{14}, B^{14}, B^{14}, will be gradually forced by the cams, B^1, B^1, B^1, towards their common centre, and closed upon any finished bolt or other object to which it may be desired to set them. Whilst the wheels are in this position, the adjustable stop, B^{12}, is moved so as to bring the projection on B^{12} near its centre, in contact with a similar projection on the wheel B^6, when the set bolt, B^{18}, may be secured. If now the wheel B^{11} be turned in the same direction as before, it will carry with it the wheel B^6, and cams, B^1, B^1, B^1; and as the dies, B^{14}, B^{14}, B^{14}, Fig. 7, and cams, B^1, B^1, B^1, will then be moving in the same direction, and at the same velocity, there will be no further movement of the dies in a radial direction. To open the dies for the purpose of withdrawing the bolt, the pulley shaft, C, being in motion, the pinion, B^7, must be moved by means of the loose collar, C^4, and handle, C^9, so as to tighten the friction clutch, B^9, which will give an equal velocity to the two pinions, B^8 and B^7, but as the pinion B^7 is larger than the pinion B^8, the wheel B^6, which is driven by B^7, will move faster than the wheel B^{11}, which is driven by B^8, thereby causing the cam box to move around the dies in the opposite direction to that first described, and the cams, 6, 6, 6, will force the dies from the centre. To close the dies preparatory to or during the operation of cutting the bolt, the pulley shaft, C, being in motion, the pinion, B^7, is forced by means of the loose collar, C^4, and handle, C^9, against the leg, A^1, which is here covered by a piece of leather, so as to create friction. This will cause the pinion to stop, together with the wheel B^6, and cams, B^1, B^1, B^1. The motion of the wheel, B^{11}, and die-box, B^8, will, however, continue, thus causing the dies, B^{14}, B^{14}, B^{14}, to revolve as first described, thereby closing the dies until the projection on the stop, B^{12}, comes in contact with the similar projection on the wheel, B^6, compelling it to move at the same velocity, when no further movement of the dies, in a radial direction, will take place. The cutting edge of the dies may be formed so as to cut a full thread by once passing over the bolt. This is accomplished by first cutting a thread in the dies perfectly straight and cylindrical, of such size as to fit the bolt when it has the thread cut upon it. The tops of the threads on the dies are then dressed off, commencing at the base of the thread where the bolt is entered, and terminating at the top of the thread, about four threads from the point of entrance. They are at the same time to be dressed back from the cutting edge, so as to give a clearance. Each thread of the dies will thus form a cutting edge, and the thread upon the bolt will be formed by a series of cuts, each one deeper than its predecessor, until the perfect thread is developed. To fit the machine for tapping nuts, a cylindrical piece of metal is provided, which is turned to fit accurately in the hollow spindle, B^4, and die-box, B^9, having a square hole through its centre, to receive the shank of the tap. In its outer surface, cut a square recess of the same width as the grooves in the die-box, B^9, which receive the dies, B^{14}, B^{14}, B^{14}. Having removed the dies, insert in place of one of them, a piece of metal of the shape and size of the die, having the end nearest the centre of the die-box made square, so as to fit in the recess of the cylindrical piece, and which serves as a key to prevent the piece from turning in its place. The piece in the die being forced into the recess by the operation of the cam, B^1, as described in the process for cutting bolts, the square shank of the tap is slipped into the hole, of the same shape, in the piece which is put in the hollow spindle, B^4, and the nut to be operated upon by the tap may be held by any of the numerous devices now in use for that purpose.

The nature of this invention, which is the most perfect of its kind that has been brought to our attention, consists partly in causing a difference of velocity between the revolving die-box and its surrounding cam-box, for the purpose of opening or closing revolving dies, having their cutting edge cut as described. It is evident that this may be performed in a variety of ways. For instance, in place of shifting the handle as described, it may be held in one position, either so as to keep the friction clutch in gear, or to hold the pinion against the leather washer on the leg, thereby reversing the motion of the dies, by which they will be opened or closed, as well as revolved. Again, the die-box alone may be driven, and by applying friction, or by holding the cam-box by the band, and reversing the motion of the dies, they will be opened or closed, as well as revolved. When the bolt is cut, in order to withdraw it without running the dies backward, the machine may be stopped, and by turning the cam-box backward, the dies will be withdrawn, but they will leave the mark of the cut upon the terminal thread. Neither of these methods is recommended, for reasons already given, but a modification may be adopted, by which the die-box and cam-box can be stopped alternately, accomplishing the same result as in machine first described.

Let both pinions be made loose on the cone pulley shaft, and in place of the friction clutch, provide a positive clutch, which will be driven constantly by the pulley shaft. This clutch can be arranged, as is well known to machinists, so as to throw into gear with either pinion, thus making either pinion a driver, whilst the other would run free. The same movement which throws the clutch into gear with one pinion, may bring a slight friction upon the other, so as to retard or stop it entirely, until it is driven by the projection on the adjustable stop before described. Then, by arranging the cams so that the dies will open when the cam-box is held still, and the dies revolved, we have an arrangement in which the motion of the dies is never reversed, and they are withdrawn from the bolt whilst they are revolving, so as to leave no mark upon the thread.

Again, dispense with the teeth on the wheels, B^{11} and B^6, and in their places put flanges on the wheels, so as to form rims of equal diameter, suitable for a belt to run upon; then provide a belt shifter and break, so arranged that when the belt is thrown upon either rim, the break may bear upon the other, and retard it. Then, arrange the cams as last described; that is, so that the dies will open when the cam-box is held still, and the dies revolved; we then have another method of giving motion to the die-box and dies, in which the motion of the dies is never reversed, and they are withdrawn from the bolt whilst they are revolving, so as to leave no mark upon the thread.

We have thus described several methods of operating this arrangement of dies; but the principal features in this machine, which stamp it as the work of a man of genius and a thorough mechanic, are: first, the use of rotating dies, having their cutting edges formed, as described, in combination with cams or their equivalents, when both are so arranged as to be capable of revolving about a common centre, at different velocities, for the purpose of opening and closing the dies as described; also, the arrangement of cams, with the open spaces between them, in combination with the die-box and dies, to facilitate the changing of the dies; and the mode of attaching the tap-holder to the revolving die-box.

ENGINES

REGULATED BY SELF-REGULATING CUT-OFFS.

Plate XXXVII. commences a series of drawings illustrating the means of adjusting the expansion of the steam, so as to use only sufficient at each stroke to overcome the resistance, without the intervention of a throttle valve. Until a quite recent period, the common method of regulating has been by the throttle valve—a kind of "damper"—in the steam pipe, which is turned as the speed increases, and chokes off the supply. This is now, as applied to large engines, considered by many but a relic of barbarism. An engine throttled is in a condition somewhat like that of a horse restrained by a brake applied to the wheels, and compelled to exert more strength than is necessary. The new system removes the brake, and puts the bit in his mouth instead. The throttling system spends the fury of the imprisoned vapor in a violent struggle to pass through a narrow opening, allowing only a fraction of its power to be expended in driving the engine. The new takes at each stroke a large or small quantity of steam, according as the work requires, and allows it to act with all its vigor. We shall give details of several of the best varieties of engines which involve this feature.

CORLISS' PATENT VARIABLE CUT-OFF.

Plate XXXVII.—The present illustration is a horizontal steam engine. Fig. 1 is a side elevation; Fig. 2, a plan of the engine; and Fig. 3, a top view of the valve motion fixtures of the self-regulating variable cut-off. This method of operating and controlling the valves is the invention of Mr. G. H. Corliss, of Providence, R. I., and is justly admired by all who have had an opportunity of testing the effect of the engine, for economy of fuel and regularity of motion.

A is the cylinder, or rather the wooden casing or lagging which surrounds the same. A^2 is the steam pipe bringing the steam from the boiler. A^3 is a branch of the same, descending vertically to the stop valve chest, A^4, in which is a suitable valve, turned by the handle, A^5, to control the admission of steam to the steam valve chests. These chests, or cavities, like those for the exhaust valves, are cast in the cylinder, and are inclosed with it in the lagging, A.

It will be observed that the engine is directly connected to the crank; that the frame is peculiar in form; that the slides are in the same vertical plane with the piston rod; and that the cross-head stands, of course, necessarily in an upright, in lieu of a horizontal position. Also, that there are, in this engine, four valves, each distinct from the other. The construction and arrangement of such valves, though somewhat differently operated, are shown in section in the details of another engine by the same makers, shown in Fig. 7, Plate XL.

It may be further observed that the upper are steam valves, to admit steam to the cylinder; the lower are exhaust valves, to allow its escape; and that while the peculiarity of the motions of the steam valves is most marked and important, the motion of the exhaust valves is somewhat different from that observed in engines where the exhaust valves are directly connected to the eccentric.

The eccentric, B^1, and the governor, C, play the most important parts: the former, to give motion to the steam and exhaust valves, and the latter to regulate the liberation of the steam valves, allowing them to close by the gravity of weights suspended on the rods, B^{22}. The governor receives its motion through the pulley, C^0, on the main shaft of the engine, and transmits its revolving motion by a belt to the pulley, C^{17}, which is fixed on the shaft, C^{14}. This shaft, C^{14}, extends through the cast-iron bearing, C^6, and revolves in it; on the opposite end, it has a bevel gear, operating another bevel gear, C^{15}, to rotate the governor. To the shaft of the bevel wheel, C^{15}, are connected the governor ball arms, C^7; and on the lower end of the arms are fixed the balls, C. Rods, C^8, connect the arms, C^7, with the loose collar, C^9, which is free to turn on the rod, C^1, but is prevented from rising or sinking thereon by the collars, C^{10}, which are firmly fixed on C^1. By the ordinary centrifugal action of the governor in lifting or lowering the loose collar, C^9, the rod, C^1, is lifted or lowered without being revolved. The lower collar, C^{10}, has two jaws, or knuckles, from which two connecting links, C^{11}, C^{12}, extend, and connect to what we shall term the cut-off regulators, C^{13}, C^{18}. These cut-off regulators are mounted loosely on the valve spindles, B^{21}, B^{21}, and each carries on the side opposite to the arm, C^{13}, a projection shown in dotted lines, which, although of small size and so located as to be almost unobserved, performs one of the most important functions of the apparatus, to wit, the releasing of the corresponding steam valve at the instant desired. Before describing the effect of the changes of position of the governor balls, in changing the period of the closing of the steam valves, it will be necessary to trace, from the eccentric to the several valves, the motion which opens them.

The eccentric rod, B^4, connects the eccentric strap, B^1, to an arm on one extremity of the rocking shaft, B^5. This shaft is supported in the cast-iron bearing, B^0, and carries on its opposite end the arm, B^6, with its pin, B^7. B^{II} is a fixed pin, or projection on the side of the cylinder, A. On this, as a centre, is mounted the wheel, or circular plate, B, so as to be free to rotate, or oscillate thereon. B^{10} is a pin, fixed in one side of the outer face of B. B^8, B^{18}, is a rod connected to B^7, and having a hook, or bent part, B^9, adapted to fit upon the pin, B^{10}. It follows, from this connection of the parts, that the throw of the eccentric, B^1, communicates a rocking or oscillating motion to B when the hook, B^9, is allowed to take hold of the pin, B^{10}. This motion of B can be stopped at any moment by lifting the handle, B^{18}, so as to disconnect B^9 from B^{10}, and when this has been done, the plate, B, which is termed the wrist plate, may be turned by hand into any position desired, by grasping the handle, 13; but when the connection to the eccentric is maintained as represented by the hook and pin, B^9, B^{10}, the wrist plate rocks once forward and backward with each revolution of the engine.

Four pins, designated each by the number 12, are fixed at unequal distances, in the inner face of B. The lowermost two serve to operate the two exhaust valve stems, B^{13}; the uppermost two of these pins serve to operate the steam valve stems, B^{21}. The motion of the exhaust valves is simplest, and will be first described.

On each stem, B^{13}, is fixed an arm, as represented. These arms are each connected to one of the pins, 12, in the wrist-plate, by means of the connections represented each by B^{11}, B^{14}. The middle part, B^{14}, of each of these connections is provided with a right-hand screw thread at one end, and with a left-hand screw thread at the other end, and is tapped into corresponding holes in the end parts, B^{11}, so that by turning B^{14} the connection may be shortened and lengthened at pleasure, so as to adjust the set of the exhaust valves very accurately, each being set independently of the other.

It will be readily seen that the exhaust valves, which are analogous in effect to cocks, are alternately opened and closed at each oscillation of the wrist plate, each remaining open during the whole or nearly the whole of the corresponding stroke of the piston.

On the stem, B^{21}, of each steam valve, is firmly fixed an arm, B^{22}, having a branch projecting therefrom, nearly at a right angle. To each of these branches is connected, by a rod, B^{23}, a heavy weight, not represented. In the inner face of each of the arms, B^{22}, is fixed a swiveling socket, which supports one end of a sliding rod, B^{15}, B^{16}, the other end of each of said rods being made to embrace one of the uppermost of the pins, 12, so as to receive motion from the wrist plate. These rods are not directly connected to the arms, B^{22}, but on the contrary are free to slide through the sockets thereon, without giving any motion thereto, except through the aid of additional parts, now about to be described.

A crooked piece, B^{17}, B^{18}, is loosely hinged to each of the rods, B^{15}, B^{16}, by the pin, B^{20}. The lower part, B^{18}, carries a small hook adapted to catch under a suitable projection on the arm, B^{22}, and is pressed up to seize the same by the force of a spring not represented. The upper part, B^{17}, of the same piece, is bent as indicated in dotted lines, and stands very near to the cut-off regulator of the corresponding valve. The rocking motion of the wrist plate, B, by giving a corresponding motion to the sliding rods described, causes the bent part of B^{17}, which is solid to B^{18}, to traverse back and forth under the corresponding cut-off regulator. The parts are so proportioned that when, in thus traversing across the bent part of B^{17}, it passes under the swelled part before described of the cut-off regulator, it invariably touches it, and is depressed. At each vibration of the wrist plate, the hook on B^{18} seizes B^{22} at the right moment, and commences to pull it, thereby opening the valve, which is of the character shown in section in Fig. 7, Plate XL., and allows the steam to enter that end of the cylinder, and act on the piston with its full force in the early part of the stroke. But the weight depending from the branch arm by the rod, B^{23}, tends to shut the valve, and is ready to do so as soon as any slight force shall disconnect the hook on B^{18} from the arm, B^{22}. Such force is found in the depression of the hook by the contact of B^{17} with the swelled part of the cut-off regulator. The moment the bent part of B^{17} touches the swell on the cut-off regulator, and is depressed slightly, it releases the arm, B^{22}, and then the gravity of the weight immediately turns the arm back to its original position, so as to close the valve.

Now it will be evident from this, that the time of the closing of the valve, in each stroke, will be affected by the position of the swell on the cut-off regulator, and as the arm, C^{18}, on each cut-off regulator is connected to the governor as above shown, and is turned in one direction or the other with each elevation or depression of the governor balls, the contrivance is a most sensitive means of closing the valve at a period varying with each variation in the velocity of the engine. The moment the engine is, from any cause, checked in its motion, and the governor balls are less affected by centrifugal force, they sink slightly, thereby depressing the collar, C^9, and pushing upon the arms, C^{18}. This movement turns both the cut-off regulators, so that the swell on each is not touched by B^{17} so soon as before in each stroke, and consequently the steam valves are held open longer, and more steam is received at each stroke than before, which restores the speed of the engine. If, on the other hand, the engine commences, from any cause, to run too fast, the balls rise, carrying up the collar, C^9, and pulling both the arms, C^{18}, so that the swells on their respective cut-off regulators come earlier into contact with B^{17}, and the valve is shut earlier. If the speed is very slow, the cut-off regulators turn so far that the arm, B^{17}, is not affected, and the hook is not detached. In such extreme case, the steam valve remains open until the wrist plate oscillates back, which requires nearly the whole stroke of the engine. This mode of working uses much more steam than usual, and produces more power, but at a greater expense of steam, in proportion to the power, than usual. In practice, the engine does not work in such manner, except for a few strokes, in starting from rest, or when a heavy resistance is suddenly encountered by the machinery. In all ordinary conditions, the parts being properly proportioned, the engine maintains an almost absolutely uniform speed, and cuts off at a point varying between the very commencement of the stroke and half stroke, according as the resistance is changed, or the pressure of the steam is allowed to sink or rise.

The disagreeable effects which might ensue from the dropping of the weights at each stroke, are avoided by causing the momentum thereof to be absorbed by a cushion of air. The engine operates without any indication of jarring, or any sound from such source.

The cylinder, C^5, standing upon the floor, contains a quantity of oil or other fluid, and a disk somewhat like a churn dash, which is connected to the governor by the rod, C^6, and serves to check any tendency to extraordinary or oscillatory motions in the rising and sinking of the governor.

CORLISS' IMPROVED PATENT VARIABLE CUT-OFF.

Plates XXXIX., XL., and a part of XLI.—The mechanical combination of the cut-off arrangement which is the subject of the present illustration, varies in many respects from Plate XXXVII. A very rapid closing of the valves is produced by the peculiar application of steel springs instead of weights, and many other valuable improvements make themselves apparent in the several parts. Fig. 4 is a side elevation, Fig. 5 a top view, and Fig. 6 an elevation of the opposite side.

In this engine the wrist plate, A^{30}, is mounted in a different position, and is operated directly by the eccentric rod, and hook, A^{41}, A^{48}. The valves are located as before, but the arm of one of the exhaust valves, B^8, extends upwards in lieu of downwards, to adapt it to the new arrangement of the wrist plate and connections. The same effect as before is produced in the working of the exhaust valves. The steam valves are not closed by weights, neither are they detached in the same manner as before, although the means are to a certain extent analogous. The air cushions to destroy the momentum of the parts without shock or noise, when the effect of closing a valve has been produced, are in this engine fixed on the machine in plain sight, instead of being concealed beneath the floor. The valves are detached at a time which is controlled by the elevation of the governor balls as sensitively as before, and the shutting of the valve is performed much more rapidly. It may seem, at the first glance, that the shutting of the valves of the engine shown in Plate XXXVII. is performed instantaneously, but it will be evident on reflection that a certain amount of time must be consumed by the weight in starting from rest, or from a lifting motion, and dropping down several inches so as to close the valve. The effect of a strong spring acting upon a few light connections, as in this engine, is sensibly more rapid; and when it is recollected that during whatever time in each stroke the valve is closing, the steam is "wiredrawn," and used at a disadvantage,—that in an engine working as most small stationaries do, at a rate as rapid as sixty revolutions per minute, only one half of a second is consumed in an entire stroke of the piston,—and that at or near half stroke, when the valve is usually released, the piston is moving at its most rapid rate,—it is evident that even a very short period of time consumed in the closing of the valve is worthy of economizing.

The connections to the steam valves in this latest engine are effected through two sliding rods, A^{18}, which run in the guides, A^{24}. The pins on the wrist plate are one on the outer and the other on the inner face, which holds the links out of line with each other, or not in the same vertical plane, but more nearly side by side. Two levers, A^{16}, are mounted side by side on the frame so as to be free to oscillate independently, and the links from the wrist plate are connected to the lower ends thereof. Hooks, A^{19}, are connected at one end to the upper ends of these levers. The other end of each hook is allowed to rest on a flat part of A^{18}, and to drop into a slight recess therein at a proper moment so as to pull open the valve. A stout spring, composed of several strips of steel plate, is bolted upon each of the levers, A^{16}, and is connected to the corresponding rod, A^{18}, by the link, A^{15}. These springs tend, like the weights in Plate XXXVII., to keep the valves always closed, each driving its rod, A^{18}, home very rapidly, and consequently turning its lever, A^{6}, so as to shut its valve with almost lightning speed, the moment its hook, A^{19}, is lifted so as to release it. The slide, or plunger rod, A^{18}, carries a small piston, not represented, which fits into a small cylinder, A^{25}, produced on the guide, A^{24}, and as the parts are driven home by the spring, the air in the interior of this cylinder cushions the piston and brings the parts gently to rest.

The means by which the hooks are lifted are peculiar and very admirable. The bell crank lever, A^{21}, is connected to the governor by the rod, D^{18}, and is weighted by the ball, A^{22}. It turns on a fixed centre, which latter is supported by the standard, A^{23}. To each of the hooks, A^{19}, is loosely connected a trip lever, A^{20}, which extends upward and is loosely connected to a corresponding link, 8, which latter is connected to A^{21}, as represented. The upper ends of the trip lever, A^{20}, are consequently moved forward and backward with each elevation and depression of the governor balls, D^{8}. The lower end of each trip lever, A^{20}, is armed on the side nearest A^{25} with a steel piece, not designated by letter, but which is plainly shown in the detail drawing, Fig. 47, Plate XLI. This piece is accurately adjusted in position on A^{20} by screws, and as the hook, A^{19}, is drawn back, and the lever, A^{20}, becomes correspondingly inclined, this piece projects below the surface of A^{19}, and pressing on the flat part of A^{18}, lifts the hook and releases A^{18}, and allows it to fly back. It is evident that the period at which this disconnection is effected will depend on the position of the top of A^{20}, and consequently on the elevation of the governor balls.

Plate XL. shows more in full many of the details above described. Fig. 7 is a vertical section of the cylinder, and an outline of much of the novel part of the valve motion. Fig. 8 is a top view of the cylinder, A^{5} being the long opening through which the steam is admitted to the steam valve chests or chambers. Fig. 9 is an end view, showing in dotted lines the outlines of the ports and valve chambers. Fig. 11° is a cross section of the steam valve, with its stem and chamber. Fig. 11 is a plan, and Fig. 12 a side elevation of a steam valve and stem. Fig. 13°

is a corresponding cross-section of an exhaust valve, with its stem and chamber. Figs. 13 and 14 are other views of the exhaust valve and stem. Figs. 15 and 16 are views of the wrist plate and rock arm. Figs. 17 and 18 are views of the guide, A^{24} and A^{26}, and of the pivot, A^{27}, which supports the wrist plate. Figs. 19 and 20 are views of the bearing for the bell crank lever, A^{21}. Figs. 21 and 22 are views of the said bell crank, A^{21}. Figs. 23, 24, 25, and 26 are views of the eccentric and eccentric strap. Figs. 27 to 34 are the governor and its connections. Figs. 35 and 36 are the eccentric hook. Figs. 37 to 40 are the connections to the levers in the valve stems. Figs. 41 and 42 are views of one of these levers. Fig. 43 is a side elevation of one of the sliding rods, A^{18}, and of the corresponding hook, A^{19}, and links, A^{15}. Fig. 44 is a plan view of the same parts, except that the hook, A^{19}, and the flat part of A^{18}, on which the hooks are supported, are represented as removed. Fig. 45 is a plan view of the hook, A^{19}. Fig. 46 is a plunger or piston, shown in Fig. 43, and which serves the important purpose of confining or cushioning the air to check the motion. Figs. 47 and 48 are the trip levers, A^{20}, showing the small adjustable piece which, by contact with the flat part of A^{18}, lifts the hook and releases the valve. Figs. 49 and 50 are views of the stop valve and chest, *E*. Figs. 51 and 52 are views of the starting lever for the stop valve. Figs. 53, 54, and 55 show the cross-head and end of the piston rod. Figs. 56 and 57 show the same kind of an adjustable cut-off as those now explained, but applied to flat sliding valves in lieu of the rolling valves now preferred for this purpose.

[For a large portion of the description of the Corliss' Engine, as also of that of the engine of the steamship *Knoxville*, and of the locomotive "Talisman," we are indebted to Thomas D. Stetson, Esq., the well-known Mechanical Engineer and Patent Attorney.]

REYNOLDS' PATENT VARIABLE CUT-OFF.

The present illustration is a variable cut-off applied to a horizontal steam engine. Fig. 1 shows the cut-off arrangement and governor in a side elevation. Fig. 2 a section through the cut-off and cut-off valve, and also through the cylinder and slide valve. Fig. 3 is a front view of the cut-off motion, and Fig. 4 a top view of the same.

In this invention the cut-off arrangement operates a separate valve, which excludes the steam from the entire valve chest proper when the piston has performed the desired part of its stroke. This exclusion of the steam is effected by a valve which is perfectly balanced, so as to be unaffected by the pressure of the steam. This valve is lifted by positive mechanism, and is returned to its closed position by springs. The period at which it is thus detached and closed is controlled by the governor. The valve is lifted and dropped twice during each revolution: first, to serve while the piston is making its stroke in one direction; and second, to accomplish the same for the return stroke.

A is the cylinder, A^1 the piston, and *S*, *S*, the cylinder ports, which serve equally for the induction and eduction of the steam. *B B* is the valve chest, *C*, *C*, the valves, and *E*, *E*, the exhaust ports. C^3 is the valve stem, which is operated through the agency of the eccentric, in the ordinary manner, the steam, at each stroke, being admitted through *S*, to drive the piston, and allowed to escape from *S* into *E*, through the cavity or hollow throat of the valve during the return stroke.

F is the cut-off valve chamber. *D*, *D*, are two disks of metal, which are adapted to slide tightly through corresponding openings in F, and are connected together by the guiding pieces, D^9. D^1 is a stem fixed thereto, and passing upward through the stuffing-box, F^5. It carries at its top a cross-head, D^{16}, which, in turn, carries the two vertical rods, D^{10}. Spiral springs, D^{11}, are fitted around these rods, which press down the cut-off valve. A shaft, D^8, is mounted in the bearing, D^{15}, in the framing, so that it is free to rock or oscillate. An arm, D^{13}, is keyed thereon, carrying an adjustable pin, D^{14}. On the valve stem, C^2, is firmly fixed an arm, D^4. From the upper extremity, D^5, of this arm, a rod, D^6, D^7, shown partly in dotted lines, communicates motion to the pin, D^{14}, causing the shaft, D^8, to rock with each motion of the slide valves, C.

On the inner end of D^8, are two short arms, as represented in Fig. 2, which are adapted to alternately lift the cut-off valve through the agency of the cross-head, D^8, and the hanging pieces, D^9, as represented. The releasing and dropping of the valve is effected by spreading the hanging pieces, D^9. This operation of spreading these parts is performed by a wedge-shaped casting, 3, which is connected to the governor rod, G^7, and is elevated and depressed by the governor, according to the speed of the engine, as is plainly shown by the drawing,—the short arm, G^8, of the governor being caused to raise or lower the rod, G^5, and through the connections, G^6, G^7, to give a corresponding motion to the casting, 3, with each elevation or depression of the governor balls, G^9.

The spreading open of D^9, and the consequent release and closing of the cut-off valve, is directly effected through the agency of pins, 9, which project from the sides of D^9, and come in contact with the inclined faces of 3, at some period in each lift of the cut-off valve; the time of such contact, and consequently of the release and descent of D, must, it is evident, depend on the elevation of the governor balls, and this changes with each change of velocity, so as to supply the steam through F^3, in exactly the quantities desired.

SETH BOYDEN'S VARIABLE CUT-OFF.

The nature of this invention consists in the application of a slide valve, with two cut-off plates lying on the top surface of it, to regulate the admission of steam into the cylinder. These cut-off plates, marked E^6, E^6, are regulated by the action of the governor, through their respective connections, and, as heretofore explained, such a quantity of steam is admitted into the cylinder, as is requisite or necessary for the amount of work to be performed by the engine, to keep up a regularity of speed. Fig. 5 is a sectional view of governor. Fig. 6 an outside view of steam chest, cylinder, and a part of framing. Fig. 7 a section through steam chest, valves, and cut-off apparatus in position when steam is admitted half-stroke. Fig. 8 is a section of the same with position of valve and cut-off fixtures when steam follows whole stroke. Fig. 9 is a transverse section through the valve. Fig. 10 a top view of steam and exhaust ports. Fig. 11 a horizontal section of steam-chest, and top view of cut-off parts. Fig. 12, top view of cut-off plates; and Fig. 14 a side view of the same. Fig. 13 is a horizontal section through pedestal of governor stand.

The governor column is screwed into the pedestal. The top part of the column has a recess for the collar C^5, which is screwed to the tube, C^4, and on the lower extremity of the tube is screwed a bevel wheel, C^8, operated by another bevel wheel, C^2, shaft, C^1, and pulley, C. The governor ball arms, C^{13}, C^{12}, are free to swing on their centres, 2, 2, and rotate with the tube, C^4, and top piece, C^5. It is a common practice to let an engine run at a certain number of revolutions per minute, but when a change takes place, and more or less work is thrown on or from the engine, the speed of the engine will be changed for an instant, until the motion of the governor throws its balls in or out, and through the short levers, 3, 3, of the governor ball arms, C^{12}, lifts or lowers the rod, C^8. The rod, C^8, has an attachment forged solid to it with a transverse slot, to communicate the motion of the rod, C^8, to the lever shown in dotted lines, and fastened on the inside to the spindle, C^{14} (see Fig. 13), which spindle carries the longer lever, C^{15}, and gives its motion to the connection, C^{16}, Fig. 6, and lever, C^{18}, and thus to the spindle, C^{20}. To the spindle, C^{20}, is fastened a double lever, C^{20}, Figs. 7 and 11, connecting the opposite ends by two links, C^{21}, to the regulating pieces, C^{23}, C^{23}, which come in contact with the inside projections, E^6, and consequently move the cut-off plates, E^6, over their ports, E, to shut off the supply of steam to the cylinder, at such a point of the stroke as has been explained before. The casting, D^4, for the cut-off fixtures, is laid on its seat without further fastening, and can be removed with little delay, in case of repairs to the cut-off arrangement. E^5 is the valve stem, which holds the valve by the valve strap, or bridle, designated also by E^5. D is the steam-chest casing, with a cover, D, on the top of it. D^2 is a flange for stuffing-box. F is a light casing, to cover up the bolts for holding to its place the steam-chest cover, D. E^7, E^7, are the induction and eduction channels for the steam. E^8 is the exhaust pipe.

N. T. GREENE'S PATENT VARIABLE CUT-OFF AND VALVE GEAR.

Plate XLIII.—Fig. 1 is a side elevation; Fig. 2 a plan; Fig. 3 a side view of cut-off arrangement, drawn on two-inch scale; Fig. 4 a transverse section through bed plate, slide holder, and slides.

Most prominent in this engine, to regulate the admission of steam into the cylinder, is the simplicity of motion resulting from the combination of the rectilinear and curvilinear motions. This combination enables the third agent, which is otherwise invariably used to effect the disconnection in detached cut-offs, to be dispensed with. This cut-off has also a great range of variability, on account of the eccentric being set very nearly with the crank, only sufficient advance being given it to take up a small lap on the valves, and the lost motion of the valve gear. In speaking of the merits of this engine, the most skillful engineers have expressed themselves in very flattering terms, and assure us if the engines are put up in good order, they will run with great regularity of speed and with a considerable saving of fuel. With reference to the several parts of the machine, the following will be a literal reference:

A is the steam chest, containing steam valves, S^{14}, moving over steam ports, S^0. These valves are connected by links, S^{18}, to arms, S^{16}, on rock shafts, S^{10}, which pass through stuffing boxes on the side of the steam chest, with weight arms, S^{17}, and rock levers, S^{11}, keyed on them. The spring tappets, S^8, carried on the sliding bar, S^6, give motion alternately to the rock levers. Motion is imparted to the sliding bar by an eccentric on the main shaft. The tappets have a lip on one side, upon which rests a gauge bar, G^5, connected by a rod, G^8, to the governor. The motion of the tappet being rectilinear, and that of the rock lever curvilinear, a separation must consequently take place between them, when the weight, S^{18}, which has been raised by the motion of the rock lever, will close the valve, and thereby effect the cut-off. The point of separation of the tappet and rock lever is determined by the amount of lap of the former upon the latter, at the commencement of their united motion, and this lap is varied by raising or depressing the gauge bar (the springs, S^{19}, keeping the tappets pressed constantly against it). The operation of raising or depressing the gauge bar is performed by the governor thus: if the speed of the engine increases, the balls expand, and depress the gauge bar, thereby giving less lap of the tappet, and a quick cut-off. On the contrary, if the balls contract, the gauge bar is raised, the tappets have greater lap, and a longer cut-off is the result. It must be borne in mind that the only period when the governor can act on the tappets is at the period when bôth are disconnected from the rock levers, but so sensitive is its action that the greatest possible variations of resistance are regulated with the utmost perfection.

The exhaust valves are operated through the eccentric, E; rod, E^4; rock arm, E^5; rock shaft, E^8; rock lever, E^6; and valve rods, E^{12}. The governor balls, G^1, G^1, are operated by two rope pulleys, one placed on the main shaft, marked G^6, and the other fastened to the shaft, G^7. On the same shaft is a bevel gear, operating another bevel gear, secured to the governor head, G.

The details of this engine are represented in Plates XLIV. and XLV. Fig. 4 is a vertical longitudinal section through the cylinder, showing the steam chest up, and the exhaust chest on the lowest part of it. The exhaust valves, E^{11}, E^{11}, are connected by a rod, E^{12}, to the arms, E^{10}, E^{10}, and held firmly to the rod, E^{12}, by two collars, with set bolts for each valve. The exhaust valves are faced on their lower surface, the steam in the cylinder pressing upon the upper surface of the valve, when they cover their ports. E^{14}, E^{14}, are covers for the exhaust chest. The remaining details of this engine are described on the various figures.

WRIGHT'S PATENT SELF-REGULATING CUT-OFF.

Plate XLVI.—Fig. 1 is a side elevation of the engine, with improved cut-off arrangement. Fig. 2 plan; Fig. 3 a transverse section through the bed plate; Fig. 4 a horizontal section through the cylinder, steam chests, and cam motion; Fig. 5 is a transverse section through cylinder, steam and exhaust chests; Fig. 6 an outside view of cam and shell, showing upper and lower parts of inside piece, and of bevel gear; Fig. 7 is a top view of cam and shell; Fig. 8 is a feather; Fig. 9 inside piece of cam and shell; Fig. 10 governor spindle; Fig. 11 a section view through shell and cam; Fig. 12 a top view; Fig. 13 an outside view of the same; Fig. 14 is a horizontal section through cam, valves, and valve rods, showing cam in three positions.

The advantages claimed for this cut-off, and the nature of its construction and operation, are briefly enumerated here.

It is simply a cam, D (see Fig. 12), and capable of a range between nothing and full stroke, if required. It is composed of two pieces. The outside piece, or shell, C^8, being a hollow cylinder, with part of one end of it, D, shaped to the form of a cam, and having a spiral groove, G, cut on the inside, running the whole length of the piece. The inside piece, H, is a hub, which exactly fits the internal diameter of the shell, C^8, but having a hole, I, through it to receive the governor spindle, which is eccentric to both pieces. On one end of the inside piece, H, is a cylindrical ring, or collar, K (see Fig. 6), which is also eccentric to both pieces, but concentric with the hole, I, through it, and with the governor spindle; both pieces, C^8 and H, are connected together by a feather, L, one side of which is of a spiral form, and the other side straight, or rectangular, and respectively fit and work in the spiral groove, G, of the shell,

C^8, and the rectangular groove, N, of the inside piece, H. The feather is connected with the governor spindle.

It is operated directly from the governor, which is placed on a stand above the cam and in the centre between the steam chests, as shown in Fig. 1, the spindle of which passes down through the governor column and is attached to the feather. The raising and depressing of this feather regulates the lift of the cam.

When any change takes place in the engine load, and the governor balls leave their legitimate plane or range, the variation is transmitted through the spindle to the feather of the cam, instantly altering the lift of the cam, and consequently that of the valves, and regulating quickly and precisely the amount of steam to be admitted to the cylinder to move the load at the proper speed.

The simplicity of this "Cut-off," the facility with which it can be adjusted—the parts being but few and always in sight and accessible—its action, so quick and perfect, and no throttling of the steam, but permitting it to enter the cylinder at boiler pressure, all combine to make it a most perfect cut-off. The governor is operated by gearings, C, C^1, C^2, C^3, and a shaft, C^4; the latter is placed on the side of the engine bed-plate, to communicate its rotary motion by two bevel wheels, C^5, C^6, to the governor shaft, C^{11}. The bevel wheel, C^9, is operated by the shaft, C^4, and engages with another bevel wheel, secured to a shaft placed across the engine bed-plate to communicate its motion, by an eccentric, to the exhaust valves.

PUTNAM MACHINE COMPANY.

PATENT VARIABLE CUT-OFF. PATENTED JANUARY 15, 1856.

Plate XLVII.—Fig. 1 is a side elevation of the engine with improved attachment; Fig. 2, a plan of the same; Fig. 3, a vertical section through the steam and exhaust valves; Fig. 4, a horizontal section of the steam and exhaust chests and valves; Fig. 5, exhaust valves; Fig. 6, steam valves; Figs. 7, 8, and 9, sections through steam and exhaust chests; Fig. 10, a transverse view of the engine.

The invention consists in a new and peculiar method of connecting the balance cut-off valves of steam engines with the governor, by which the steam may be cut off at any point of the stroke, according to the amount of work upon the engine. The parts for effecting this are simple of construction, and not liable to get out of order or to require repairs. In the accompanying drawing, Plate XLVII., A is the cylinder; A^1, cylinder cover; A^2, the steam chest; A^2, A^3, covers to get access to the steam and exhaust valves; A^4, screw valve; A^5, steam pipe; A^6, elbow of steam pipe; A^7, wheel for screw valve; A^8, oil cups; A^{10}, stuffing box; A^{11}, gland; A^{12}, cross-head; A^{13}, piston rod; A^{14}, connecting rod; A^{15}, crank; A^{16}, crank shaft. The crank shaft carries a cog-wheel, C, which engages with a gear, C^1, upon a short shaft, C^4; C^2, is a bevel wheel upon the shaft, C^4, which gears with the wheel, C^8, upon the shaft, C^5, which latter carries the cams that actuate the steam and exhaust valves. The connections between the governor and the apparatus which actuates the valves, will now be described. The governor spindle is moving through the centre of the governor column, G, with a bevel gear, G^2, on its lowest extremity operated by the bevel gear, G^1; A, small rod actuated from the extreme ends of the governor ball arms, connects to the angular lever which moves the lever C^{20}. The small rod connected to the ends of the governor arms, moving in a vertical direction, moves the angular lever, and the angular lever, the lever C^{20}, which is fixed to the shaft, C^{18}, vibrating in the bearings, C^8, C^8, C^8. Projecting from the frame-work of the steam engine, C^{19}, C^{19}, are arms attached to the shaft, C^{18}, and vibrating with it. C^{11}, C^{11}, are bent levers of the form represented in Fig. 10, which are pivoted to the arms, C^{11}; the extremities of these levers are allowed to play freely in slots at the lower end of the valve rod, C^{14}; as the rod connected to the governor now rises and falls, the shaft, C^{18}, is caused to vibrate in one direction or the other, and the levers, C^{11}, are withdrawn more or less from the slots in the stems of the induction or steam valves, for a purpose which will now be explained. C^{10}, C^{10}, C^{10}, C^{10}, are cams upon opposite sides of the shaft, C^7, and directly beneath the levers, C^{11}, upon which they impinge as the shaft, C^7, revolves, and thus when the valves are raised from their seats and steam is admitted to the cylinder, in proportion as the balls of the governor diverge, the rod is raised, and the ends of the levers, C^{11}, are withdrawn from the slots in the bottom of the valve stems, C^{14}, thus carrying the shoulders of the bent levers farther from the circles of revolution of the cams, C^{10}. The levers, C^{11}, are so arranged with reference to the governor and the cams, C^{10}, that when the balls are at the highest point, the shoulders of the levers are withdrawn so that the cams revolve without touching the levers, and the inlet valves are not opened. When, however, the balls are in any other position, the shoulders of the levers, C^{11}, lie more or less over the cams, C^{10}, and as the latter revolve, the levers, C^{11}, are raised, and the inlet valves are kept open for a longer or shorter space of time. The steam is thus cut off, sooner or later, in proportion as the balls are more or less distended. The levers, C^{11}, are curved upon the under side, so that they may be eased down gradually as the cams revolve. It is evident, if the cams which open the valves be allowed to make as many revolutions as the main shaft, that they will revolve 180° whilst the piston moves once through the cylinder, and that with such an arrangement, it would not be practicable to regulate the operation of the valves; it therefore becomes necessary to revolve the shaft, C^6, a less number of times than the main shaft, in which case, more than one cam will be required. If the shaft, C^6, revolve half as fast as the main shaft, two cams will be necessary, as in the case represented in the accompanying drawings. If the main shaft revolve three times as fast as the shaft, C^6, the latter will require three cams, and so on. It will be perceived, that by means of the arrangement above described, the valves may be made to feel the slightest change of velocity in the governor balls; while, at the same time, no work is thrown upon the governor, as even the weight of the rod and the parts connected therewith, may be balanced by a weighted arm projecting from the opposite side of the shaft, C^{18}; if in any case it be found desirable. In practice, however, they have not found this to be necessary. The exhaust valves, E, are operated similarly to steam valves, except that they are allowed to open and close without variation. The levers, C^{16}, C^{16}, are pivoted to the bearings, C^8, and they are raised by the cams, C^{10}, which are upon the shaft, C^7. As these levers are stationary, the levers, C^{16}, are so formed with reference to the cams, C^{10}, that the exhaust valves shall be held open during the whole passage of the piston, as required. The construction and arrangement of the valves and steam passages will now be explained.

Figs. 3 and 4.—Within the steam chest, A^2, are two other boxes, F, F; upon these secondary boxes are the upper and lower seats of the steam valves, S, S. S^1 and S^2, is the spindle of the valve, and is guided above by the projection, S^3, cast to the cover, and below, by the hollow screw, C^{15}. The boxes, F, F, being entirely surrounded with steam, when the valves, S, S, are are raised, the steam passes in through the open rings, through the steam port, S^4, into the cylinder. The exhaust valves close the eduction passage when resting upon their seats. E^3, E^3, is the spindle of the valve, and is guided above by the cover, D, and below by the hollow screw, C^{13}. The seats of the exhaust valves are upon the interior box, the space within which is in constant communication with the exhaust passage, E^8, and is entirely shut off from the interior of the boxes, F, F, except through the valves, E, E, when the latter are raised from their seats. The operation of these valves is as follows: when the steam valves, S, S, are raised, steam is allowed to enter from the steam chest into the box, F, and through the passage, S^4, into the cylinder; as the piston commences its return stroke (the valves, S, S, having been previously closed,) the exhaust valves, E, E, are raised, and the steam passes from the cylinder through the opening, S^4, into the box, F, through the exhaust valves, E, E, into the interior box, thence through the eduction passage, E^8, into the conductor. This arrangement of valves and steam passages is economical, compact, and of simple construction. The claims of the inventors, Charles H. Brown, and Charles Burleigh, are: operating the valves by means of the revolving cams, C^{10}, in combination with the bent levers, C^{11}, and their connection with the governor, in the manner and for the purpose substantially as herein set forth.

SEVEN FEET ENGINE LATHE.

Plate XLVIII.—contains illustrations of a seven-feet engine lathe, built by Bement & Dougherty, Philadelphia, designed and constructed for all kinds of heavy work, and for which the frame, gearing, and all its parts are well arranged, in regard to strength, convenience, and facility. The slide-rest is self-acting. The speed of both the lathe and the slide-rest may be increased or diminished by arrangements hereafter described. In addition to which, there is a small drilling or boring machine fixed upon the movable head-stock, G, so that a crank, driving wheel, or any similar object, can be drilled or bored outside the centre, whilst secured to the face-plate, or resting between face-plate and head-stock, G. This device is particularly serviceable, as by these arrangements, a crank may be bored, faced, and fitted to receive the crank pin, without being subject to removal. The upper part of the head-stock, G^1, may be moved at right angles to the centre line of lathe, by means of the handle and screw, G^5, if required.

Fig. 1, is a side elevation; Fig. 2, a longitudinal section view, showing the method of varying the gearing attached to the carriage, C, C, and slide-rest, C^1; Fig 3, is a plan or top view of lathe, also showing the drilling or boring apparatus; Fig. 4 is a back view; Fig. 5, a front view; Fig. 6, a top view of the gearing to connect the slide-rest; Fig. 7, an end view of the same. A, is the bed-plate; A^1, are ribs, cast solid with the bed-plate, to give strength, as shown in Fig. 3; A^2, A^2, A^2, are projections of the bed-plate, to secure it to the floor by bolts, as shown in Figs. 2 and 3; A^3, A^3, are the slides upon which rests the loose head-stock, G; A^4, is the rack; A^5, the screw; B, is the fast head-stock, B^1, the main spindle, carrying the face-plate, B^2, to which is secured the other part, B^3, and is held down by the caps, B^4, B^4, and bolts, 1, 1; B^5, shows the hub of the face-plate; B^6, is a projection from B, to carry the shaft, B^9; B^7, B^7, are caps to secure the shaft, B^9, as shown; at B^8, is placed a small pinion to gear into, B^3; B^{10}, are pulleys, to which is secured the pinion, B^{13}, conveying motion to B^{14}, upon the shaft, B^9; to the other end is secured the pinion, B^{12}, gearing into B^{11}, upon the shaft, B^9; B^{15}, is a short spindle, as seen in Fig. 2, keyed into the main spindle and carrying with it the toothed wheels, D, D^1, D^2. C, is the carriage; C^1, the slide-rest; C^2, a plate to secure the tool, C^4, in the tool-holder, C^5, by means of the screw, 2; C^7, is a hand-wheel, which is connected to the rack, A^4; C^8, is another hand-wheel, connected by means of the wheels, C^9, C^{10}, and C^{11}, to the screw acting upon the slide-rest, best shown in Figs. 6 and 7. To obtain the motion of the slide-rest, the cone, C^{12}, must be secured to the wheel, C^{11}, by means of the handles K, (see Fig. 7) so that as the shaft, D^{16}, revolves, it gives motion to C^{19}, (Fig. 6,) and through the wheel, C^{18}, to the shaft, C^{17}, carrying with it the bevel-wheel, C^{20}, and so gives motion through the wheel, C^{14}, to the shaft, C^{15}, upon which the cone, C^{12}, is secured. If required to move the slide-rest by hand, it is only necessary to loosen the handle K, thus allowing the cone, C^{12}, to revolve freely in the wheel, C^{11}; the motion of the slide-rest is reversed by means of the handle, C^{13}, being raised or depressed. D, D^1, D^2, are three wheels which gear into three corresponding wheels, D^4, D^5, D^6, any one of which may be secured by the rod, D^1, having inserted in the other end a key, I, projecting on either side, so that any one of the wheels may be driven separately, together with the pinion, D^0, which gears into D^7, D^7 into D^8, and D^8 into another wheel secured to the back of D^9; D^9 gears into the wheel, D^{14}, (see Figs. 1 and 4), secured to A^7, and which is coupled to the screw at A^6, (Fig. 3.) From D^9, the motion is continued through D^{11}, to D^{13} and D^{15}, which are placed upon the shaft, D^{16}, (see Fig. 3), having a pulley, D^{17}, secured to it by means of a key, sliding freely in a slot made for that purpose, in D^{17}. G, (Fig. 2), is the lower part of the loose head-stock, having projections which fit into corresponding grooves in the upper part, at right angles to the centre line of lathe; G^2, is a separate part, which is screwed into G^1, as shown in section; G^3, is a pulley secured upon the screw by a nut at G^7; G^4, G^4, are set screws to secure G^8; G^6, a handle placed in the wheel, G^8; G^0, is the centre; H^0, is the table projecting from the head-stock, G; H, is a spindle driven by the toothed wheel, H^1, geared into the pinion, H^2; H^3, is a screw; H^4, a spindle upon one end of which is secured the pinion, H^2, upon the other end is the pulley, H^5, which is driven by a belt running from the movable pulley, D^{17}; H^6, is the base, secured to the head-stock by the bolts, 9, 9, having two projections, H^6, H^6, to carry the shaft, H; H^7, is a hand-wheel, having an internal thread to suit the screw, H^3, the wheel being retained between the bracket and the bearing, H^8.

SPEED OF THE LATHE.

The lathe can be worked with single or double gear: the speed can be changed with single gear five times, and with double gear five times, which will make ten changes of speed. The slowest speed of the cone pulleys on the lathe marked B^{10}, is 40 revolutions per minute, and the fastest speed, 330 revolutions per minute; which would turn the face-plate, B^2, if single-geared, 4½ the slowest, and 36½ the fastest speed per minute, and if double-geared, the face-plate would make, in three minutes, 1 revolution the slowest speed, and 3 revolutions per minute the fastest speed. The slowest speed of the pinion, B^8, which drives the face-plate, is 3 revolutions per minute. The feed for the slide-rest carriage can be changed six times, to advance from $\frac{1}{16}$ of an inch to $\frac{1}{4}$ of an inch per revolution of face-plate.

The main shafting from which the motion is taken, makes 44 revolutions per minute, and the pulley fastened on said shafting, to operate the intermediate shaft and pulleys for the lathe, is 4 feet in diameter, and has 11 inches face. The intermediate driving and loose pulleys are 19 inches diameter and 6 inches face; on the same shaft are cone pulleys of the same size as the cone pulleys on the lathe. The intermediate pulleys and shaft make 110 revolutions per minute.

www.ingramcontent.com/pod-product-compliance
Lightning Source LLC
Chambersburg PA
CBHW021704210326
41599CB00013B/1513